THE EU AND NEIGHBORS

A Geography of Europe in the Modern World

THE EU AND NEIGHBORS

A Geography of Europe in the Modern World

BRIAN W. BLOUET
College of William and Mary

JOHN WILEY & SONS, INC.

Vice President and Publisher   *Jay O'Callaghan*
Executive Editor   *Ryan Flahive*
Production Editor   *Nicole Repasky*
Executive Marketing Manager   *Jeffrey Rucker*
Senior Designer   *Kevin Murphy*
Senior Photo Editor   *Lisa Gee*
Photo Researcher   *Ramón Rivera Moret*
Production Management Services   *Pine Tree Composition, Inc.*
Editorial Assistant   *Courtney Nelson*
Media Editor   *Lynn Pearlman*
Cover Photo   *age fotostock/SuperStock*
Cover Design   *David Levy*
Wiley 200th Anniversary Logo Design   *Richard J. Pacifico*

This book was set in Times Roman by Laserwords Private Limited, Chennai, India and printed and bound by Donnelley/Crawfordsville. The cover was printed by Phoenix Color.

The book is printed on acid-free paper. ∞

To order books or for customer service please, call 1-800-CALL WILEY (225-5945).

Library of Congress Cataloging-in-Publication Data

Blouet, Brian W., 1936–
   The EU and neighbors : a geography of Europe in the modern world / Brian W. Blouet.
      p. cm.
   ISBN-13: 978-0-471-65554-1 (pbk.)
   1.   Europe—Economic integration.   2.   Europe.   I. Title.
   HC241.B545 2008
   341.242′2—dc22
                                                                              2007008126

Printed in the United States of America

10 9 8 7 6 5 4 3 2 1

PREFACE

▶ TO THE READER

This book is an entry-level, introductory, systemic, and regional survey of Europe, which uses the European Union as an organizing focus for the text. It is not assumed that students have already taken a geography course and as geographic concepts are introduced they are explained.

The book is divided into two parts. The systematic chapters cover the environment, cultural geography, population, settlement systems, economies, and political geography. The regional chapters examine Europe on a country-by-country basis, but countries are grouped in the order they entered the European community. In broad terms the sequence is Western Europe, Southern Europe, Northern Europe, and Eastern Europe. Turkey and the countries of Eastern Europe, still awaiting entry, are covered as are states emerging from the former Soviet Socialist Republics, which are not yet candidates for EU entry.

Place names are usually in the form used in *Goode's World Atlas*, 21st edition, but the *National Geographic Society College Atlas of the World* (2007) was made use of, as was Saul Cohen (ed) *Columbia Gazeteer of The World*. New York, Columbia University Press, 1998.

▶ ACKNOWLEDGMENTS

This book results from conversations with Ryan Flahive, who deserves much credit for his contribution in shaping the project. Ryan was supported by able assistants including Courtney Nelson who has seen the book through the production phase at John Wiley.

High quality reviewing was provided by William Berentsen (University of Connecticut), Renate Holub (University of California, Berkeley), Cub Kahn (Oregon State University), Jon Kilpinen (Valparaiso University), Darla Munroe (Ohio State University), Ann Oberhauser (West Virginia University), Erik Prout (Texas A&M University), and Donald Zeigler (Old Dominion University). Thanks to the reviewers for valued inputs.

At John Wiley thanks go to: Jay O'Callaghan (VP and Publisher), Ryan Flahive (Executive Editor), Jeffrey Rucker (Marketing Manager), Nicole Repasky (Production Editor), Lisa Gee (Senior Photo Editor), Sandra Rigby (Senior Illustration Editor), Kevin Murphy (Senior Designer) and Courtney Nelson (Editorial Program Assistant).

Academic debts are extensive. H.R. Wilkinson, my supervisor at Hull University, was an authority on the Balkans and the former Yugoslavia. At the University of Sheffield I worked with Alice Garnet, Charles Fisher, Bryan Coates, David Grigg, and Malcolm Lewis. John Sibbons taught me much about France. At the University of Nebraska-Lincoln influential colleagues included Leslie Hewes, David Wishart, Dean Rugg, Bob Stoddard, and Richard Lonsdale. At Texas A&M University Peter Hugill, Robert Bednarz, and

Clarissa Kimber shared much knowledge. At William and Mary, where I hold appointments in the School of Education and the Department of Government, I have benefited from interaction with James Bill, Mike Tierney, Paula Pickering, Rani Mullins, Maria Ivanova, Clay Clemens, Alan Ward, and Laurie Koloski.

On research leaves I have been aided extensively in the library of the School of Geography, Oxford, and enjoyed visits to Hertford College Senior Common Room. Jean Gottmann, John House, Andrew Goudie, Paul Coones, Ian Scargill, Gordon Smith, and Andrew Warren all helped.

Patty Donovan, Pine Tree Composition, got us to deadlines on time and was a pleasure to work with. Ramón Rivera Moret did a fine job of photo researching. Alice and Will Theide of Cartographics made sketches into maps. Some maps from William H. Berentsen, *Contemporary Europe,* John Wiley and Sons, 1997, were adapted for the present work. Laura Wyatt and Nicole Hopkins, in the School of Education, have been excellent editorial assistants, Nicole seeing the book through the production phase. Nicole, Courtney, and Olwyn helped with the selection of photos. My wife, Dr. Olwyn Blouet, made many contributions and encouraged the project at every stage.

To all of the above, and all those I should have recognized, thank you.

Brian W. Blouet
Huby Professor of Geography
and International Education
College of William and Mary

Contents

INTRODUCTION: THE EUROPEAN UNION

Europe, a region rich in resources, has been continually divided by wars and partitioned by boundaries. Since World War II the countries of the region have moved to share geographic and economic space and removed many impediments to cross-border movement of goods, people, and capital.

This book examines the origin and evolution of the European Union within the physical, cultural, and political framework of the geographic region of Europe. The European community emerged after World War II, as a partnership between France and Germany to prevent conflict between them, at the same time creating an enlarged market to promote economic growth.

▶ THE COAL AND STEEL COMMUNITY (1951) AND THE EUROPEAN ECONOMIC COMMUNITY (1957)

The first institution involving the original six members of what became the European Union was the Coal and Steel Community, signed into being in the Treaty of Paris (1951) by France, Germany, Italy, Belgium, The Netherlands, and Luxembourg. The same six countries, at the core of western Europe, signed the Treaty of Rome (1957) establishing the European Economic Community (EEC) to promote, over a period of time, the free movement of goods, labor, and capital (Figure I.1). The movement of goods among countries was encouraged by the removal of tariffs on manufactured products originating in the territory of member states. The enlarged market of the six member states was protected by a common external tariff. Imported cars, for example, would pay the same common tariff whether they entered the territory of the six at the port of Hamburg in Germany, Rotterdam in The Netherlands, Antwerp in Belgium, Bordeaux in France, or Genoa in Italy.

The six adopted a Common Agricultural Policy (CAP). Farmers were paid above world market prices to produce agricultural staples. Cheap grain from North America was brought up to the CAP price by the imposition of dues at the port of entry. The merits and disadvantages of the still existing Common Agricultural Policy are discussed in later chapters.

▶ EFTA: THE EUROPEAN FREE TRADE ASSOCIATION (1959)

Not all European countries favored the EEC economic model. Neutral countries like Switzerland and Sweden thought the EEC compromised neutrality because aspects of sovereignty, for example control of borders, were merged. Britain, which imported food

Figure I.1 Countries of Europe.

from the United States, Canada, Australia, and New Zealand, did not want to increase food prices by adopting the CAP. Britain suggested the creation of a European Free Trade Association (EFTA) in which countries would agree to remove tariffs on goods passing between member states. Each country would be free to trade with non-EFTA states on terms of their own choosing. In 1959, the United Kingdom, Sweden, Denmark, Norway, Switzerland, Austria, and Portugal joined EFTA.

EFTA and the European Economic Community (EEC) were not opposed to each other. They were different models for increasing regional trade. In 1992, the organizations created a European Economic Area that allowed free trade between the two sets of countries.

► ENLARGEMENT OF THE EUROPEAN ECONOMIC COMMUNITY (EEC)

The European Economic Community was the larger entity in terms of GDP and the Common External Tariff disadvantaged EFTA countries like Britain and Denmark in trade with the six. In the 1960s as France, under General de Gaulle, established leadership over the six, applications to join the EEC from Britain and Denmark were twice rejected. With the exit of de Gaulle, enlargement became possible and the process followed a logical geographical pattern.

In 1973, Britain, Denmark, and Ireland were admitted. Denmark traded heavily with Germany, and Britain had strong trade links with Belgium, The Netherlands, and Germany. Ireland was closely linked economically with Britain. The 1980s saw membership for southern European countries: Greece (1981), Spain (1986), and Portugal (1986) all joined partly as a reward for establishing democracy after long periods of military or dictatorial rule.

► EUROPEAN UNION (1993)

The countries of Europe moved to a greater merging of sovereignty with the signing of the Treaty on European Union in the Dutch town of Maastricht in 1991. Some countries, including Britain and Denmark, wanted opt-out clauses and the treaty was not ratified by all 12 members until 1993.

The collapse of the Soviet Union changed the geopolitical scene, allowing formerly strictly neutral countries to merge into the European Union. Finland, Sweden, and Austria became members in 1995. The dissolution of the Soviet "Near Abroad" empire and the Warsaw Pact resulted in countries of east Europe joining NATO and applying for EU membership. The Baltics, Poland, Czech Republic, Slovakia, Slovenia, and Hungary joined in 2004, along with Greek Cyprus and Malta. Romania and Bulgaria entered in 2007. Croatia, Bosnia Herzegovina, Serbia, Macedonia, and Montenegro are in talks about applying, while Turkey, with nearly 80 million inhabitants that are mostly Muslim, is seen as difficult to absorb.

► ORIGINS OF THE EUROPEAN UNION

Since the Napoleonic wars, the great powers of Europe—France, Germany, the Russian Empire, and Austria-Hungary—fought to control space, people, and resources. The geopolitical competition for territory quickened in the decades before World War I. The Swedish geopolitician Rudolf Kjellén advocated a northern super state dominated by Germany. The political geographer Friedrich Ratzel spelled out the mechanism of *Lebensraum* (living space) by which strong states would grow in size at the expense of weaker states. World War I was a fight to control territory and economic space (Blouet, 2001).

The war resulted in the disintegration of Austria-Hungary, a reduction in territory controlled by Germany, the collapse of the Ottoman Empire, the appearance of the Soviet Union, and the creation of new states. The peace treaties of 1919–1920 did not repudiate geopolitical competition. France, Britain, and the United States just wanted to make it more difficult for Germany to expand. Events showed that breaking Europe into smaller pieces made it easier for Germany, and then the Soviet Union, to take over the smaller states of eastern Europe, such as Czechoslovakia.

In the 1920s an influential minority discussed a possible European Union. The idea came to the League of Nations before being buried by the protectionism of the Great Depression and the rise of totalitarian states. Nazi Germany and the Soviet Union wanted to create self-sufficient economies to avoid international trade and cooperation.

▶ WORLD WAR II

Trench fighting in World War I involved huge loss of life, but, outside war zones, civilian casualties were low. After World War I, the treaties signed in Paris in 1919–1920 constituted a victor's peace, "... the peace to end all peace." (Macmillan, 2001) World War II was a total war. Civilians were blitzed, shelled, starved, deported, and massacred. The war was so terrible that at the end of it, in western Europe, reconciliation overcame retribution. World War II brought western Europe to an understanding of the need for cooperation.

After World War II church men and women, from many denominations in many countries, called for reconciliation. A group of Christian Democrat politicians revived the ideas of international cooperation discussed between the wars. The politicians included the Rhinelander Konrad Adenauer, who became the first Chancellor of West Germany; Paul-Henri Spaak, the Belgium premier; Robert Schuman of France; and the Italian Alcide de Gasperi. It was Spaak who, exiled to London in World War II, told people "... we must unite Europe. We cannot afford more civil wars among our nations or we will destroy civilization." These men, along with the French economic planner Jean Monnet, worked on the formulae of economic integration, leading to the eventual European Union. The politicians were bonded by common experiences of defeat, occupation, and resistance for the "germ of the European Movement, as it developed after the Second World War, is to be found in the resistance movements against Nazi domination" (Mowat, 1973, p. 21).

Britain lay outside Europe across the English Channel, which had saved the country from occupation in World War II. Britain had large debts, both monetary and moral, to commonwealth countries like Canada, which had sustained the war effort and provided fighting units. Winston Churchill, when he called for a United States of Europe in 1946, envisaged a British Empire/Commonwealth cooperating with, but separate from, Europe.

▶ WESTERN EUROPEAN DEFENSE NEEDS

Coming from the east, dark forces quickly reentered Europe after World War II. The Red Army, posing first as liberator, occupied eastern Europe. Poland, Czechoslovakia, Hungary, Bulgaria, and Romania came under Moscow control. The Baltics, absorbed by the Soviet Union in 1940, did not reemerge. The USSR had occupation forces in East Germany and Vienna.

At Westminster College, at Fulton, Missouri, on March 5, 1946, with President Harry Truman on the platform, Winston Churchill delivered his Iron Curtain speech: "From Stettin in the Baltic to Trieste an iron curtain has descended across the continent [of Europe]."

▶ THE TRUMAN DOCTRINE, THE MARSHALL PLAN, AND NATO

Would communist regimes be established west of the Iron Curtain? In World War II, France and Italy were under fascist control and occupation. Communist parties strengthened. Postwar the largest political groupings in France and Italy were the communist parties. A civil war between royalists and communists divided Greece. The Truman Doctrine (1947) gave aid to Greece and was followed by the Marshall Plan (1948) promoting economic reconstruction in western Europe. Countries in the Soviet bloc were told by Moscow not to take Marshall Plan funds. The Soviets saw the Marshall Plan as a scheme to promote the dollar as the currency of world trade.

Fears for the security of western Europe led, in 1948, to a Treaty of Mutual Assistance between Britain, France, and the Benelux countries, to be followed in 1949 by NATO. The Cold War, dominated by the U.S. and the USSR, became a force for European unification, with the Marshall Plan funding economic cooperation and the Soviet Union fostering fear and collective security.

After World War II, west European economic recovery was rapid. Prewar production levels were surpassed by 1949 and in the 1950s economic growth was sustained. The Coal and Steel Community was operating and the need for a wider European economic framework was discussed.

▶ AGREEING ON A EUROPEAN FRAMEWORK

For the countries of western Europe to merge aspects of sovereignty, the right political climate and an agreed framework had to exist. Governments had to stop thinking of their economies as discreet entities, in competition with adjoining states, and merge economic space to allow goods to move easily in an enlarged western European market. In regard to some economic activities, boundaries between states would be removed.

What model would an enlarged western European market follow? Would it be a neoliberal economic model with the removal of barriers to trade? The Dutch had pioneered free trade in the seventeenth century and Britain had created a widely spread free trade system in the decades before World War I.

Or would the economic model be mercantilist and employ a *zollverein* system, bringing all member states into a customs union with a common external tariff? The French had developed a closed mercantilist system in the seventeenth century trying to ensure that imports to France came from French colonies in French ships. The Germans used a *zollverein* customs system long before the German states unified into the German empire (1871). If Britain was an influential part of the debate on the European economic framework, free trade concepts would be strongly advanced. Germany and France were more protectionist in their thinking. Now events intervened to drive France and Britain apart and bring Germany and France together.

► THE SUEZ CANAL CRISIS

An episode in 1956, hardly noticed in North American texts, helped shape modern Europe and the Middle East. Colonel Nasser, the leader of Egypt, wanted an Aswan High Dam to impound the Nile and provide more irrigation water. The World Bank would finance the project. Then in the summer of 1956 Nasser signed an arms deal with the Soviet bloc. John Foster Dulles, U.S. Secretary of State, was outraged and withdrew World Bank support for the Aswan project. Colonel Nasser legally took over (nationalized) the Suez Canal, owned by a French company in which the British government had a large shareholding. Negotiations to establish a canal users' association to maintain access to all shipping were unwelcome in Egypt and undermined by Dulles.

In collusion with France and Britain, Israel attacked Egypt. Britain and France, by treaty, intervened to "protect" the canal. The U.S. joined the USSR in calling for a halt to the attacks. Britain caved in first because of economic pressure from the United States. Prime Minister Anthony Eden phoned France to say the campaign was over. The French Prime Minister, Guy Mollet, was with the West German Chancellor Konrad Adenauer struggling to reach agreement over Europe. Mollet could not contain his anger at the British collapse. Wise Adenauer, taking his moment, said to Mollet make a unified Europe your revenge on the United States and its client state, Britain. France and Germany came together. The Franco-German economic model prevailed; Britain was on the outside of EEC and helped create EFTA (1959). French suspicion of Britain's relationship with the U.S. persists to the present.

In the Middle East Western influence collapsed. Within a year Arab nationalism had destabilized the region. Countries including Iraq and Lebanon have never recovered.

► JEAN MONNET

The French international economist, Jean Monnet, is thought of as the framer of the European Community. However, he was not developing concepts in isolation and drew widely on European experience pulling together common threads of thought that France and Germany shared. Further, the Benelux customs union, created by Belgium, The Netherlands, and Luxembourg in 1947, provided a basic model that, if followed, would ensure that the three countries would join the European community.

The European model Monnet produced was protectionist; with a common external tariff on goods entering Europe from nonmember countries. The model was "statist." There was to be a central authority in Brussels with the power to regulate aspects of economic activity and the funds to establish communitywide programs, such as the Common Agricultural Policy (CAP).

► THE COAL AND STEEL COMMUNITY

Protectionist and statist characteristics were on display in the first major instrument of European economic integration—the Coal and Steel Community. During World War II the coal, iron, and steel industries of the Ruhr, the Saar, Luxembourg, Belgium,

Photo I.1 Strasbourg, France, UNESCO-World Heritage Site showing a central square and cathedral.

Source: ©Knoll/AgeFotostock.

and northern France worked together under German domination. Prior to World War I, corporations had made cross-border investments in an iron and steel region, which interlocked in terms of resources and markets. For example, Luxembourg possessed a large iron ore field, but lacked coal, and the home market was too small to absorb all the steel produced. Luxembourg had to import coal and sell steel into neighboring markets.

Integrating the coal mines, blast furnaces, and steel plants of western Europe into one market with equal access to the resources of coal and ore meant that national governments could no longer protect home industries, and countries with resources could no longer rig the market to make coal, for example, more expensive to plants in adjoining countries. In economic terms, French Lorraine iron ore and German coalfields were in the same space. France and Germany had fought three wars over border-area coal and ore.

If governments could no longer protect home coal and steel producers from European competition, older mines and uneconomic iron works would close. Belgium is one of the oldest industrial countries in the world. In the Sambre-Meuse valley, coal and ore beds were exposed, and, in the nineteenth century, complexes of coal mines, blast furnaces, and steel works developed. By the second half of the twentieth century the ore beds were exhausted and the best coal mined. With the creation of the Coal and Steel Community, Belgian steel makers got access to cheaper Ruhr coal. The Coal and Steel Community made funds available to phase out production at high-cost Belgium coal mines and to

Photo I.2 Old Strasbourg is sited on an island in the river Ill two miles before the Ill joins the Rhine.

Source: SGM/AgeFotostock.

retrain miners for other work. Politically, the Coal and Steel Community would not have been accepted in countries with older plants without provision for structural adjustment funds. Structural adjustment funds are a continuing part of EU programs. East European countries, with older industrial sectors, have access to the funds on becoming members. The Schuman Plan, as the coal and steel proposal was called, after the French politician, was devised by Monnet with input from the eventual members. Britain was not consulted. The plan was announced and Britain told to take it or leave it. For political reasons Britain could not take it (see Chapter 8), but this should not obscure the fact that the Coal and Steel Community was a Franco-German construct to reduce rivalry in the area of heavy industry.

The Coal and Steel Community rules now apply to all EU economic space. The plan has produced a relatively efficient distribution of iron and steel industries and did allow Italy to build several large integrated iron and steel plants at port locations using imported coal and iron ore.

▶ ROME TO MAASTRICHT

The Treaty of Rome (1957) created the European Economic Community (EEC) to allow the free movement of goods, labor, and capital among the original six members: France, Germany, Italy, and the Benelux countries. Brussels and Strasbourg became major centers of administration and the European Parliament (Photos I.1, I.2, I.3).

Creating shared economic space was a slow process. It took until 1968 to remove tariffs on goods, originating in the community, moving between member countries. Even with the removal of tariffs, paperwork barriers remained.

Photo I.3   European Parliament building, Strasbourg, beside the navigable river Ill.

Source: Ingolf Pompe/Alamy Images.

The free movement of capital took decades to achieve and is not fully complete. Labor mobility has increased but cultural factors, including language, cause impediments. The establishment of a community sharing economic space was always seen as a step toward the creation of common business conditions and further merging of sovereignty. In 1986 the single European Act removed bureaucratic barriers to trade and harmonized many regulations.

The treaty of Maastricht, coming into force in 1993, created the European Union, and made provision for a common currency—the Euro.

▶ EU EXPANSION AND EXPANSION FATIGUE

The largest EU expansion began in 2004 when ten countries, including a tier of eight east European states stretching from the Baltics to Slovenia at the head of the Adriatic Sea, became members. (Figure I.1) Borders have opened, goods are moving in east–west directions, structural readjustment funds are helping phase out old industries, and investment from the west is pulled east by the lower costs of a workforce possessing industrial skills.

There are signs of expansion fatigue. When the ten countries joined in 2004 only three existing members (Ireland, Britain, and Sweden) agreed immediately to accept the free movement of labor from the new member states. In 2005 an EU constitution was to be ratified. Some countries approved by legislative action; other countries had referenda. Voters in France said "non," followed by the Dutch saying "nee." The constitution would have advanced the merging of sovereignty. The rejection of the constitution in two EU core countries suggests that voters are wary of deeper EU involvement. Bulgaria and

Romania entered EU in 2007. Full membership for Turkey has become problematic, and there is a general sense that expansion should slow for a time.

▶ EU ISSUES

EU protectionism has been reduced in many areas by World Trade Organization (WTO) rules. The Common Agricultural Policy, vehemently guarded by French farmers, takes 40 percent of the EU budget and restricts market access for agricultural products grown elsewhere.

The EU has not eradicated economic nationalism. Several countries resist the takeover of flagship corporations by firms from other EU countries. Governments use nontariff barriers to favor indigenous corporations and government contracts often go to home firms. In the service sector, Europe's economic space is not fully open to competition. This issue will require additional negotiations.

▶ THE EU TODAY

The EU is a remarkable achievement born from World War II devastation. France and Germany live in peace. Authoritarian regimes in Greece, Spain, and Portugal have gone and the countries absorbed into Europe. Eastern Europe, the battleground between Germany and Russia, is becoming part of the European Union. There is large-scale movement of goods, people, and capital across the region.

Whether or not the EU can evolve from being a successful regional organization to embrace a globalizing world provides material for discussion. Countries like Britain, The Netherlands, and Spain, with expansive, imperial pasts, naturally look outward. France and Germany, both with a history of trying to dominate Europe, have a more continental viewpoint. France, in particular, discusses the need for the EU countries to develop common foreign and defense policies. France sees Europe as a counterweight to the United States in international affairs.

The relationship of the EU with Russia is another question for debate. Even in the Soviet era, Europe bought oil and gas from Russian oil fields. Fearing Soviet control of Europe's energy supply, President Ronald Reagan refused to allow U.S. corporations to help build a major pipeline from the east into Europe. Energy dependence on Russia has grown, raising fears that Russian rulers will use oil and gas as a control mechanism. In the winter of 2005/2006, Russian gas supplies to the Ukraine were interrupted, which had the effect of curtailing supplies to Italy.

▶ A MULTICULTURAL EU

As a result of economic growth, labor migration, and decolonization, peoples from other cultural realms have settled in Europe. Britain has communities drawn from the Indian subcontinent and the Afro Caribbean, Germany has Turks, France has many Muslims from former North African colonies. Increasingly, legal and illegal migrants find a way to Europe from West Africa. European law protects human rights but the absorption of people from other cultures is often slow and prone to tension as France discovered recently when the unemployed sons, and some daughters, of North African migrants disrupted life in many cities. Radical Islam has found a foothold in the United Kingdom, as illustrated by London bombings in 2005 and plots, in the summer of 2006, to blow up planes bound for the United States.

► THE EUROPEAN UNION AND NORTH AMERICA

Aspects of the European Union have North American roots. The German economist Friedrich List (1789–1846), after visiting the United States, saw the need to unify economic space within a single tariff system as described in his *National System of Political Economy* (1846).

More immediately we can say, without the intervention of Canada, other Commonwealth countries, and the United States, Europe would not have been liberated in 1945. Britain alone did not have the resources to invade fortress Europe. Without North American intervention, Europe would have been dominated by a totalitarian regime, Fascist or Communist, or uneasily divided between the two. Both the Fascists and the Communists wanted autarky, self-sufficient economic systems largely cut off from world trade.

The Marshall Plan and the creation of NATO were important steps promoting European unity. When the EEC was created in 1957, one of its aims was to encourage investment, by multinational corporations, within the shared economic space of the community. Much of the multinational investment came from North American corporations who built production facilities within the common external tariff and enjoyed access to the enlarged European market.

There have been conflicts between Europe and the United States. The U.S. has imposed, from time to time, protective tariffs in violation of WTO rules. The U.S. imposed tariffs on imported steel products, disadvantaging European exporters. For its part, Europe has kept out meat products, where hormones have been used to speed the growth of animals. The European Union resists the import of GM (genetically modified) crops. These trade disputes are usually settled by negotiation, although there are, on both sides of the Atlantic, protectionists who want trade wars.

There are European leaders who want to see the EU adopt common defense and foreign policies with the aim of making Europe a counterweight to the United States in international affairs. France is the leader of this movement. It is difficult to see how the present 25 members of EU could agree on a common stance in world affairs. Several members are neutral in military matters. Viewpoints on international relations contrast widely, with nations having differing histories and locations. The countries of eastern Europe are more concerned with Russian relations than the future of the Atlantic world.

In 1989, Canada and the United States entered into a free trade agreement. In 1992, with the addition of Mexico, NAFTA (North American Free Trade Agreement) was created. It is tempting to think that a free trade agreement could be made between the EU and NAFTA, but it is unlikely because of the difficulties of harmonizing the complex regulations of the parties. Even if agreement was reached, it would not apply to agriculture. North Atlantic free trade in agriculture would destroy the heavily subsidized Common Agricultural Policy and result in the reduction of North American farm subsidies. Too many vested interests on both sides of the Atlantic would resist.

The Development of the European Union, 1945–2007

1947	**The Benelux Customs Union**
	Belgium, The Netherlands, and Luxembourg
1951	**The Coal and Steel Community,** Treaty of Paris, France, Germany, Italy, Belgium,
	The Netherlands, and Luxembourg
1957	**European Economic Community (EEC),** Treaty of Rome, France, Germany, Italy,
	Belgium, The Netherlands, and Luxembourg
1973	Britain, Denmark, and Ireland join EEC
1980s	Greece (1981), Spain (1986), and Portugal (1986) join EEC
1993	**European Union (EU)**, Treaty of Maastricht
1995	Finland, Sweden, and Austria join EU
2004	Estonia, Latvia, Lithuania, Poland, Czech Republic, Slovakia, Hungary, Slovenia,
	Greek Cyprus, and Malta join EU
2007	Bulgaria and Romania join EU
2007	Turkey involved in talks to join EU

▶ ORGANIZATION OF THE TEXT

The following six systematic chapters introduce the physical, cultural, population, economic, settlement, and political geography of Europe. The subsequent regional chapters cover the countries of EU, largely in the order of joining. Some states rejected EU membership. They are covered along with neighboring EU countries. For example, Norway, which did not join EU, is described in the same chapter as Sweden, which did.

Countries waiting to join EU, including the new states emerging from defunct Yugoslavia and Turkey, are all covered. Former Soviet Socialist Republics—Russia, Belarus, and the Ukraine—are unlikely to join EU soon. However, economic stability in Russia is important to the European community because the region needs access to Russian resources.

The book closes with a chapter on Europe and the wider world. A glossary of terms used in the text is also provided.

▶ FURTHER READING

Bainbridge, T. *The Penguin Companion to European Union*. 3rd edition. New York: Penguin Putnam, 2002.

Berentsen, W.H. *Contemporary Europe: A Geographic Analysis*. 7th Edition. New York: Wiley, 1997.

de Blij, H. *Why Geography Matters*. New York: Oxford University Press, 2005. (see Chapter 10, "European Superpower.")

Blouet B.W. *Geopolitics and Globalization in the Twentieth Century*. London: Reaktion Books, 2001.

Coudenhove-Kalergi, R.N. *Pan Europe*. New York: A.A. Knopf, 1926.

Dinan, D. *Even Closer Union: An Introduction to European Integration*. 3rd Edition. Boulder, CO: Lynne Rienner, 2005.

Heffernan, M. *The Meaning of Europe: Geography and Geopolitics*. London: Arnold, 1998.

Herriot, E. *The United States of Europe*. London: G.G. Harrap, 1930. (Translated from French)

Leonard, D. *Guide to the European Union*. 9th Edition. London: Profile Books, 2005.

Macmillan, M. *Paris 1919: Six Months that Changed the World*. New York: Random House, 2001.

Mowat, R.C. *Creating the European Community*. New York: Harper Row, 1973.

Ostergren, R.C., and J.G. Rice. *The Europeans: A Geography of People, Culture, and Environment*. New York: Guilford Press, 2004.

Spaak, P. *The Continuing Battle: Memoirs of a European*. Boston: Little, Brown, 1971.

Western, J. Neighbors or Strangers? Binational and transnational identies in Strasbourg. *Annals of the Association of American Geographers*, March 2007, 97(1): 158–181.

PART

I

SYSTEMATIC SURVEY

1

PHYSICAL ENVIRONMENTS

Physically Europe is not separated from Asia. Traditionally the physical geography boundary between Europe and Asia has been the Ural Mountains. The Ural physical boundary is to the east of many recognized boundaries of European cultural traits.

In the north Europe is defined by Arctic seas, to the east by the Urals, in the South by the Mediterranean, and to the west it is open to the Atlantic Ocean. The Mediterranean is connected to the Atlantic by the Strait of Gibraltar.

► CLIMATE AND WEATHER

There are four major influences on the climate and weather of Europe: the Arctic Seas to the north, the landmass of Eurasia to the east, the Sahara desert and Mediterranean Sea to the south and the Atlantic to the west. The zones are source regions for the air masses that produce Europe's climate and weather (Figure 1.1). (Photo 1.1)

Maritime Polar (mP) air acquires characteristics over Arctic Seas. This cool, moist air arrives over Western and Central Europe from the north and northwest. The air is unstable, and upon reaching land and being pushed upward quickly, produces rain showers, hail, or flurries of snow depending on the season. Maritime Polar air is most frequently encountered in the winter months when the polar front (the leading edge of the cool Arctic air) is well to the south; in a wet, cool summer, mP air can penetrate northwest Europe and drop temperatures into the 50s.

Continental Polar (cP) air forms in a zone of high pressure over Siberia in the long, continental winter. The air has similar characteristics to the pool of cold Canadian air that forms over North America in winter, but the Siberian air is colder. Much of Siberia is frozen for more than half the year and in European Russia, west of the Urals, rivers, land, and lake surfaces are frozen for three to four months. There are plenty of cold surfaces over which cP air can form. In the winter months the high pressure cP Siberian air flows outward, bringing bitterly cold air to Manchuria and Japan in the east, and below freezing weather to central Europe. Every few years the Siberian air extends westward across The

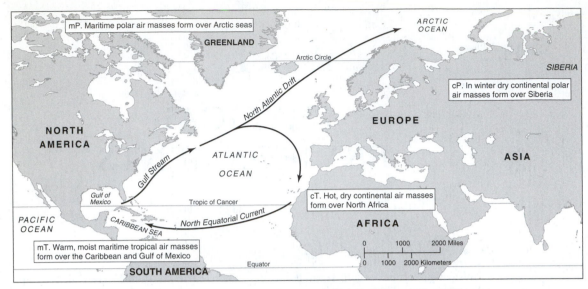

Figure 1.1 Air masses and ocean currents influencing the weather of Europe.

Photo 1.1 Artist, Constable, John (1776–1837) Cumulonimbos clouds are forming in maritime tropical air over Highgate Hill, north London. Later, in the afternoon, there was rain.

Source: View from Highgate Hill (oil on canvas) by Constable, John (1776–1837) ©Private Collection/©Ackerman and Johnson Ltd, London.

Netherlands, Belgium, northern France, and the North Sea to Britain. Abnormally cold conditions prevail and patterns of life are disrupted by the extreme cold, with daytime temperatures not rising much above freezing.

Continental Tropical (cT) air has its source over the Sahara desert. The air mass is a strong influence in the Mediterranean and southern Europe where cT air contributes to the long, hot, dry summers.

The westerly winds are the prime weather maker for western Europe, with the oceanic influence lessening as we go eastward into continental Europe. Maritime Tropical (mT) air acquires its characteristic warmth and moisture over the Gulf of Mexico. Maritime Tropical air travels across the Atlantic in a northeasterly direction, along with the Gulf Stream, to bring mild, moist conditions to western Europe, and, to a lesser degree, central Europe. Of course, the mT air and the sea currents cool as they travel on a northeasterly path to Europe but are still relatively warm on arrival in western Europe, where ports are ice free in winter. Follow the line of latitude of Bergen, in Norway, back to North America and you are in Labrador, Canada, with a long, icy winter. The extensive Norwegian coast is ice free during winter, while Greenland and Labrador, at similar latitudes, have frozen coasts.

North Atlantic weather is seasonal. In winter, along the boundary between the mT air and the mP air, depressions form and track westward, bringing rain and in a cold winter, snow, to Europe. In the summer months the leading edge of the mP air retreats to the north and the Azores' high pressure cell establishes sunny, stable, warm conditions over western Europe. Depressions still form but as the mP air has retreated toward the pole, the low pressure systems follow a more northerly track.

That is how it is on average. In a cool summer the Azores' high may not become well established and depressions still pass over western Europe, bringing clouds and rain. For example, in late June 2004, a depression tracked southwest to northeast across England accompanied by high winds and heavy rain. Temperatures did not exceed 60°F. There was no play at Wimbledon, winds gusted to 65 mph in the English Channel, and in Portugal, well to the south of Britain, at the 2004 Euro soccer competition, matches were played in heavy rain. The weather did not improve. The jet stream was well south of its normal summer path and continued to steer depressions, with cool, cloudy, rainy weather into western Europe (Figure 1.2). However, the summer of 2004 did not break records for coolness and the low temperatures were not near the worst on record. In 1845 cold, wet weather dominated from July to September. The grain harvest failed in western Europe; the Irish potato blight took hold, with catastrophic results.

In contrast to the cool, wet 2004 summer the preceding summer of 2003 was hot and dry. The grass in London's green parks went gray. In France, the death rate surged as thousands died in the heat. In the French wine-producing region of Burgundy, grapes are usually picked in September. In the summer of 2003, by mid-August, Pinot Noir grapes for red Burgundy were ripening to raisins. The Chardonnay grapes, for white burgundy, were shriveling. The grape harvest was early.

If the summer of 2004 was wet in western Europe, the fall was dry. In the spring of 2005 rainfall was below average. By July 2005 water use restrictions were imposed in many parts of France. Then in a southern Europe heat wave, Portugal, Spain, and southern France experienced exceptionally high temperatures. Cattle died in Portugal's drought. In Spain fires erupted around Guadalajara, in the center of the country. Dry

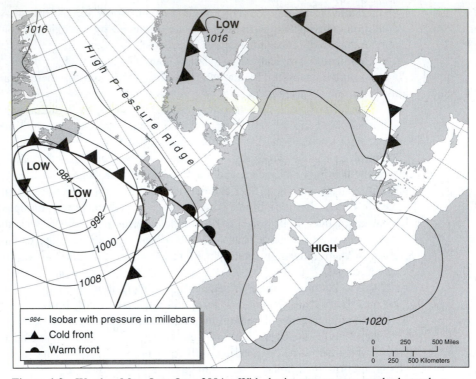

Figure 1.2 Weather Map, Late June 2004—With the jet stream on a southerly track, a depression approaches western Europe.

grassland and scrub ignited in the summer heat and fire fighters were killed trying to contain the fire. Crops failed, livestock died, and in Portugal, tourist resorts were supplied with water by tanker.

While fires burned in Iberia, heavy rain in Germany, Austria, Croatia, Bulgaria, and Romania caused flooding. Commentators thought the jet stream was following an unusual path. To underline weather variability, at the end of July 2005, rain came to Spain. There were flash floods in southwest England and at Birmingham a tornado tore roofs from houses. By early August the Azores' high had strengthened, the highs in Iberia were around 100°F, and fires were burning again in Portugal and southern France.

Clearly the climate of Europe is variable from year to year. June 2004 was cool and wet. In June 2006 the Azores' high pressure cell was strongly established and weather was settled and temperatures high (Figure 1.3). Britain had the hottest June for thirty years and soccer players at the World Cup in Germany experienced heat exhaustion. However, at Wimbledon there was little play on days one and two (June 25 and 26) as the westerly winds briefly penetrated the British Isles, bringing clouds and rain. In Germany, the World Cup weather remained hot and fine, but to the south, in the Black Forest, on June 27, massive thunderstorms dropped tennis ball–size hail. Snow ploughs cleared the roads.

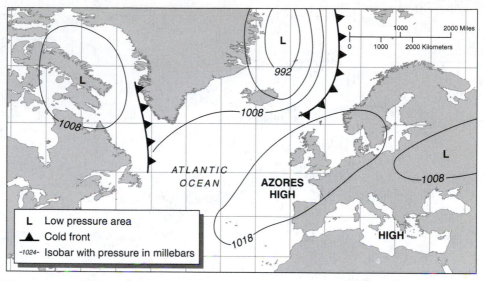

Figure 1.3 Weather Map, High Pressure Summer 2006—The Azores high is well established and high pressure dominates western Europe.

General Climate Considerations

When we draw maps of climate regions (Figure 1.4) we use data averaging conditions over decades. The averages smooth away unusual events. The climate of Europe, particularly in the west and the northwest, is highly variable, severe storms appear in summer months, hurricane-force winds occur in winter, and unexpectedly heavy snowfall can disrupt road and rail networks.

Western Europe has a maritime climate. Precipitation is plentiful with rain in all months, though on average, more in winter than in summer (Figure 1.5). Temperatures, in both winter and summer, are moderated by the surrounding seas. In the west average January temperatures are above freezing. East of Germany average January temperatures are below freezing.

Central and eastern Europe experience continental influences. Winter temperatures are cold. In summer the land mass warms up and so does the air above. Warm, expanding air is less dense and as the warm air rises, it cools; clouds form and thunderstorms result. At Moscow there is more precipitation in summer than in winter. The mean monthly Moscow temperature for January is below freezing and in July it is 68°F (Figure 1.6).

Maritime influences and precipitation diminish as we go eastward into the continent. The oceans are the major source of moisture in the atmosphere. In Western Europe droughts are rare but do occur. In the Ukraine rainfall is adequate to support grain farming in an average year. Below-average precipitation will result in a poor harvest. Going further east, to the semi-arid lands at the north of the Caspian Sea, farming, other than wheat, requires irrigation.

Southern Europe—the Iberian peninsula, southern France, Italy, and Greece—has a wet, mild winter, although it can be cold in upland areas such as the Abruzzi uplands of

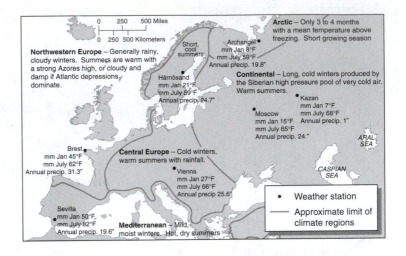

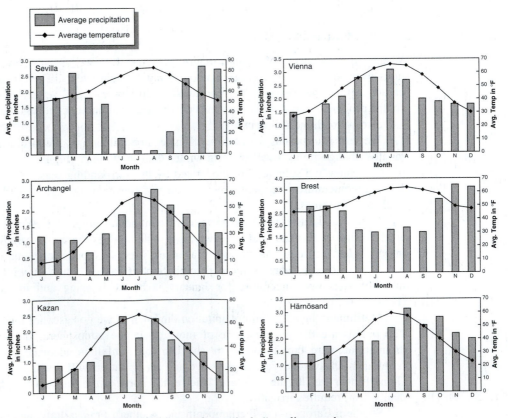

Figure 1.4 Generalized climate regions—including climographs.

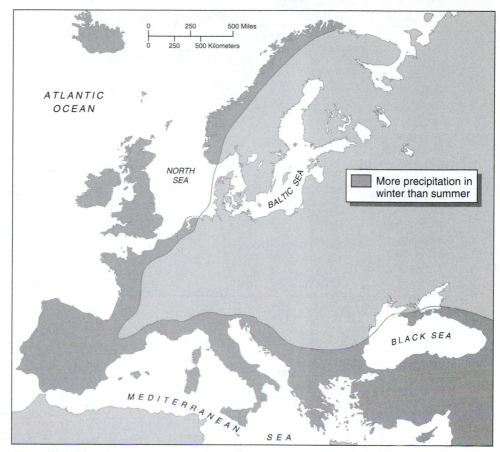

Figure 1.5 European areas with more precipitation in winter than in summer.

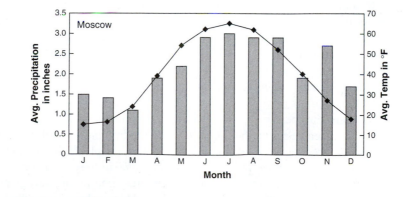

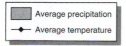

Figure 1.6 Climograph, Moscow.

TABLE 1.1   Climate Data for Selected Cities

	MM Jan temp. °F	MM July temp. °F	Precipitation inches
Reykjavik (Iceland)	34	52	31
Glasgow (UK)	39	59	38
Dublin (Ireland)	40	59	30
London (UK)	39	64	25
Stockholm (Sweden)	27	62	22
St. Petersburg (Russia)	17	64	19
Paris (France)	37	64	23
Berlin (Germany)	31	66	23
Warsaw (Poland)	24	65	22
Moscow (Russia)	15	65	25
Lisbon (Portugal)	51	71	27
Madrid (Spain)	41	77	17
Naples (Italy)	50	77	35
Milan (Italy)	36	75	38
Athens (Greece)	48	81	16
Istanbul (Turkey)	42	74	27

MM, Mean monthly temperature °F

central Italy and the mountains of Greece. Summer in southern Europe is generally hot and dry, with southern Spain, parts of Italy, and Greece having a July average temperature of over 75°F (Table 1.1).

In Table 1.1, notice how mild winters are in Reykjavik, Glasgow, and Dublin. Moving east from Paris through Berlin, Warsaw, and Moscow, winters get colder and summers slightly warmer, reflecting continental influences. Winter cooling is also apparent going from Stockholm to Petersburg.

In southern Europe, traveling from west to east, Lisbon and Naples have mild January mean temperatures, around 50°F. Madrid is much colder in winter, reflecting the continental influence of the Iberian peninsula. Northern Italy also displays continentality. Milan has a January mean of 36°F, lower than Glasgow much further north but in the westerly air stream for much of the year. Athens is hot and dry in the summer. Istanbul is cooler than Athens in both winter and summer, reflecting the influence of the Black Sea.

Southern Europe can experience a southerly summer wind from the Sahara. Crossing the Mediterranean Sea, the hot air picks up humidity and by the time the wind reaches southern Europe it can be so hot and humid that people feel lethargic and sometimes nauseous. In southern Spain the wind is the *Leveche,* in Italy the *Sirocco,* in Malta *Xlokk.*

As we go up in the atmosphere, temperatures drop at an average rate of 3.7°F for every thousand feet ascended. This loss of temperature has a big effect in regions like the Highlands of Scotland and the mountains of Norway, where growing seasons are shortened. The Alps and the Pyrenees have mountain climates. In Switzerland mountain sides are often covered with conifers but higher up the growing season is too short for trees and woodland gives way to alpine grassland. The grasses grow rapidly in the

Photo 1.2   The Dolomites, Italian Alps, Italy. In mountain environments trees give way to grassland with altitude. The grassland is used as summer pasture. Note chalets for summer habitation.

Source: Creatas/SUPERSTOCK.

summer and traditionally cattle are driven up to graze. In this example of transhumance, part of the farming family moves uphill with the livestock, lives in a chalet, and makes butter and cheese. (Photo 1.2)

In northern Europe days are long in the summer and short in the winter. The long hours of summer daylight allow crops to grow rapidly. In the Arctic circle the sun barely sets in June and cruise boats nudge up through ice flows to allow tourists a view of the midnight sun. At midwinter there is a twilight for a few hours. (Photo 1.3)

► NATURAL VEGETATION

In its natural state, before the arrival of human activities, the vegetation of Europe was predominately woodland. In the far north trees give way to tundra. The Danubian plains and the Steppes of the Ukraine were natural grasslands. The sequence from northern Norway to the Mediterranean Sea via the Danubian plains was tundra, coniferous forest, mixed deciduous and coniferous forest, broadleaf deciduous forest, grassland, and in the Mediterranean region, evergreen broadleaf and deciduous woodland (Figure 1.7). (Photo 1.4)

The tundra, with a growing season lasting from late May to early August, cannot support tree growth. Tundra vegetation consists of grasses, sedges, mosses, lichens, and

Photo 1.3 Baltic Sea, Ice Flows. In the north, the Baltic freezes and closes to shipping in winter.

Source: Kai Honkanen/AgeFotostock.

a few tough, low-growing scrubs. Tundra thaws in the short summer to become a water-logged breeding ground for clouds of midges and mosquitoes. The coniferous forests that stretch from Scandinavia to Siberia still exist. Most of the forests have been repeatedly cut over to provide timber, timber products, and wood pulp, as is the case in Norway, Sweden, Finland, and Russia. The trees are replanted or regenerate naturally and are regularly harvested.

Landscapes in western and central Europe look well wooded on the basis of trees lining field boundaries and roads. Woodland is often maintained on poorer soils and on estates as landscape features and cover for wildlife that may be hunted or shot for sport. There are extensive remnants of famous forests like Sherwood Forest and the Black Forest, but most woodland has been cleared for agriculture and what remains has been modified by human action. The natural grasslands of Danubia and the Ukraine have been plowed into farmland.

In the Mediterranean, woodland was cleared in antiquity or before. An ancient Greek commentator noted that there were hillsides in Attica that supported nothing but bees. The woodland has been cut, the thin limestone soils eroded exposing the underlying rock. On the rocky surfaces the characteristic vegetation of Mediterranean limestone lands appeared—low-growing scrubs, succulents, wild flowers, and herbs. This vegetation type is known as *matorral* in Spain, *macchia* in Italy, and *maquis* in France. (Photo 1.5)

As we go eastward into Europe a shorter growing season and decreasing precipitation impact vegetation patterns. The coniferous forest-belt expands, and deciduous forest decreases. From the Hungarian plain into the Ukraine and Russia grassland becomes the dominant natural vegetation type.

The rich, black earth soil (*chernozem*) of the Steppes were ploughed up in the nineteenth century to grow wheat that was exported through Black Sea ports like Odesa. Further east, in the semi-arid lands around the Caspian and Aral Seas, irrigated cotton acreage

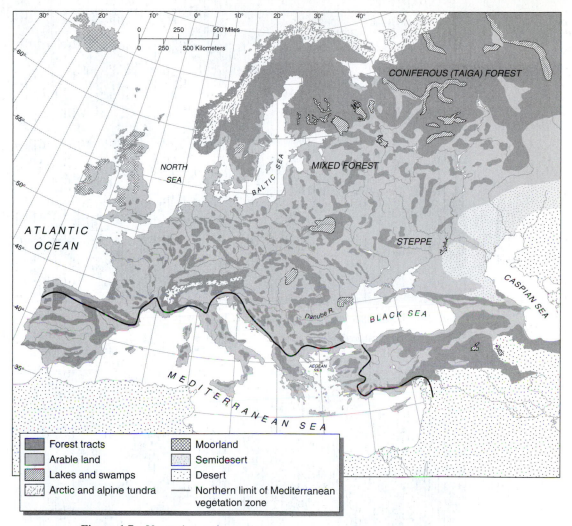

Figure 1.7 Vegetation regions.

was greatly increased in the twentieth century. Diverting water from rivers feeding the Aral Sea, into irrigation canals, led to the shrinking of the sea and the creation of desert.

► TOPOGRAPHY: MOUNTAINS, PLATEAUX, RIVERS, PLAINS, AND COASTS

Caledonian Mountains

The oldest mountains in Europe, dating back at least to the Cambrian, are the Caledonian mountains, stretching from Scandinavia to the highlands of Scotland and the Welsh upland (Figure 1.8). The mountains are the stumps of higher peaks that were eroded away

Photo 1.4 Coniferous Forest on the Atna river in Norway.

Source: Blickwinkle/Alamy Images.

Photo 1.5 Mediterranean scrub at Es Port, castle. Cabrera National park. Balearic Islands, Illa de Cabrera, Spain.

Source: Age fotostock/SUPERSTOCK.

Figure 1.8 Physiographic features.

long ago. The surviving surfaces are blocks of igneous and metamorphic rocks that formed batholiths (deep-seated igneous intrusions) at the core of much higher mountains. The highest points in Scotland (Ben Nevis, 4,408 feet), Wales (Snowdon, 3,560 feet), and Norway (Galdhøpiggen, 8,097 feet) are impressive but the great peaks capped with sedimentary rocks have gone. There is much bare granitic rock in Norway, Scotland, and Wales and the thin soils in the damp upland climate are acidic. The vegetation consists of grasses, sedges, mosses, lichens, and a few stunted shrubs on steep hillsides. Conifers grow at lower levels in the uplands. (Photo 1.6)

Hercynian Mountains

The uplands of Hercynian age, equivalent to the Appalachians of North America, are indicated in Figure 1.8. Moving from west to east, the uplands of Hercynian age are

Photo 1.6 The lake of Tyin looking to Jotunheimen mountains, Norway. Notice the building materials.

Source: Staffan Anersson/AgeFotostock.

the Spanish Meseta, the Brittany (Breton) peninsula, the Massif Central of France, the Ardennes, the Vosges Mountains to the west of the Rhine, the Black Forest, the Hartz Mountains, Bohemia, and the Urals.

Mountains of Alpine Age

In a roughly west-to-east direction run chains of mountains of Alpine age, displaying high peaks, deep-cut valleys, steep slopes, glaciers, and snowfields. (Photo 1.7) The Alpine system starts in the west with the Cantabrians, followed by the Pyrenees, along which runs the border between France and Spain. In the south of the Iberian peninsula, still running roughly west to east, is the Sierra Nevada of Andulucía. The highest peak in the peninsula is found in the Sierra Nevada (11,411 feet). The Pyrenees rise to 11,168 feet at Pico de Aneto.

East of the Pyrenees, beyond the Rhône valley, is the Jura and the Alps from which the most recent age of mountain building takes its name. In the Americas the Andes and the Rockies are products of the Alpine age of mountain building. From the Alps, in Switzerland and Austria, branches of the mountain system run south, forming the Apennines of Italy, and the Dinaric Alps that run through the Balkans, parallel to the Adriatic Sea, to Greece. To the east of the Danube, Alpine-age mountains reappear in the Carpathians, which run through southeastern Europe to the Black Sea. The west-to-east orientation of the mountains of Alpine age results in much cold air being shut out of the Mediterranean basin. But notice the gaps through the mountains, such as the Rhône valley. In winter, if there is low pressure over the western Mediterranean, cold northern winds can blow south through the gaps to the sea.

Photo 1.7 Jungfrau Mountain, Switzerland. Notice the rail line serving the settlement.

Source: Yoshio Tomll/SUPERSTOCK.

Glaciation

Europe has glaciers in Scandinavia, the Alps, and the Carpathians (Figure 1.9). In the last Ice Age glaciers advanced down mountain valleys and a great ice sheet spread out from Scandinavia to the North Sea, the Baltic, and the European plain from the Jutland peninsula to Russia. (Photo 1.8)

When the glaciers retreated they left behind extensive deposits of glacial drift, sometimes called boulder clay. The drift was a mixture of clays and bedrock scoured up by the glaciers as they moved forward. As a result, the north European plain has heavy clay soils. The rock in the clay could be used for building.

Out of the snouts of glaciers came streams of meltwater carrying silt, sands, and gravels. The west side of the Jutland peninsula, lying on the margin of the Scandinavian glacier, is composed of sands and gravels. To the east the peninsula is composed of glacial clays. South of the zone of glaciation is an area in which windborne silt (loess) was deposited on the north European plain, the Danubian lands, and the steppes of the Ukraine.

As glaciers retreated lakes were left in valleys that had been deepened by glacial scouring. Southern Finland is a land of lakes left among uneven glacial deposits. The Lake District of England has a group of lakes, similar to the Finger Lakes of upper New York State, which fill the floors of valleys deepened by glaciers. In some areas, after the glacial retreat, drainage was never integrated. The Pripet marshes of eastern Europe are an example of a large wetland created at the end of the glacial period. In the Soviet era part of the Pripet region was artificially drained to make farmland.

In the glacial retreat, rivers worked to reestablish courses to the sea over terrain that had been blocked by ice and glacial deposits. Often huge lakes formed at the edge of

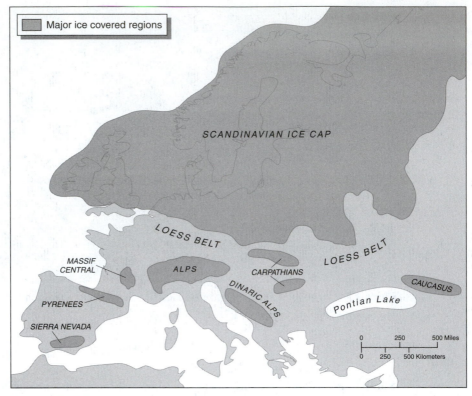

Figure 1.9 Maximum extent of glaciation.

glaciers until water spilled over a moraine or hilly ground to cut a passage to the sea. The rivers crossing the north European plain, like the Elbe and the Vistula, display diversions and meanders that emerged as the ice retreated.

As the ice age came to an end, water that had been held in the ice caps and valley glaciers was released. Sea levels rose, drowning the mouths of many rivers and forming deep estuaries. Along the coasts of Norway and Scotland the sea invaded valleys formerly filled with ice, creating fjords and firths. These inlets of the sea have been utilized as at Glasgow on the Firth of Clyde and in the fjords that house the sheltered cities of Bergen, Oslo, and Trondheim.

The glacial retreat was not permanent; at times glaciers readvanced. In the seventeenth-century Little Ice Age, Alpine glaciers advanced down the valleys. In the cooler and wetter climate agriculture retreated from higher ground and crop yields declined. Over the last century, climate has become warmer, glaciers have retreated, and the area covered by Arctic Sea ice has shrunk.

Oceans, Seas, Estuaries, and Rivers

The relative mildness of the climates of western Europe (Venice, Genoa, and Bordeaux, roughly the same latitude as Minneapolis) is mainly due to the westerly winds and

Photo 1.8   Switzerland. Swiss valley with active glacier.

Source: Prisma/SUPERSTOCK.

the North Atlantic drift, with the former being the most important (Seager, 2006). The northeast-flowing Gulf Stream has cooled by the time it reaches North Cape, well inside the Arctic Circle in Norway, but the water is still warm enough to keep the Norwegian ports ice free in winter.

Over the North Atlantic in winter, deep depressions can form, which generate gale-force winds. Deep depressions accompanied by high winds are a threat to shipping and coastal communities. In early 1953 a depression in the North Sea accompanied by high winds flooded the fenland of eastern England, breached the Dutch dykes, and caused widespread flooding in Belgium, The Netherlands, southern Denmark, and adjoining areas of Germany.

The sea, particularly in winter, can be a threat. Fortunately, the coasts of Europe form an intricate pattern of gulfs, bays, fjords, inlets, and estuaries that offer shelter from the sea and allow waterborne coastal communication. Many rivers with extensive estuaries allow inland navigation. The Elbe, which flows to the North Sea at the base of the Jutland peninsula, is navigable inland via the ports of Hamburg, Magdeburg, and Dresden into the Czech lands.

Western and central Europe possess many rivers that arise deep in the landmass and are navigable far upstream. The Rhine is navigable through The Netherlands, Germany, and into Switzerland, above the port of Basle. The Danube is navigable from Ulm and Regensburg in Germany through Austria, Hungary, Serbia, Bulgaria, and Romania, to the Black Sea. Many of the major cities of western and central Europe are on navigable water. Examples include London on the Thames, Paris on the Seine, Koblenz and Cologne on the Rhine, together with Vienna, Bratislava, Budapest, and Belgrade on the Danube. (Photo 1.9) (Photo 1.10)

In the Mediterranean region few rivers are navigable far inland, although human made improvements to the channels have allowed vessels to use the Rhône. The Po, in northern Italy, is little used for navigation, and the Arno, which in the Middle Ages allowed vessels to sail upstream from the Mediterranean to Pisa and Florence, became silted and less useful.

The Mediterranean Sea, while mild and friendly to sailors in the summer, was notoriously dangerous to oar powered vessels from classical times until sails replaced oars. Even in the present the winter north winds—the Mistral, the Tramontana, and the Bora—can disrupt shipping in the Gulf of Lion, the Tyrrhenian Sea, and the Adriatic.

Plains and Coasts

In southern Europe and the Mediterranean lands coastal plains are not extensive and areas of cultivatable lands are limited. On the Caledonian mountain coasts of Wales, Scotland, and Scandinavia rugged headlands, separated by deep firths and fjords, are characteristic. Again, agricultural land is limited to narrow strips, for example, at the head of Norwegian fjords.

Photo 1.9 The confluence of the Rhine and Mosel rivers, Koblenz. Germany.

Source: Age fotostock/SUPERSTOCK.

Photo 1.10 The Danube at Regensburg, Germany. The relatively low arches on the stone bridge indicate that only smaller vessels procede further upstream.

Source: Werner Otto/AgeFotostock.

The major plains in Europe include:

The European Plain

An extensive belt of country runs from the basin of Aquitaine, in southwest France, through the Paris basin, across the glaciated north European plain including Germany, Poland, Belarus, and the Baltics, to Russia. The European plain is not homogenous, as many rivers cross it creating flood plains, river terraces, and interfluves. In the Paris basin escarpments break the country into rolling landscapes with limestone outcrops. As the plain extends from France to Russia there are variations in soils, drainage, climate, and vegetation but the plain and the rivers that cross it have facilitated the movement of people, goods, armies, and ideas.

The Danubian Plain

The plain begins just below Vienna where the Danube flows out from between the Alps and the mountains of Bohemia onto a lowland that extends from eastern Austria, through southern Slovakia, Hungary, and Serbia, to Bulgaria, Romania, and the Black Sea. The plain is composed of riverborne deposits laid down by the Danube and its tributaries over which were laid deposits of windblown loess at the end of the glacial period. The plain is a natural grassland, converted to a rich agricultural region by human hand.

The Danube river, and the surrounding plains, have been an important routeway in European prehistory and history. The routeway does not stop where the plain ends, for the Danube valley penetrates into southern Germany and gives access to north-running lines of communication.

The Steppes

The Steppes extend across the Ukraine and Southern Russia into Kazakhstan, becoming drier toward the east. The natural vegetation is grassland and there are few trees except along water courses. In the Ukraine rainfall is sufficient to allow cultivation without irrigation. To the east it becomes necessary to irrigate farmland to obtain a crop.

The richness of agriculture in the Ukraine reflects the widespread distribution of *Chernozem* soils, the "black earths" of the Steppe, which had their origin as deposits of loess in the glacial period.

▶ HUMAN IMPACT ON THE ENVIRONMENT

The industrial revolution resulted in environmental degradation. Spoil heaps from mines, the waste from metal smelting, and waste from textile and chemical industries polluted soil and water and many rivers ceased to support fish. Atmospheric pollution, the result of burning coal for domestic heat and industrial power, released soot, particulate matter, and sulphur dioxide into the atmosphere. The sulphur dioxide became a weak sulphuric acid that returned to earth as acid rain. Pollution from the large industrial areas of Britain, Germany, France, and Belgium blew to Scandinavia where it damaged coniferous forests and lakes. However, coal burning in power stations decreased with the discovery of the Groningen gas field in The Netherlands and North Sea oil and gas. In recent decades there has been a marked decline in coal usage. In addition, countries introduced regulations to prevent the burning of solid fuels for domestic heating. The European Union (EU) has tightened regulations on the burning of coal in power stations and further reduced pollution.

The impact of European community policies on the environment has been mixed. In the last decades of the twentieth century the Common Agricultural Policy encouraged large-scale agriculture, which involved increased mechanization and high inputs of fertilizers and pesticides. Field sizes increased with the loss of hedgerows and other wildlife habitats. It is now recognized that the Community overproduces agricultural products, and policies are moving toward the protection of rural environments. Under the Community Fisheries Policy the fishing resources of the region were pooled and quotas allocated to the fleets of member countries. Policies encouraged the use of larger, industrial-type fishing vessels that depleted stocks of fish, such as cod. Now the EU has cut the quotas for some species of fish and traditional fishing communities are angry. Their quotas have been cut too, but traditional fishermen feel the big, modern vessels overfished and depleted stocks.

Environments have been altered by activities that are not controlled by EU policies. For decades there has been a boom of tourism in southern Europe. Large sections of the coasts of Portugal, Spain, the south of France, and particularly Italy have been built over with hotels, villas, apartments, and resorts. Pollution of the Mediterranean Sea has increased. Labor has left the land in southern Europe; marginal land has gone out of cultivation and become scrubland. Such vegetation is a fire hazard, particularly in the hot, dry southern summer.

EU and the Environment

The Treaty of Rome of 1957 did not directly address environmental issues. The Single European Act (1985) did propose to preserve and protect the environment and promote the prudent use of natural resources. In the 1990s the European Environment Agency was established in Copenhagen. Most EU members have signed on to international environmental conventions including the Kyoto Treaty, which aims to reduce emissions of greenhouse gases. European countries have accepted targets to lower the output of the gases and are actively reducing emissions.

The countries that were most active, initially, in promoting a green EU agenda were Germany, The Netherlands, and Denmark. The Netherlands has immediate contact with the environment, with much of the country below sea level and vulnerable, should global warming produce a sea-level rise. Germany became worried about atmospheric pollution when, a few decades ago, coniferous forests began to die as a result of acid rain and automobile exhaust emissions. Denmark won a landmark case at the European Court of Justice defending the right of the country to insist that beer and soft drink containers be reusable. Some exporters of drinks to Denmark argued that the rule interfered with EU free trade. The court upheld the right of countries to protect the environment.

When Sweden and Finland joined EU in 1995, they pressed for more environmental protection measures. Both countries had good environmental records but suffered from exported pollution produced by heavy industry in Britain, Germany, and the former Soviet Union.

Non-EU members can join the European Environment Agency, and Iceland, Norway, and Liechtenstein have done so. There is now a cluster of countries in the Baltic region and northwest Europe that are particularly concerned with environmental issues, including Iceland, Norway, Sweden, Denmark, Finland, Germany, and The Netherlands. The countries of southern Europe tend to resist policies that protect the environment. Much of southern Europe has a tradition of Roman law in which property is seen as the owner's, to use and abuse as they wish. Regulation is often resisted or ignored.

▶ FURTHER READING

An excellent supplement to this chapter is an atlas such as *Goode's World Atlas.* Chicago: Rand McNally, 2005, or *College World Atlas,* Washington D.C.: National Geographic Society, 2007. A comprehensive atlas will contain maps of landforms, climate, soils, vegetation, and land use. An atlas is invaluable in following places, regions, and features named in the text.

Brown, N. *History and Climate Change: A Eurocentric Perspective.* New York: Routledge, 2001.

Grove, A.T., and O. Rackham. *The Nature of Mediterranean Europe: An Ecological History.* New Haven and London: Yale University Press, 2001.

Gupta, J., and M. Grubb. *Climate Change and European Leadership: A Sustainable Role for Europe.* Boston: Kluwer Academic, 2000.

Jones, P.D, A.E.J. Ogilivie, T.D. Davies, and K.R. Briffa (eds.). *History and Climate: Memories of the Future.* New York: Kluwer Academic/Plenum, 2001.

Jordan, A. *Environmental Policy in the European Union.* Sterling, VA: Earthscan, 2005.

King, R., L. Proudfoot, and B. Smith (eds). *The Mediterranean: Environment and Society.* London: Arnold, 1997.

Koster, E.A. (ed). *The Physical Geography of Western Europe.* New York: Oxford University Press, 2005.

Seager, R. "The Source of Europe's Mild Climate." *American Scientist* 94 (July–August 2006): 334–341.

Thomas, J.B., and J. Wainwright. *Environmental Issues in the Mediterranean.* New York: Routledge, 2002.

2

CULTURAL GEOGRAPHY

▶ INTRODUCTION

Europe remains a mosaic of cultures, nationalities, and regional identities even as it integrates aspects of economic and political life. Like any major culture realm Europe has a large number of cultural traits derived from numerous sources. When we begin to analyze origins we quickly enter debates concerning the diffusion of cultural traits versus the independent origin of cultural traits. We can all agree that Europe's major religions originated in the Middle East and diffused into the region, and we understand that the Indo-European language family developed in the region adjoining the Black Sea. Other traits are more difficult to establish in terms of origin. The Greeks had a system of democratic government, but it is difficult to link that system directly with modern European democracies. Between the collapse of classical civilizations and the emergence of the modern world, there were long periods in which government was disorganized or control was exercised by tyrants. When the barons sat King John down at Runnymede, beside the Thames in 1215, and demanded basic rights like habeas corpus, in Magna Carta, they were not necessarily calling for something new but the restitution of rights that had previously existed. The origin of those rights may be found in the organization of Anglo-Saxon tribes that invaded Britain after the collapse of the Roman Empire. The evolution of a universal franchise, something not present in Ancient Greece, is a product of the last two hundred years. We look to the American Revolution and the French Revolution for the development of modern democratic institutions.

Here we examine the development of two major cultural traits related to language and religion.

▶ LANGUAGES

With a few exceptions, including the Basques, Finns, Estonians, and Hungarians, the peoples of Europe speak a tongue in the Indo-European language family. The main groups within the family are the Celtic, Germanic, Slavic, and Romance languages. Greek, Armenian, and Albanian are also Indo-European.

The proto Indo-European language probably developed in a region adjoining the Black Sea. As populations grew and occupied new lands on agricultural frontiers, culture groups became detached; new words, phrases, and pronunciations emerged and separate languages evolved. The process was not simply differentiation by isolation, as migrations and conquests influenced language distribution.

Celtic Languages

The Celts were formerly widely distributed in the British Isles, France, Central Europe, northern Italy, parts of the Iberian peninsula, southeastern Europe, and Turkey. With the rise of the Roman Empire many of these regions were conquered by Rome, and with the fall of the empire, Germanic tribes and their languages came to dominate central Europe and lowland Britain. The Celts were confined to the peninsulas of Brittany, Cornwall, and Wales, the highlands and islands of Scotland, Ireland, and the Isle of Man. In Brittany, Breton is spoken, Welsh is the first language in central Wales, and Gaelic is spoken in western Ireland and the highlands and islands of Scotland. The last men and women who spoke Cornish, as a mother tongue, died early in the nineteenth century. The language is preserved as a trait of Cornish identity—*kernow bys vykken* (Cornwall forever!) On the Isle of Man, Celtic Manx was widely spoken but few speak the language today.

Place names used in Celtic times remain widespread. In England river names like Ouse, Avon, Esk, and Thames have been inherited from Celtic speakers, pushed out in Roman times and the later invasions of Anglo-Saxons speaking Germanic languages. Julius Caesar conquered Celtic-speaking Gaul (which included the territory of France, Belgium, and Luxembourg) between 58 and 51 B.C. The Gauls eventually adopted Latin, out of which the Romance language of French evolved. Some Celtic words for common everyday articles were incorporated into French, and the names of the tribes of Gaul are retained in the names of French towns and cities. Paris records the name of the Parisii tribe that controlled the area before the Romans. Rheims memorializes the Remi, Amiens the Ambiani, Nantes the Namnetes, Tours the Turnones, and Poitiers the Pictavi tribe (Elcock, 1975, p. 207).

The Celtic word *dun* means a fortified settlement. There are numerous examples of the place name element in Ireland and Scotland. Dun is still a part of place names in northern Spain, France, and elsewhere, a pertinent example being the French fortress town of Verdun.

The Celts possessed metallurgy, making adornments, coins, weapons, and armor with iron by the time of the Roman contact. The Celts were warlike. Celtic subjugation by Rome was the result of Roman military organization, not inferior weaponry or will on the battlefield. Before the Celts, Europe was occupied by people with stone age cultures—Paleolithic, Mesolithic, and Neolithic. Neolithic farmers were tool makers using bone, wood, leather, and stone. There was trade in materials like flint and obsidian. Neolithic farmers cleared woodland by girdling trees. Using rollers, levers, and rafts, rock was quarried and moved distances to construct large stone (megalithic) monuments, of which Stonehenge is the best known. Megalithic monuments and burial mounds are common in the landscapes of Europe. (Photo 2.1) (Photo 2.2)

Neolithic people, and later those with copper and then bronze age technology, must have given names to river valleys, hills, routeways, and raw material sources. To what degree these names were incorporated into Celtic toponyms is a debated issue.

Photo 2.1   In addition to Stonehenge Britain contains many standing stone monuments including Avebury, Wiltshire, England, pictured above.

Source: Robert Harding Picture Library/Alamy Images.

Photo 2.2   Tarxien Temple, Malta. Maltese Neolithic Temples are older than Stonehenge.

Source: Eddie Gerard/Alamy Images.

Phoenicians and the Alphabet

The Phoenicians developed their manufacturing and trading culture in the area of modern Lebanon and by the beginning of the first millennium B.C. were sailing, settling, and trading across the Mediterranean to the Atlantic. The name Phoenician is probably derived from the Greek word for red, possibly after the dye the Phoenicians traded in the Mediterranean. The Phoenicians spoke a Semitic language and called themselves *k'an'ai* and they are known to us as the Canaanites of the Hebrew bible. Homer referred to them as Sidones, for many sailed from the east Mediterranean port of Sidon (modern Saydal) but other ports were involved including the Papyrus port of Biblos near modern Jubayl. From Byblos the Greeks derived the term *book* and thus the word *bible*. The Phoenicians developed a 22-symbol alphabet, with one sign representing one sound, which diffused to Greece where the five vowel sounds were added. Our alphabet derives from the Phoenicians, via Greece and Rome.

Phoenician trade was based on luxury craft goods, glassware, metals including bronze, papyrus, spices, textiles, timber (cut from the Cedars of Lebanon), wine, weapons, ivory, and a purple dye extracted from the Murex mollusc. By the eighth century B.C. the Phoenicians established port settlements in the Mediterranean on Sicily, Malta, Sardinia, the Balearic islands, and the Iberian peninsula. Later, as the Persians pushed the Phoenicians from the east Mediterranean coast, Carthage in North Africa near the modern city of Tunis, rose to preeminence. Phoenician foundations include Palermo in Sicily, Barcelona, Málaga, and Cádiz, on the coast of Spain. On the Atlantic coast of Morocco, Larache is a Phoenician foundation and the Phoenicians sailed extensively on the coasts of Africa (Figure 2.1). There is no truth to the idea that Phoenicians traded for tin in Cornwall, in southwest England; this myth was made up by a school teacher in the reign of Queen Elizabeth I.

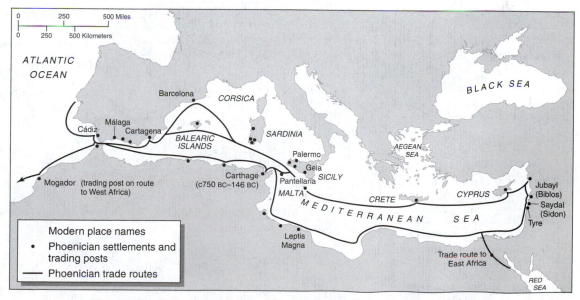

Figure 2.1 Phoenician—Carthaginian settlements and trade routes.

The Greeks and Phoenicians were hostile to each other. With the rise of Rome the Romans and the Phoenicians/Carthaginians came into conflict in the period 264–146 B.C. At the end of the Punic wars Carthage had been devastated and abandoned. Europe probably owes more culturally to the Phoenicians than is recognized. Much of what we know about them was written by hostile Greeks and Romans.

Greece

By the seventh century B.C. the Greeks had established settlements around the Aegean, along the shores of the Black Sea, in Sicily, southern Italy, and the Mediterranean coast of Spain and France (Figure 2.2). Marseille (*Massilia*) is a Greek foundation as are the adjoining places of Antibes, Monaco, and Nice. In southern Italy Naples (*Neapolis*), Taranto (*Taras*), and Paestum (*Poseidonia*) are examples of Greek foundations and at Paestum, which was totally abandoned later, there are fine examples of Greek architecture. (Photo 2.3) In Sicily, Siracusa, Catania, and Arrgigento are a few of many Greek city foundings. On the Black Sea, Odesa, Sevastopol, and nearby Yalta were all established by the ancient Greeks.

In everyday speech we use Greek words including alpha, beta, gamma, and delta. Most Greek words were introduced into languages by educated elites with a classical education. However, some Greek words were incorporated into Romance languages long ago, possibly from Greek colonies in the western Mediterranean. The Greek word for stone, *petra,* survives in Spanish as *piedra* and the Greek word *platea* is the origin of the Italian *piazza,* the Spanish *plaza,* and the French *place.* The word for poet, in several languages, is derived from Greek. The Greeks had important influences on art, architecture, philosophy, education, sports, and theater.

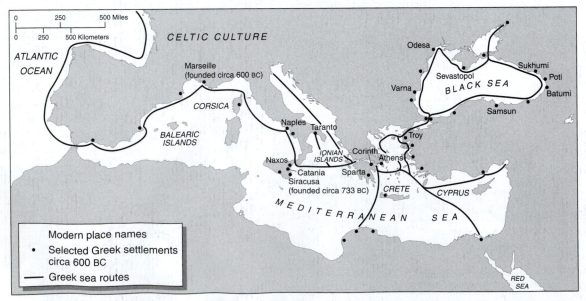

Figure 2.2   Greek foundations c. 600 B.C.

Photo 2.3   Paestum in southern Italy is an abandoned classical Greek settlement.

Source: Wojtek Buss/AgeFotostock.

Rome and Romance Languages

The Roman Empire had a *lingua franca,* Latin, and the language survived, after the collapse of the empire in the fifth century A.D., to evolve into French, Spanish, Portuguese, Provençal, Catalan, Italian, and Sicilian, together with regional dialects. Latin did not die with the empire because it was the language of Christianity, a religion legalized by Rome in 313 A.D. Theological works were published in Latin, sermons were preached in Latin, and the Church kept records in Latin. Up until half a century ago Roman Catholic churches conducted services in Latin. Ancient European universities such as Oxford to the present day conduct ceremonies, including graduation, in Latin. When Christianity reached England in the seventh century A.D., hundreds of new words from Latin, Greek, and eastern languages were introduced including bishop, monk, disciple, and shrine from Latin, pope, and apostle from Greek.

When European scientists published treatises, they wrote in Latin. The Pole, Nicolaus Copernicus, published on the *Revolutions of Heavenly Bodies* (1543) in Latin as did other scientists of the sixteenth, seventeenth, and eighteenth centuries. Scientists no longer write papers in Latin but all plants and animals are classified using Latin names.

Roman law became the basis of many European legal systems and even where Roman law was not the foundation of the system, as in England, legal terms such as *subjudice* are still conveyed in Latin phrases. The source of many concepts in civil law is Justinian's Code organized between 528 and 534 A.D.

The Roman Empire, in Europe, was largely defined by the Rhine and the Danube (Figure 2.3). At maximum extent, around 120 A.D., the empire did not securely hold lands east of the Rhine and, by 200 A.D. the Germanic tribes east of the river were a threat. In central and southeast Europe, Roman control did not extend north of the Danube with the important exception of Dacia, which occupied territory now in Romania. Romanian is a Romance language. Romans settled around the Black Sea and in Turkey.

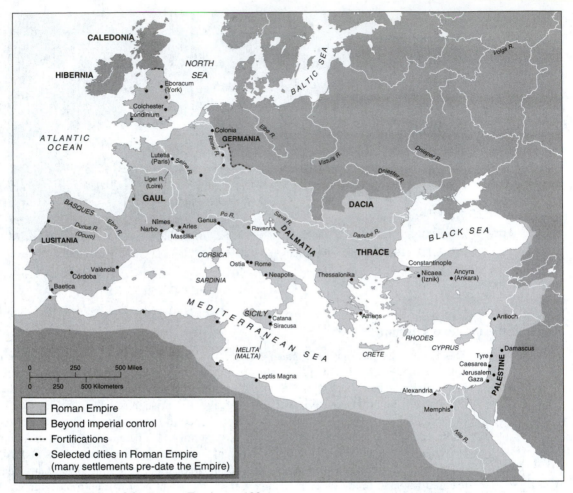

Figure 2.3 Roman Empire, c. 130 A.D.

In 330 A.D. Emperor Constantine moved his capital to Constantinople at the entrance to the Black Sea. The western empire, based on Rome, was prone to political rivalries and the Germanic tribes developed from a threat to a force capable of overrunning western Europe. In 410 A.D. the Visigoths looted Rome. The Vandals came across the Rhine, moved through France to the Iberian peninsula and into North Africa where there were Roman settlements. The Vandals successfully attacked Rome in 455 A.D. The last Roman emperor was deposed in 476 A.D. by a Germanic force. The eastern Byzantine empire, based on Constantinople, continued and often sent forces west without being able to reestablish the western empire. Constantinople (present-day Istanbul) did not fall until 1453 A.D. when the Turks took the city as they expanded into southeastern Europe.

Rome in Cultural Landscapes
Roman civilization has a continuing presence in the culture and cultural landscapes of Europe. English is a Germanic language, the result of the Angles and Saxons settling

Photo 2.4

Source: View of the Campo Vaccino, Rome by Claude Lorrain (Claude Gellee) (1600–82) ©Louvre, Paris, France/Lauros/Giraudon/.

in Britain as Rome weakened and withdrew, but English has many words of Latin origin including *normal, laureate, fibula,* and *feline*. Many words, particularly scientific terms, were brought into English by scholars in the sixteenth and seventeenth centuries, including *exaggerate* and *habitual* from Latin and *monopoly* from Greek via Latin.

The place names of western Europe still record the Roman presence as at Cologne, based on the Latin *coloni,* and London (*Londinium.*) We use innumerable Latin words, symbols, and abbreviations in everyday writing and speech, including A.M. (*ante meridiem*), i.e. (*id est*), *de facto, non sequitor, post mortem, vice versa, prima facie, de jure,* and *per se.*

Roman Towns, Roads, and Place Names

Everyone knows that the Romans built roads and planned towns (Photo 2.4). What is less well known is that many sites on which the Romans laid out towns had already been occupied by settlements housing Celts or other peoples brought into the Roman Empire. The Romans wanted to bring their rule and their law to conquered territory and deliberately built towns at or near existing concentrations of population. (Photo 2.5)

Photo 2.5 Roman building at the center of modern Nîmes, France. Nîmes also possesses a Roman Coliseum in which events are still staged.

Source: Walter Bibikow/AgeFotostock.

Julius Caesar in his description of the conquest of Gaul provides descriptions of Celtic life. On settlements he describes towns (*oppida*) and cities (*urbes*). *Oppida* contained a concentration of population protected by a wall and a ditch. *Urbes* were tribal centers and had a larger range of functions. The Romans tended to lay out towns at the tribal centers. Thus at Colchester, in southeast England, the capital of the Catuvellauni tribe, a new walled town was created at a place that was already a center of Celtic administration and commerce, and possessed a mint. The name Colchester includes the Celtic name for the nearby river Colne and the basically Latin word *chester,* meaning a town.

The capital of the Catuvellauni tribe had enclosed, within earthworks, a huge area of 15 square miles. The more compact Roman town, laid out shortly after the successful Roman invasion of southern England in A.D. 43, was walled rather than protected by earthworks. In general Roman town plans featured two major streets: the *Cardo,* running from the north gate to the south gate, and the *Decumanus,* running east to west. The two major thoroughfares met at the Forum, a square with civic and mercantile functions.

The laying out of a new town took territory from Celts and if a new city was a *coloni,* settled by time-expired legionaires, there was an allocation of farmland to former soldiers. Not surprisingly, laying out *coloni* caused friction as at Colchester when the surrounding Celts, led by the tribal leader Queen Boudicca, sacked Colchester and newly created London (*Londinium*) in A.D. 59–60.

Eventually Rome created a prosperous landscape of towns, cities, and villas linked by roads that facilitated trade and the interchange of ideas and cultural traits. Modern landscapes are still influenced by Roman cultural contributions to field systems, irrigation works, roadways, canals, aqueducts, fortifications, town plans, and architecture.

Photo 2.6   Roman aqueduct. Segovia, Spain.

Source: Oscar Garcia Bayerri/AgeFotostock.

(Photo 2.6) The landscapes of Italy are full of Roman remains but there are many major structures elsewhere in Europe, including Hadrian's wall, which runs across northern England from the Irish to the North Sea. You can still walk the wall today and visit the towers, towns, and camps that housed Roman soldiers and their families.

Roman building employed classical architecture, a style still used in the present and the basis of other styles including Renaissance and Baroque. Classical architecture is not confined to Europe. Washington, D.C., is full of buildings in the style and many state capitols, courthouses, and post offices in the United States reflect classical architectural traditions.

Germanic Languages

With the collapse of the Roman Empire Germanic tribes spread their influence further to the west, coming to control the North European plain, the Jutland Peninsula, England, and lowland Scotland. The Germanic tongues include the Scandinavian languages, Dutch, Frisian, Flemish, German, and English. Although Germanic tribes conquered Gaul, Italy, and the Iberian Peninsula, their presence was not strong enough to displace languages evolving from Latin.

Slavic Languages

The majority of the inhabitants of eastern Europe speak a Slavic language including Polish, Czech, Slovak, Serbian, Croatian, Slovene, Bulgarian, Macedonian, Belarusian, Russian, and Ukrainian. Czechs and Slovaks can understand each other as do Serbs and Croats, Russians and Ukraines. Romanian is a romance language. Hungarian is not a Slavic tongue, being Ural-Altaic in origin.

Uralic and Altaic Languages

The Uralic and Altaic languages probably originated, as the name implies, around the Ural mountains and were spread east and west by migrating peoples. The Hungarians, speaking a Uralic language, arrived on the Danubian plains beginning in 896 A.D. We are less clear on the arrival of Finnish, Karelian, Saami (the language of the Lapps), and Estonian.

The Altaic languages are related to the Uralic, and Turkish is an Altaic language. In Turkey most of the population speaks Turkish but the 12 million Kurds speak Kurdish, and Arabic is used in some parts of the country. Over a million Bulgarians speak Turkish, many as their first language.

Arabic

Today there are Arabic-speaking communities in European cities. Marseille, for example, has large populations speaking Algerian or Tunisian dialects of Arabic. Historically, the Arabs had widespread control of territory in southern Europe, including the Iberian peninsula, the Balearic islands, Sardinia, Sicily, and Malta (Figure 2.4). Sicily and Malta were reconquered by 1091 A.D. The *reconquista* of Iberia was completed in 1492 with

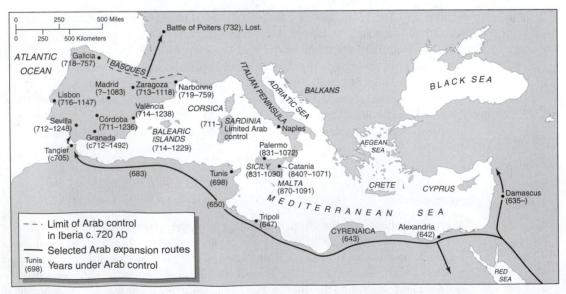

Figure 2.4 The Arab Empire in Europe.

Photo 2.7   Torre del Oro and Guadalquivir river, Sevilla, Andalucía, Spain.

Source: R. Matina/AgeFotostock.

the fall of the Kingdom of Granada but the *Moriscos* were not expelled until the first decade of the seventeenth century.

Arab influence remains. The Maltese speak a basically Arabic language to which Latin, Italian, Sicilian, French, Spanish, and English words have been added. From Arabic we derive words and concepts, including *algebra, zero, zenith, tarif, alfalfa, admiral,* and *sherif.* In the Iberian peninsula, particularly Andulucía, many place names (toponyms) are of Arabic origin. A high percentage of the names with an Arabic origin begin with *a* or *al,* including *alquería* (farmhouse), *alfolí* (granary), *aceña* (water mill), and *almazara* (oil mill). Even the *maquila* of the *maquiladora* has an Arab origin.

River names in Spain with a Guada prefix have an Arabic origin. Guada is derived from the Arabic word *Wadi,* meaning a river or a valley. Examples include Guadalajara (stony river), and Guadarrama (sandy river). The river running through Sevilla is the Guadalquivir: the big (*kibir*) river (Photo 2.7). When we speak of the Ramblas, the café-lined, promenade street in Barcelona, we are using a basically Arabic word, meaning a sandy place, although the city paved the area long ago.

In Spain, and elsewhere in southern Europe, the words used to describe elements in irrigation farming have a base in Arabic. The Spanish words *acequia* (irrigation ditch), *azud* (dam), *alberca* (reservoir), and *aljibe* (cistern), are all Arabic in origin as is the word *noria.* The *noria* lifts water from the well to the *acaquia,* which irrigates the field. (Photo 2.8)

European Languages Beyond Europe

Many European languages were spread beyond Europe by colonial and imperial systems. English and French were brought to North America by settlers from Europe. Spanish

Photo 2.8　Museo de la Huerta, Murcia, Spain. The noria lifted water to the irrigation channel.

Source: Luis Castaneda/AgeFotostock.

came to the New World with the conquistadors, and the Portuguese settlers gave Brazil its language, although Native American peoples still speak indigenous languages. In North America a number of Indian words entered English, including *canoe, caucus,* and *raccoon.* From the Indian subcontinent English acquired *pundit, curry, juggernaut, jute, jungle, bungalow,* and *verandah.* French, Portuguese, and Spanish (e.g., Huracán) all picked up words from overseas.

▶ RELIGIONS

Christianity, Islam, and some numerically smaller religions, including Judaism, Huguenots (French Protestants), and Quakers, have played an important part in economic and intellectual life.

Christianity

Christianity had its origins in the eastern Mediterranean Roman Judaea. From there the religion spread to Greece and then to the city of Rome before diffusing widely through the Roman Empire. Christian belief competed with other gods and in Rome, emperor worship. However, in 313 A.D. Constantine converted to Christianity and allowed freedom of worship. In 325 A.D. Constantine convened the first ecumenical council at which the Nicene Creed was adopted. When in 330 A.D. the emperor moved his capital east to Constantinople, the church at Rome and the church at Constantinople evolved differently. The eastern empire became the basis of the Eastern Orthodox Church, also referred to as Greek Orthodox and Russian Orthodox. The schism between the Church at Rome and the Eastern Orthodox Church created one of the lines of division between western and eastern Europe, just as the Reformation created a division, in the west,

Photo 2.9 Cathedral of Notre Dame, Reims, France.

Source: Uwe Dettmar/AgeFotostock.

between northern Europeans who were largely Protestant and southern Europeans who were Roman Catholic. (Photo 2.9) (Photo 2.10)

The eastern church, with its own rituals, vestments, monasteries, and doctrines, became the dominant denomination in the Greek lands, Russia, and the Ukraine. (Photo 2.11) In the Balkans, the Serbs were Orthodox; the Croats and Slovenes were Roman Catholic.

The Eastern Orthodox Church and the Roman Catholic Church are a powerful part of European cultural landscapes. Every city, town, and village has a church, or churches, of Catholic or Orthodox origin and later denominations have added their places of worship to the fabric of the built environment. Cities have cathedrals and are the seats of bishops to whom all the clergy of the parishes of the bishopic report.

The center of all historic towns in Europe is dominated by a church or cathedral. In the medieval era the Christian church controlled patterns of life: birth, confirmation, marriage, death. Church bells called parishioners to morning prayers and back from fields and workplaces for evensong. In rural and urban places Sunday was given over to church services and devotional activities. The major festivals, Christmas and Easter, were religious occasions and every parish celebrated its patron saint with a special day. Foodways were influenced by religion: fish on Friday and fasting during Lent. Many of these patterns survive into the modern world and contribute to the fabric of life. Certainly the places of pilgrimage visited in medieval times, including Rome, Canterbury, and Compostela, still attract millions of visitors for worship and admiration of religious monuments. (Photo 2.12)

Photo 2.10 Rear view of Gothic cathedral displaying the flying buttresses supporting the walls.
Source: Doug Scott/SUPERSTOCK.

Photo 2.11 Orthodox church. St. Petersburg. Russia.
Source: Wojtek Buss/AgeFotostock.

Photo 2.12 Santiago cathedral. Obradoiro square. Santiago de Compostela. La Coruña province. Galicia. Spain.

Source: P. Narayan/AgeFotostock.

Saint Benedict, in the sixth century, established monasteries and other orders followed the example of the Benedictines. Monasteries brought church establishments into rural areas and although places of prayer, worship, and contemplation, the religious foundations had an economic base, often developing large agricultural estates and introducing elements of commercial agriculture into rural landscapes. (Photo 2.13)

The Reformation
The Roman Catholic Church and the Eastern Orthodox Church developed rich religious rituals. In parts of the Catholic world many wanted simpler services that reflected the humble gatherings of the early Christians. Many resented the materialism of the Church hierarchy and the sale of indulgences (remission of sins). Out of this grew the Reformation and the emergence of Protestant churches, partially as the result of the preachings of figures like the Czech John Huss (1372–1415), Martin Luther (1483–1540), and John Calvin (d. 1564). In time, most of northern Europe embraced Protestant beliefs. Denmark, Norway, Sweden, and Finland are predominately Lutheran; Switzerland embraced Calvinism. All broke with the Church at Rome. In England Henry VIII exploited the growing Protestant movement to create a Church of England (1534) with the sovereign as head, to promote a sense of national identity, and, of course, to facilitate his divorces.

Photo 2.13   The Benedictine Monte Cassino Monastary was entirely rebuilt after being destroyed in World War II.

Source: Brian Harris/Alamy Images.

To raise revenue Henry dissolved the monasteries and sold off land to ambitious farmers and merchants. Although much stone and timber from monasteries went into the fabric of farmhouses, the monastic remains, such as Fountains Abbey, are an impressive part of historic landscapes in England.

Calvinists wanted more radical reform, leading to the establishment of Presbyterian and Congregationalist churches (in which ministers are of equal rank and congregations can choose their pastor). In spite of the break with Rome, Anglicans continued to follow many of the old patterns of worship in the existing churches. The Puritans wanted simpler forms of worship. Many came to New England between 1620 and 1640 to establish communities.

In general, the Scandinavian countries, Finland, northern Germany, The Netherlands, northern Belgium, England, Scotland, and Wales are Protestant. The Poles and the Irish are Roman Catholic, as are most people in France, southern Germany, the Czech lands, Spain, Portugal, Italy, Slovenia, Croatia, and the island of Malta, where St. Paul, on his way to trial and martyrdom in Rome, was shipwrecked in A.D. 60. (Figure 2.5)

Islam

After the breakup of the Roman Empire, the Eastern Orthodox Church and the Roman Catholic Church had to compete with paganism in many parts of Europe, although Christianity eventually became the dominant religion in most of the region. A more powerful competitor to Christianity than paganism was Islam.

Muhammed died in 632 A.D. but Islam expanded the area it controlled, and the number of adherents to the faith grew rapidly after his death. By 650 A.D. many formerly

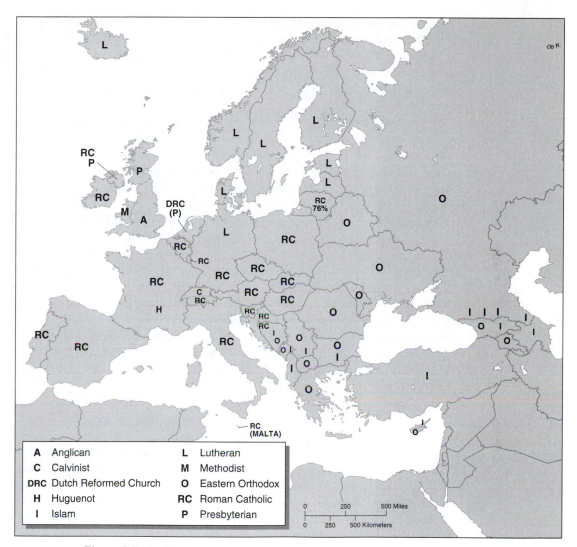

A	Anglican	L	Lutheran
C	Calvinist	M	Methodist
DRC	Dutch Reformed Church	O	Eastern Orthodox
H	Huguenot	RC	Roman Catholic
I	Islam	P	Presbyterian

Figure 2.5   Religious Affiliation by Country.

Christian centers, including Palestine, Antioch, and Alexandria, were in the Islamic world. Then Islam rapidly expanded into North Africa and parts of southern Europe. The Arabs crossed from North Africa into Iberia in 711 A.D. and by 720 controlled the peninsula apart from the Basque country. By 870 A.D. Sicily and Malta were under Arab control (Figure 2.4).

Although Sicily and Malta were restored to Christendom in the twelfth century, the reconquest of Iberia was not completed until 1492 with the fall of the Kingdom of Granada. Many Moors stayed in Andalucía and were not forced out until 1609. The Kingdom of Granada left an impressive architectural presence, including the Alhambra in Granada itself. The place names of Iberia retain many Arabic elements.

The Arabs had an impact on agriculture in southern Europe, introducing crops like cotton, citrus, and sugar cane. Sugar cane was diffused by Iberians to the Atlantic islands of Madeira, the Canaries, and São Tomé, which became stepping stones for the introduction of sugar cultivation in Brazil and the Caribbean. In Sicily, many terms relating to irrigation farming incorporate Arabic elements, indicating a contribution to irrigation technology.

As Christendom reconquered Iberia, another Islamic threat emerged in the eastern Mediterranean. The Turks spoke Turkish and were not Arabs, but they were Muslims. By the fourteenth century Turks were powerfully established in Anatolia and were commanding surrounding regions. In 1389 Turkish armies routed the powerful Serb army in Kosovo at the Field of the Blackbirds and established a presence in southeastern Europe.

Constantinople was increasingly isolated and in 1453 the Ottoman Turks took the city on the Bosporus, the waterway leading from the Black Sea to the Aegean and the Mediterranean. The great church of Sancta Sophia became a mosque. The Eastern Orthodox Church survived as an institution in Istanbul (Constantinople) because the Ottoman Empire was cosmopolitan and the city was a major trade center with ethnic communities drawn from all over Europe and the Middle East. After the fall of Constantinople, Athens came under Turkish rule in 1456 and the Ottomans had control of the Aegean Sea.

Suleiman I, who ruled from 1520 to 1566, expanded the Ottoman Empire into North Africa, Mesopotamia, and the Danube valley to include the Hungarian lands. In 1529 Vienna was besieged but successfully resisted. The Turks were still strong enough to besiege Vienna in 1683.

For expansion at sea, Suleiman developed a powerful navy. In 1522 the island of Rhodes was taken from the crusading Order of St. John of Jerusalem, which was pushed westward to Malta. Tunisia and Algeria were brought under Ottoman suzerainty, providing bases to threaten Christendom in the western Mediterranean. In 1565 a huge Turkish fleet brought an army to besiege Malta but the devoutly Christian Maltese, and the crusading Knights of St. John, withstood the nearly 4-month battle. Turkish seapower was checked at the battle of Lepanto (1571), by a combined Christian force, in the gulf of Corinth.

Turkish control of southeastern Europe did not dissolve until the nineteenth century when Greece established independence (1821–32). Romania, Bulgaria, and Serbia emerged from the Ottoman Empire in the second half of the nineteenth century. The religious influence of Turkish rule remains in the Balkans. The Muslim communities of Bosnia-Herzgovina emerged under Ottoman control when many families found they could advance in the service of the empire if they embraced Islam.

Islam today has a growing presence in Europe. Since World War II, large Muslim communities have been established in western and central Europe. As the result of migration from former colonies most British Muslims originally came from Pakistan, a part of the empire until independence in 1947. The Netherlands attracted Muslims from Indonesia (the former Dutch East Indies) and Surinam. Italy possessed Tripolitanea and Cyrenica (modern Libya) until the end of World War II. Algeria (1830–1962), Morocco (1912–1956), and Tunisia (1881–1956) were colonies of France. All major French cities have large Muslim communities, particularly in the southern port of Marseille. Spain

Photo 2.14 Mosque. Paris, France.

Source: Kevin George/Alamy Images.

ruled part of Morocco, across the Gibraltar strait, until 1956 and still, controversially, holds Ceuta and Melilla in Morocco.

All the North African colonies became sources of migrants moving to the former imperial power. As the migrants tended to cluster in the same areas of cities, spoke Arabic or Berber, and worshipped at newly established mosques, the new populations were not integrated, although some local inhabitants converted to Islam. Half a century ago mosques were rare in Europe. Today, in western Europe they are a part of urban cultural landscapes. (Photo 2.14)

Minority Religions

Europe, like North America, has a large number of minority religions as a result of admitting migrants from around the world. Historically, some minority religions have made important contributions to economic and intellectual life. These religious groups include the Huguenots, Jews, and Quakers (the Religious Society of Friends). (Photo 2.15) The Huguenots were a Protestant, Calvinist minority in Roman Catholic France. Fear of the minority grew and on August 24, 1572, Huguenots suffered the St. Batholomew's

Day massacre. Thousands were murdered. In 1598 the Edict of Nantes gave Huguenots religious freedom. When the Edict was revoked in 1685, many Huguenots left for the American colonies, Britain, and Prussia. In Britain, Huguenots contributed to the textile industry, one of the foundations of Britain's early industrial revolution.

Judaism has produced innumerable men and women who have contributed to economic and intellectual life in Europe. For example, by the early nineteenth century, the Rothschild family had created a banking network that included the head office in Frankfurt and branches in London, Paris, Vienna, and Wales. The bank was big enough to bail the French government out, after the Franco Prussian War, and to lend the funds (1875) that allowed the British prime minister Benjamin Disraeli to buy out the interest of the ruler of Egypt and gain control of the Suez Canal company for Britain. These spectacular events should not obscure the fact that innumerable professional and business enterprises were created by Jews, often after they had fled persecution, particularly in eastern Europe. (Photo 2.16)

The Society of Friends, the Quakers, was formed during the English civil war (1642–49). Quakers held that a formal church, services, and ministers were unnecessary, for God was in everyone. The Society was active in getting the British Toleration Act passed in 1689. William Penn, a prominent Quaker, founded Pennsylvania as an experiment in religious toleration.

In England, Quakers formed business networks. The first commercial railroad, the Stockton and Darlington, was organized and financed by Quaker businessmen and bankers. The candy industry in Britain was dominated by Quakers including the Cadburys and the Frys, families that built decent housing for workers and supported social reform causes, including the abolition of slavery. Elizabeth Fry (1780–1845) campaigned for prison reform. Today she is on the five-pound note issued by the Bank of England.

Photo 2.15 The Friends Meeting House blends into the English landscape.

Source: Shangara Singh/Alamy Images.

Photo 2.16 Synagogue. Malmö, Sweden.

Source: Guido Donati/AgeFotostock.

Religion and Cultural Landscapes

All across Europe churches, temples, and mosques are prominent in the landscape. The Roman Catholic Church and the Orthodox Church were dominant in the Middle Ages and churches associated with them are at the core of settlements, urban and rural. Medieval cities had three structural elements at the center: the church, the castle, and the market-place, with the church often standing beside the market. Villages all possessed a parish church around which the life of the community revolved.

When Protestant sects arose, in northern Europe, for the most part the church building was retained, even as services were simplified. Walking across the rural landscapes of Europe, often the first indication that the traveler is approaching a small settlement is the tower or spire of the village church pushing up through the trees, or standing out across the plain.

As immigrants have settled in European cities they have added their religious structures to the fabric of urban places. Mosques are easily visible in cities with Muslim communities and the temples of Sikhs and Hindus, while less frequent, are found in cities with communities large enough to erect places of worship. The Quakers and the Jews often have restrained buildings that fit quietly into the built environment.

▶ CULTURE CONTRASTS

Europe now embraces a number of different worldviews. In the north and west of the region the Scandinavian countries, Britain, and The Netherlands, with long parliamentary traditions of democracy, retain constitutional monarchies. The monarchs, as heads of state, sign acts of parliament into law, but do not enter political debates or attempt to influence policy. Iceland, a republic, formerly part of Denmark, and Finland, formerly a part of Sweden, can be placed in the northern and western group.

France, famously portrayed by both Admiral Mahan (1890) and Edward Whiting Fox (1971) as having a dual personality—part maritime, part continental—has had aspirations to create a vast overseas empire and, under Napoleon, came close to controlling the continent of Europe. Defeats at Waterloo and in World War II fine-tuned the French desire for continental control. The European community was largely a French creation, the rules would be written by French bureaucrats, and France would choose which rules applied to her. The Italian viewpoint is more limited and confined to identifying the rules that can be exploited for the benefit of Italian farmers and manufacturers. The Iberians possessed a global viewpoint from the sixteenth century and are less susceptible to seeing EU as a protective club. Germany sees the economic space of EU as a platform to support major corporations that will compete in global markets. The introduction, in 2004, of countries like Latvia, Lithuania, Poland, the Czech Republic, Slovakia, and Slovenia added a Slavic cultural element to EU perspectives. The Hungarians and the Greeks have viewpoints born of distinctive cultures.

For the most part, European Union leaders have been successful in avoiding threatening confrontations founded on differing national views of the region. There has been less success in creating environments where cultures introduced by immigrants can flourish within historic national spaces. Turkish workers were segregated and persecuted in German cities. North Africans live in Marseille ghettos. Paris is ringed by run-down public housing projects populated by immigrants who suffer rates of unemployment twice the high national average. In The Netherlands, Muslim immigrants from the former Dutch East Indies (Indonesia) and Surinam, in South America, have high rates of unemployment and are concentrated in sectors of cities. In Britain large communities originating in the West Indies and Pakistan reside in a few cities: the West Indians in parts of London, Bristol, and Liverpool, Pakistanis in textile towns of Yorkshire and Lancashire. Birmingham, as an economic growth center, attracted migrants from many source areas and cultures.

It is not difficult to highlight the contributions immigrant groups make to literature, education, business, popular culture, and sport. Zidane, the son of Algerian immigrants living in Marseille, scored the goals that helped France win the World Cup in 1998. Nasser Hussein, son of a family from the Indian subcontinent, captained an England cricket team that contained members who were sons of immigrant families from the West Indies. Innumerable small businesses providing food and services are owned by immigrants who filled niches in the economic fabric of European cities.

At another level there is a deep-seated failure with roots going back before the arrival of the immigrants. The inability of many European countries to educate and integrate all classes in society created a situation in which there was an alienated, unskilled segment that feared immigrants. This element felt threatened as immigrants moved into the more dilapidated streets of working-class housing areas. Adjoining residents saw the migrants as competitors for scarce housing and a cheap labor source to threaten job security. That the migrants were often Muslim seemed to make the supposed threat sinister. Immigrants felt under siege from the beginning. Indians from the Indian subcontinent did not face the same difficulties as Pakistanis in Britain. The Pakistanis took low-paying work, clustered in poorer streets, and spoke Urdu. Indians tended to be better educated, spoke English, often had or acquired professional qualifications, dispersed into communities and by adoption of English ways were largely assimilated. Englishmen and Indians can always talk cricket. For many Brits India remains a fascinating place, an enchantment that did not end when the Republic of India became independent in 1947. India is a popular "gap year" destination for students.

The present-day communities of Muslim migrants, and their descendants, contain many alienated members, as was apparent in 2005 and 2006. Early in July 2005, three bombs exploded on the London Underground and one on a bus. The death toll was over fifty people. The horror was made worse by the discovery that the suicide bombers were not fanatics from overseas but second- and third-generation Muslims from communities in the north of England. Then in October and November trouble developed in the Muslim communities in the housing projects around Paris (Richburg, 2005) and spread to many other cities including Toulouse, Strasbourg, Nantes, Toulon, Orléans, and Dijon. The burning of cars was commonplace and there were attacks on police and emergency services when they arrived to extinguish fires. The violent protestors were young men who had grown up in France within immigrant communities. Community leaders came to the streets to plead for calm and the restoration of order. The younger-generation rioters ignored the elders.

As in Britain, leaders of Muslim communities in France were surprised at the hostility the young displayed to the countries the elders had happily joined to build new lives. Immigrant communities in several European countries have produced a rich literature and a theme in the writing is the alienation of many Muslims born in Europe. (Rumbelow, 2005). Hanif Kureishi's short-story film *My Son the Fanatic* (1993) portrays the bewilderment of the hardworking, taxi-driving father as his good student, cricket-playing son hungers for *Jihad*. "I love this country" says Dad, speaking of Britain, you can "do almost anything here." "That is the problem," replies the son.

The EU, Cultures, and Cultural Landscapes

The EU wants to conserve European national and regional cultural diversity while at the same time highlighting commonalities in the cultural heritage of the region. The EU provides subsidies to help restore major sites including the Doges' Palace in Venice, the Acropolis at Athens, and the city of Lisbon. The Council of Europe has promoted a number of European Cultural Routes using themes including the pilgrimage to Santiago de Compostela, the Vikings, and the silk road.

Freedom of religion is guaranteed under European law and the EU uses most of the member states' languages in translation and in the preparation of official documents. In practice the most frequently used languages are English and French but if Europeans write to an EU institution in the language of a member state they can expect a reply in that language.

► FURTHER READING

Aubet, M.E. *The Phoenicians in the West: Politics, Colonies, and Trade*. New York: Cambridge University Press, 2001.

Cartledge, P. (ed.). *Cambridge Illustrated History of Ancient Greece*. Cambridge, UK: Cambridge University Press, 1998.

Cunliffe, B. *The Ancient Celts*. New York: Penguin Putnam, 1997.

Elcock, W.D. *The Romance Languages*. London: Faber and Faber, 1975.

Fox, E.W. *History in Geographic Perspective: The Other France*. New York: Norton, 1971.

Haywood, J. *The Penguin Historical Atlas of the Vikings*. New York: Penguin Putnam, 1995.

Heather, P. *The Fall of the Roman Empire: A New History of Rome and the Barbarians*. New York: Oxford University Press, 2005.

Hobson, J.M. *The Eastern Origins of Western Civilization*. Cambridge, UK: Cambridge University Press, 2004.

Jones, G. *A History of the Vikings*. New York: Oxford University Press, 2001.

Jordan, T.G. *The European Culture Area*. New York: Harper-Collins, 2001.

Mahan, A.T. *The Influence of Seapower Upon History, 1660–1783*. Boston: Little, Brown, 1890.

Martin, G.J. *All Possible Worlds: A History of Geographical Ideas*. New York: Oxford University Press, 2005.

Matvejevic, P. *Mediterranean: A Cultural Landscape*. Berkley: University of California Press, 1999.

McCrum, R., W. Cran, and R. MacNeil. *The Story of English*. New York: Viking, 1986.

Kulikowski, M. *Late Roman Spain and its Cities*. Baltimore: Johns Hopkins University Press, 2004.

Nolan, M.L., and S. Nolan. *Religious Pilgrimage in Modern Western Europe*. Chapel Hill: University of North Carolina Press, 1989.

Ostergren, R.C., and J.G. Rice. *The Europeans: A Geography of People, Culture, and Environment*. New York: Guilford Press, 2004.

Richburg, K. "The Other France, Separate and Unhappy." *Washington Post,* November 13, 2005, p. B1.

Rumbelow, H. "Real Life is Unfolding Like Fiction." *The Times,* July 30, 2005, pp. 9–24.

Sawyer, P. (ed.). *The Oxford Illustrated History of the Vikings*. New York: Oxford University Press, 1997.

Talbert, R.J.A. (ed.). *Barrington Atlas of the Greek and Roman World*. Princeton, NJ: Princeton University Press, 2000.

Tanner, M. *The Last of the Celts*. New Haven, CT: Yale University Press, 2004.

Ward-Perkins, B. *The Fall of Rome: And the End of Civilization*. New York: Oxford University Press, 2005.

Now out of print but still useful is Max Cary, *The Geographic Background of Greek and Roman History*. Westport, CT: Greenwood Press, 1982, originally published by Oxford University Press in 1949. Professor Cary drew widely on geographical writers including Ellen Churchill Semple.

CHAPTER

3

POPULATION GROWTH, DISTRIBUTION, DENSITY, AND MIGRATION

▶ INTRODUCTION

Over two centuries ago, a country clergyman in southern England, with a mathematics degree from Cambridge University, was alarmed at the high birth rate among the poor of his parish. The laboring poor lived near subsistence level and he feared increasing numbers would depress wages and living standards. The Reverend Robert Malthus (1766–1834) wrote an *Essay on Population* (1798), postulating that while food supply increased arithmetically (1, 2, 3, 4, 5, etc.), population grew geometrically (1, 2, 4, 8, 16, etc.). Growth in numbers was checked by famine, warfare, and disease.

As a clergyman Malthus had access to data. He kept the parish registers of baptisms, marriages, and burials and could document the rising birth rate in his parish. In fact, all across England death rates were dropping; more children were surviving to adulthood, marrying, and forming families.

What Malthus, and other commentators, could not know was that Europe was beginning a marked population change. Europe was entering the dynamic second phase of the demographic transition, when death rates fall, birth rates remain high, numbers grow, and the age structure of the population becomes more youthful.

▶ DEMOGRAPHIC HISTORY

Between 1600 and the mid-eighteenth century the population of Europe had grown, on average, at a rate of 0.15 percent per annum. The average conceals the fact that in some years numbers fell as a result of famine, warfare, or disease. Around 1750 population growth quickened. In the century from 1750 to 1850 numbers grew at 0.63 percent per annum. In Britain the rate of growth was faster; in France it was slower.

Figure 3.1 shows us that Europe experienced population growth from the second half of the fifteenth century to the second half of the twentieth century, with a slowdown in the first half of the seventeenth century. During nearly five centuries the population

European Population Growth 1500–2000

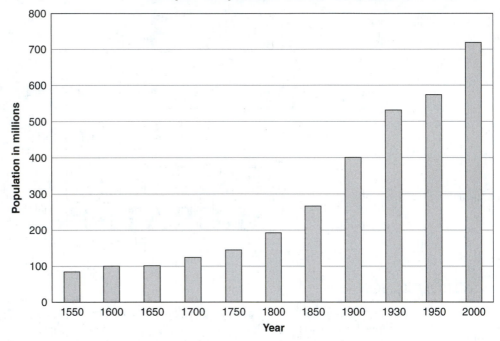

Figure 3.1   Europe: Population growth 1500–2000.

of Europe not only grew but Europeans emigrated in large numbers to the Americas, southern Africa, Australia, New Zealand, and many other overseas colonies.

In 1500 Europe had around 85 million people. By 1600 there were 110 million Europeans. Population growth slowed in the seventeenth century due to depression, warfare, and poor harvests. At the end of the century the population of Europe was 125 million. Growth then quickened, particularly in the second half of the eighteenth century. In 1800 there were nearly 200 million Europeans, and by 1900 over 400 million. In the nineteenth century the population of the region more than doubled, even though large numbers of English, Irish, Scots, Scandinavians, Germans, Bohemians, Poles, Spanish, and Italians moved overseas. In the decade 1901–1910, 3.6 million Italians, 1.0 million Spaniards, .3 million Swedes, .3 million Portuguese, and 3.1 million people from the British Isles emigrated overseas. The great outmigration from Germany was in the 1880s. After World War I European source regions produced many fewer emigrants.

Today Europe contains approximately 730 million inhabitants. In the twentieth century, even though outmigration (emigration) slowed, the population of the region did not double. World War I, starting in August 1914 and lasting until November 1918, killed millions of men and many more suffered wounds, shell-shock trauma, and the effects of gas warfare. After World War I Europe had fewer men leading full, productive lives (Table 3.1).

TABLE 3.1 Europeans Killed or Missing in World War I (in millions)

Country	Killed or MIA	Prewar population	Postwar population	Prewar BR/DR (1913)	Postwar BR/DR (1919)
France*	1.4	39.1	38.8	19/18	13/19
Germany*	2.1	65.0	63.1	27/15	20/16
Russia	1.7	170.0	(Becomes Soviet Union)		
Austria-Hungary	.94		Dissolved		
Romania	.25	7.2	15.5 (Territorial gains)	42/26	23/20
Serbia	.37	2.9	(Part of Yugoslavia)	36/22 (1911)	37/21 (1921 Yugoslavia)
Italy*	.47	35.0	36.4	32/19	21/19
UK	.93	36.0	38.0	24/14 (Eng./Wales)	19/14 (Eng./Wales)

*Numbers adjusted for postwar boundary changes.
BR–Birth Rate. DR–Death Rate.

So many men volunteered, or were drafted into armies, that birth rates in a number of countries, including France and Germany, fell below death rates during the war. When the armistice came on November 11, 1918, France had lost one man in five between the ages of 20 and 44. Many families were not reunited. Many marriages were never made as fiancées failed to return from the battlefield. France was a land of widows and spinsters. The birth rate continued to decline during the interwar years. Worldwide, the Spanish flu of 1918–1919 killed more people than had died in the war. (Photo 3.1)

World War II was a total war and the death toll was much higher as civilian populations were blitzed, shelled, and exterminated. Probably 20 million inhabitants of the Soviet Union were killed during the war. Deaths were particularly high in the Ukraine where one German war aim was to kill off Slavs to make space for Nordic agricultural colonization. Many Ukrainians died of hunger and exposure after being driven from the land. Soviet punishment squads shot anyone suspected of collaboration with German forces. In total, around 6 million Ukraines died in World War II. This death toll is similar to the number of Jews, from communities across Europe, under Nazi control, who were deported to extermination camps, largely in Poland.

During the war, civilian death rates rose as people were short of food, clothing, and shelter. The old died sooner. Infant mortality rates rose. In The Netherlands, occupied by German troops in 1940, infant mortality in 1945 was eighty deaths per thousand live births, twice the prewar figure.

European conditions did not improve quickly at the end of the war. The housing stock in major cities was depleted, food and fuel were scarce, and official rations were often below starvation levels. A new class of migrant walked the roads: the displaced person. Poles were put out of the Ukraine, Germans were displaced from Poland and

Photo 3.1 Trench warfare. A battlefield scene from "All Quiet on the Western Front".

Source: John Kobal Foundation/Getty Images.

other countries. People left alive at extermination camps like Auchswitz or concentration camps like Dachau first had to regain strength before attempting to return to homes that had been looted, squatted in, or bombed out. Poland and Germany had many refugees, displaced from farms, fields, and towns further east. Nevertheless, there was a postwar baby boom and by 1947 the population of Europe began to grow again slowly, until about 20 years ago when total fertility rates in many countries fell below replacement levels.

The war speeded the social and economic changes that brought in lower birth rates but the main factor in declining fertility was the demographic transition.

▶ THE DEMOGRAPHIC TRANSITION

It is easy to see the demographic history of Europe being shaped by great wars, epidemics, and emigration. However, during the last two centuries Europe has been through a cycle of change known as the demographic transition.

The demographic transition model describes the sequence by which countries move from high birth and death rates to low birth and death rates. In the transition, death rates fall before birth rates and the population grows until birth rates decline too and natural increase slows (Figure 3.2).

The demographic transition is usually portrayed in four phases but in the European case there are five phases:

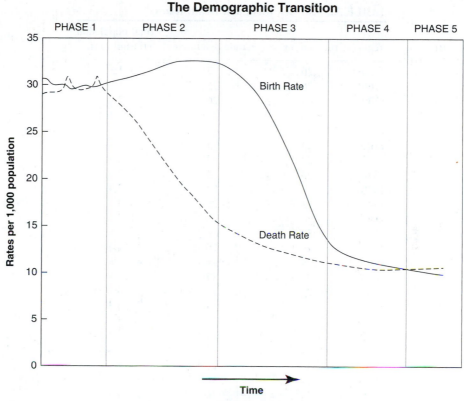

The Demographic Transition

Figure 3.2 Phases of the European demographic transition.

Phase 1   Birth and death rates are high. There is little increase in population numbers.

Phase 2   Death rates drop, birth rates remain high. Total population numbers increase and the age structure of the population becomes more youthful.

Phase 3   Birth rates drop. Death rates remain low. The population continues to increase but at a slower rate than in Phase 2.

Phase 4   Birth and death rates are low. Population growth is less than 1 percent per annum. The average age of the population increases.

Phase 5   Birth and death rates are about equal with death rates sometimes exceeding birth rates. Population numbers are stagnant or in slow decline. The percentage of the population over 65 years of age increases.

The Swedish Example

Sweden displays the patterns of the demographic transition as well as any country in Europe (see Table 3.2 and Figure 3.3). In 1740 crude birth (32) and death rates (35.5) were over thirty per thousand. Then the death rate declined slowly but crude birth rates remained high. Between the mid-eighteenth century and the mid-nineteenth century the population of Sweden increased from 1.77 million to 3.46 million. Birth rates started a

TABLE 3.2   Sweden Birth and Death Rates

Year	Crude birth rate	Crude death rate	Total population (in millions)
1750	36	27	1.8
1760	36	25	1.9
1770	33	26	2.0
1780	36	22	2.1
1790	30	30	2.2
1800	29	31	2.3
1810	33	32	2.4
1820	33	25	2.5
1830	33	24	2.9
1840	31	20	3.1
1850	32	20	3.5
1860	35	18	3.9
1870	29	20	4.2
1880	29	18	4.6
1890	28	17	4.8
1900	27	17	5.1
1910	25	14	5.5
1920	24	13	5.9
1930	15	12	6.1
1940	15	11	6.4
1950	16	10	6.7
1960	14	10	7.5
1970	14	10	8.1
1980	12	11	8.3
1990	14	11	8.6
2005	11	10	9.0

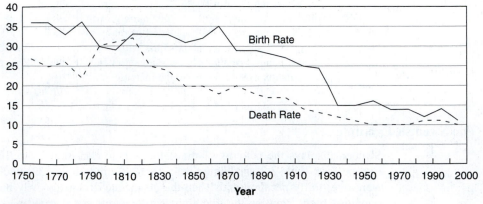

Figure 3.3   Swedish birth and death rates 1750–2000.

decline with marked drops after World War I and World War II, although neutral Sweden was not a combatant in either conflict. Since World War II the general downward trend in the birth rate has continued. Today, for every thousand Swedes there are, on average, 11 births and 10 deaths, each year. Nearly 20 percent of the population is over age 65. Sweden remains one of the world's most prosperous countries and ranks highly on quality of life indexes.

Most countries in northwestern Europe followed a pattern similar to Sweden. The countries of southern Europe, such as Italy and Spain, started the demographic transition later. In Germany, Czechoslovakia, Hungary, and Austria death rates did not decline until the last decades of the nineteenth century.

Declines in Death and Birth Rates

In the demographic transition death rates and birth rates decline. Deaths decline first, to be followed by births. In the period where deaths decline but birth rates remain high, population numbers grow. The declines in mortality have allowed world population to grow in the last two centuries.

In 1750, for every thousand inhabitants in France there were, on average, 38 deaths. The crude birth rate was 39 per thousand and population growth was slow. By 1816, at the end of the Napoleonic wars, the crude death rate was 26 and the birth rate 34. Population numbers grew. By 1850 the crude death rate was 21 and the birth rate was 27. The population growth rate had slowed since 1816 but numbers were still growing. At the end of the century birth and death rates were about equal and there was little growth in population numbers, by natural increase. Between the beginning of the twentieth century and 1946, due to wars and a low birth rate, total population numbers in France were stagnant. After World War II the birth rate rose relative to death rates and there was a baby boom. Population also increased by migration from former French colonies. (Table 3.3).

Reducing Mortality

What factors started a decline in death rates during the second half of the eighteenth century? In the Enlightenment, sometimes called the Age of Reason, people became less fatalistic and more inclined to listen to reasonable explanations. There were modest improvements in sanitation, people boiled water to kill germs, and innoculation was accepted as a precaution against smallpox. Food supply systems improved and regional shortages became less of a problem as canals and better roads allowed surpluses to be moved more easily. In parts of western Europe agricultural innovations including crop rotations, manuring of fields, better breeds of livestock, and new crops from the Americas, like the potato, produced more calories per acre than grains, improving agricultural output. Overseas trade brought in sugar from Brazil and the Caribbean. The mortality declines in the nineteenth century were slow, partly because of urbanization. Cities did not cope with population growth in terms of providing clean drinking water and sewage disposal. Dysentery and related diseases were common and outbreaks of cholera, as in North American cities, could result in surging urban death rates.

Prosperity, water, and sewage treatment plants brought west European death rates down below twenty per thousand by the early twentieth century. World War I increased

TABLE 3.3 France: Birth and Death Rates, 1750–2006

Year	Crude birth rate	Crude death rate	Total population
1750	39	38	
1816	34	26	30,449,000
1851	27	22	35,783,000
1896	22	20	38,269,000
1911	19	20	39,192,000
1916	9.5	17	
1921	21	18	38,798,000
1931	18	16	41,228,000
1938	15	15	
1946	21	13	39,848,000
1962	18	11	45,500,000
1970	17	12	
1982	15	10	54,273,000
2006	13	9	60,700,000

Sources: Population Reference Bureau. B.R. Mitchell, *International Historical Statistics: Europe 1750–2000,* New York: Stockton Press, 2002.

death rates but the slow, downward trend in mortality resumed in the interwar years before reversal in World War II.

Since the end of World War II crude death rates across Europe have fallen and now average around 12 deaths per thousand of the population. There are exceptions. Death rates in Russia and the Ukraine have increased since the breakup of the Soviet Union and the smoking Scots have higher death rates than the average for the United Kingdom. Today, crude death rates (12) per thousand in Europe are above birth rates (10) per thousand; population numbers will slowly decrease without immigration.

What Brought European Birth Rates Down?

It is easy to explain death rate declines: food supply becomes assured and sanitation, public health facilities, and medical services are improved. No one has to be persuaded not to die. Birth rates are different. The age of marriage, family size, and structure are influenced by social, economic, and cultural factors. The technology of contraception was not fully developed and available until the second half of the twentieth century.

The early phases of industrialization did not immediately bring down the birth rate. Urbanization allowed young people to leave rural areas where the age of marriage was often relatively high because the institution was not entered into until the male had ensured access to farmland and the means of subsistence. But, over the long term, urbanization represented a fundamental change. Urban areas were in the cash economy. Rural areas had many subsistence elements in which people and families provided directly for their own needs. In urban areas food, accommodation, and clothing were purchased and a larger family meant greater cash demands. Eventually the industrial city saw family size decrease. In the modern world we are all familiar with the factors that delay the arrival of children. A significant proportion of the population goes to university and many proceed to graduate school. Then there are careers to establish and an expensive lifestyle to maintain. In Europe the modern woman has 1.4 children, in Canada the figure is 1.5, and in the United States it is 2.0 largely due to the contribution of Hispanics, other minorities, and immigrants.

Total Fertility Rate

The total fertility rate (TFR) is the average number of children born to the average woman during her lifetime. TFR is an indicator of family size and changing household patterns. To replace itself a population needs a TFR slightly over two.

In England and Wales, in 1850, the average woman produced five children. In France the figure was 3.3. Britain's demographic transition had characteristics that differed from France. In prosperous, nineteenth-century Britain, which had an urban majority by 1851, birth rates stayed high longer than in France. In France the proportion of the population living in rural areas remained high. In Britain, as young people moved to waged work in industrial cities, the age of marriage fell in many areas and the number of children per household went up.

By 1900 the TFR in England and Wales was down to 3.4. France had declined slightly to 3.0. In the 1930s, in both countries, as a result of the losses of men in World War I and the Great Depression, the TFR fell below two but rebounded after World War II. Now in the United Kingdom (England, Scotland, Wales, and Northern Ireland) the TFR is 1.8. In France the figure is 1.9, reflecting a higher birth rate in the larger immigrant population, and government policies to encourage more births.

Today, the average European woman produces 1.4 children. The countries near replacement level include Iceland (2.1), Albania (1.9), and Ireland (1.9). The countries of eastern Europe and Russia have a TFR of 1.3 or less. In southern Europe the Spanish and Italian TFR is 1.3. Explanations of what used to be high southern European birth rates involving hot climates, Latin lovers, and the Church at Rome are now redundant!

Infant Mortality Rates

Infant mortality is an indicator of quality of life and standard of living. Poor countries with low living standards have high rates of infant mortality. Some countries in Africa have infant mortality rates in excess of one hundred; for every thousand live births, one hundred children do not survive to their first birthday. One child in ten dies in the first year of life.

A century ago infant mortality rates were high in Europe and North America. In Austria and Germany more than 20 percent, one child in five, did not survive the first year. The death of children before their fifth birthday was commonplace; a majority of families experienced the trauma. (Table 3.4).

► DISTRIBUTION OF POPULATION IN EUROPE

A variety of environmental, economic, and locational factors influence the distribution and density of population (Figure 3.4). Here are some observations to consider:

1. Population densities decline in northern Europe where harsher environments, with shorter growing seasons, are encountered. Similarly in upland regions steep slopes

TABLE 3.4   Infant Mortality: Selected Countries, 1900–2006

	1900	1915	1938	1950	2006
Austria	231	218	80	66	4.1
France	162	111	66	52	3.6
Germany	229	164	60	55	3.9
England and Wales	154	110	53	33	5.1 (UK)
Italy	174	148	106	64	4.1
Netherlands	151	87	37	27	4.9
U.S.		181 (b) 99 (w)	79 (b) 47 (w)	44 (b) 27 (w)	6.7
Canada	187		64	42	5.3

Sources: Population Reference Bureau. B.R. Mitchell, *International Historical Statistics: The Americas 1750–1993*, New York: Stockton Press, 1998. *Statistical Abstract of the United States*. Washington D.C.: U.S. Government Printing Office, 1915.

and shorter growing seasons, induced by cooler temperatures at higher altitudes, reduce agricultural productivity and population densities. On the population distribution map the highlands of Scotland, Iceland, the Scandinavian upland, northern Russia, and central Wales are regions of low population density. (Photo 3.2)

In the southern half of Europe we can pick out areas of low density associated with highlands. In the Iberian peninsula, the Cantabrian mountains, the Pyrenees, the Sierra Morena and the Sierra Nevada have low population densities. In France the Massif Central has generally low population densities. The Alps and the Apennines (the backbone of peninsula Italy), the Dinaric Alps (paralleling the Adriatic), and the arc of the Carpathians are areas of relatively low population densities.

2.   We must not conclude that the physical environment and agricultural potential determine population distribution and density. Notice how many capital and major cities have acquired a locational momentum that has drawn migrants to them and their size far exceeds the ability of the surrounding agricultural region to support the inhabitants. London and Paris owe their origins to being important bridging points on the Thames and Seine, respectively. The Thames basin and the Paris basin are major agricultural regions but now the populations of London and Paris are fed by food supplies drawn from across Europe and the globe. The imports are paid for by incomes derived from service industries concentrated in the capitals, including the huge governmental administrations. (Photo 3.3)

3.   Madrid became the administrative center of Spain in 1561 because of a central location in the Iberian peninsula. Summers are hot, winters are cool, and agriculture, in the surrounding region, is not rich. But capital status, jobs in the bureaucracy, and the presence of the court of Philip II drew the wealthy, and those seeking work, to a city that grew to be the largest in Spain, overtaking regional centers like Barcelona in size and influence.

4.   In general overland routeways sustain high population densities, at least in the cities that are nodes in the communication and exchange system. Ports are located at the end of overland routeways. In the Mediterranean region are the major ports of Barcelona, Marseille, Genoa, Naples, Palermo, and Piraeus (Greece) together with innumerable small

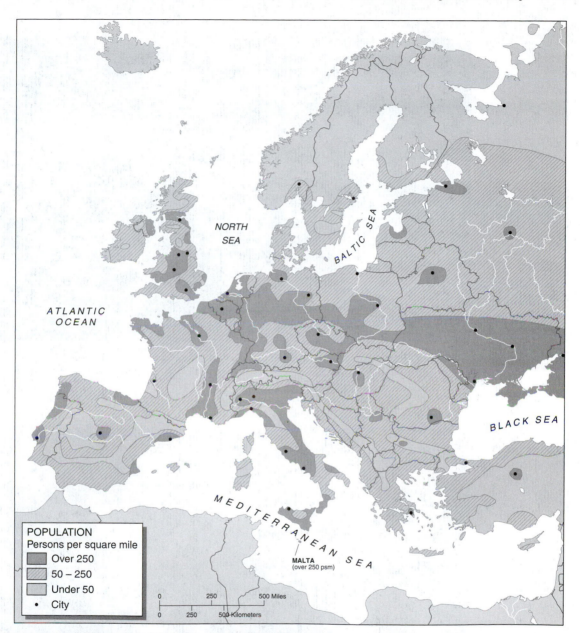

Figure 3.4 Distribution of population.

shipping and fishing places. On Atlantic coasts ports tend to be on tidal rivers, estuaries, and fjords: Bristol and Bordeaux, Liverpool and Lisbon, Hull and Hamburg, Southhampton and Sevilla, Antwerp and Oslo. Inland are the river ports on the Rhine: Cologne, Mannheim, and Basle and, on the Danube, Vienna, Budapest, and Belgrade. The ports of Europe have sustained high population densities in the city and in the hinterland.

Photo 3.2 Low densities of population are usual in upland regions of Scotland.

Source: Garry Black/Masterfile.

Photo 3.3 Paris. Capital cities display high densities of population.

Source: Dan Porges/AgeFotostock.

When raw materials are unloaded it is often cheaper to process them at the port rather than transport them to an inland plant. Ports sustain shipbuilding and repairing, ships chandlers who provision vessels, merchants, bankers, and historically, innumerable jobs heaving and hauling cargoes on and off quays and lighters. Inland routeway cities, like Lyon, owe population growth to the emergence of economic activity at converging lines of communication.

5. Soil fertility influences population density in rural areas. In the Paris basin, an extensive area of good agricultural land drained by the river Seine and its tributaries, population densities can be one or two hundred people per square mile based on agriculture including viticulture. In the Massif Central, with thin soils and much grazing land, the density of inhabitants is lower.

6. The great European industrial regions, developed in the nineteenth century on coal fields, acquired an economic momentum that has carried into the twenty-first century. Population densities remain high. Many coal mines and steel mills have closed, but other forms of work in manufacturing, and particularly services, provide employment. Population densities remain higher in the Ruhr than on the North German Plain.

Urban Populations

Approximately 75 percent of Europeans live in urban areas but there are large variations from country to country. The countries of northern (82%) and western Europe (80%) have high rates of urbanization. Smaller percentages of the population are urbanized in eastern (68%) and southern Europe (74%). In northern Europe, Iceland (94%) and the United Kingdom (89%) have particularly high rates of urbanization. In western Europe, Belgium is 97 percent urban! (Photo 3.4) Further east, Albania and Moldova still have rural majorities. Romania and Slovakia are just over 50 percent urban and have large rural populations. In southern Europe, Italy is 90 percent urban but Spain (76%), Portugal (53%), and Greece (60%) have substantial rural populations.

The trend will be for the major metropolitan areas to contain an increasing proportion of the population and to control social and economic life in their hinterlands. With a huge commuter network, London dominates much of southern and eastern England, as does Paris in northern France (Table 3.5).

Density of Population

The density of population is the average number of people occupying a square mile, or square kilometer, of a country or other unit of calculation. In the United States there are about eighty inhabitants for every square mile of national territory. In Canada, with huge northern, arctic, and tundra regions that are lightly inhabited, the average population density is less than ten people per square mile. In Europe there are 82 people per square mile of the region and great density contrasts from place to place (Figure 3.5). There are approximately 730 million people living in Europe, compared to 300 million in the United States and 33 million in Canada. The highest population densities, over 400 people per square mile, are found in western Europe in countries like The Netherlands (1,033 per sq. mile), Belgium (883 per sq. mile), and Germany (600 per sq. mile). France, with many scantily populated regions, averages less than 300 for each square mile, although

Photo 3.4 Groenplaats Antwerp. Belgium is heavily urbanized.

Source: Age fotostock/SUPERSTOCK.

TABLE 3.5   Primate Cities: London and Paris

UK		France	
City	Population in millions	City	Population in millions
London	7.2*	Paris	9.1*
Greater Birmingham	2.6*	Lyon	1.4
Greater Manchester	2.5*	Marseille (with Aix)	1.3*
Leeds, West Yorkshire	2.1*	Lille	1.0*
Liverpool, Merseyside	1.4*	Nice	0.9
Sheffield, South Yorkshire	1.3*	Toulouse	0.77
Tyne (Newcastle) and Wear (Sunderland)	1.1*	Bordeaux	0.75
Glasgow	0.6	Nantes	0.55
Edinburgh	0.5		

*Signify conurbations.

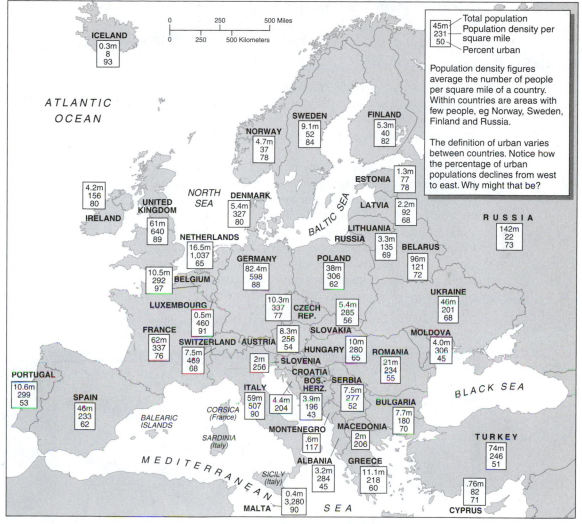

Figure 3.5 Total population, population density per square mile, and percent urban.

in the Paris metropolitan area there are thousands of French men and women in each square kilometer. The tiny principality of Monaco has a density of 44,000 people per square mile.

Population densities in eastern Europe average less than 100 people per square mile. In the west the average is 435 per square mile. In northern Europe, because of lightly settled upland, arctic, and tundra environments, overall densities, at 142 per square mile, are relatively low even if concentrations of population stand out around Stockholm, Oslo, Helsinki, and Reykjavik. Southern Europe, averaging 300 people per square mile, is a region of contrasts. Population densities are low on the Spanish *meseta*, the Italian Apennines, and the uplands of Greece. The Mediterranean coasts of Spain and Italy are largely built over. Athens and the port of Piraeus sprawl along the coast of Greece. The hundred and twenty square miles of Malta, on average, each have over three thousand inhabitants.

Population Policies

There are two major forms of population policy: antinatal and pronatal. Some antinatal policies are well known such as the Eugenic Protection Act passed by Japan in 1948, and the one family one child policy of China adopted in 1981 (effective for a time but now women in China average 1.6 children).

Pronatal policies take many forms. At a basic level a tax code that gives tax allowances for children is mildly pronatal. In Europe most countries provide tax breaks, child allowances in cash to supplement family income, childcare services, and paid maternity leave. Sweden was the country to first implement the policies but the actions did not prevent falls in the birth rate. Today Sweden has a total fertility rate of 1.8, which is above the European average of 1.4, so the pronatal policies may have reduced the declines. In the modern European world, taking a paid maternity leave for a year eliminates income loss associated with childbearing, but the year away from work may be perceived, by the recipient, as a possible impediment to advancement. Household costs and the time commitment to childrearing can be seen as an obstacle to enjoying modern lifestyles in the consumer society. (Photo 3.5)

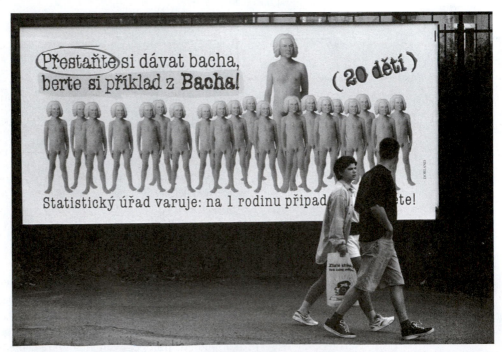

Photo 3.5 Billboard in Prague, Czech Republic. "Stop being so careful, follow the example of Bach! (20 children)."

Source: Photo by Sean Gallup/Newsmakers/NewsCom.

As long as the total fertility rate in European countries is below 2.0, the population of the region will age and decline in number. The average age of the population could be lowered and the total fertility rate increased by admitting more immigrants. Most immigrants are young adults. From a policy perspective, this course of action is problematic for previous influxes of migrants—for example, Algerians to France, Pakistanis to Britain, Turks to Germany—have not been assimilated. In France unemployment rates in immigrant communities are twice the national average. Most European populations have an anti-immigrant bias. Further, there is unemployment in Europe and no aggregate demand for more workers. New immigrants are willing to take on the jobs others will not do but their sons and daughters are not, even if unemployed. (Photo 3.6)

The admittance of migrants is not fully controlled by EU policies. Illegal migrants and persons claiming asylum find a way in. We are used to stories of asylum seekers crossing Caribbean waters to the Gulf coast or the Florida peninsula. In Europe, everyday there are stories of boats from North Africa making a landfall in southern Europe. Sometimes the boats fail and there are deaths by drowning.

The smuggling of people gains publicity as the result of tragedy as when Chinese, who had paid to be brought illegally to Britain, died of suffocation in a truck when the ventilation failed, as the vehicle made the journey from The Netherlands to Britain. The abduction and smuggling of women from non-EU eastern Europe to western Europe for prostitution is common.

Photo 3.6   Beauty shop, immigrants, Paris. France.

Source: Berndt Fischer/AgeFotostock.

▶ POPULATION TRENDS IN THE EU

Unless more migrants are admitted to Europe, we can predict, overall, numbers will decline slowly and the average age of populations will increase. As unemployment is high in many European countries, it is difficult to justify higher rates of immigration. The admittance to EU, in 2004, of a tier of eastern European countries from the Baltics to the Black Sea will result in workers moving from eastern to western Europe. This migration will slightly improve fertility rates in the west and slightly slow the aging of the population. In the east the migration will accentuate population loss.

The population of Europe is aging (Photo 3.7) and, in some countries, shrinking in size. The figures illustrate the story. (Table 3.6). In Europe, as a whole, the birth rate, the number of live births per thousand of the population, per annum, is 10. The death rate, the number of deaths each year per thousand of the population, is 12. The difference between the birth rate and the death rate is referred to as natural increase. In Europe, overall there is no increase. Although some countries, including Norway, The Netherlands, United Kingdom, Spain, and Portugal, do have slow population increases, other countries, including Germany and Hungary, have no natural increase or a minus figure on the natural increase account. From Estonia through the Baltics to Belarus, and

Photo 3.7 Older populations require additional facilities and support.

Source: Alamy Images.

TABLE 3.6   Percentage of Population over Age 65 in 2006

Country or region	Percentage over 65 years of age
Italy	19.0
Greece	18.0
Germany	19.0
Spain	17.0
Sweden	17.0
Switzerland	16.0
Belgium	17.0
France	16.0
United Kingdom	16.0
North America	12.0
Latin America and the Caribbean	6.0
Asia	6.0
Africa	3.0

Source: Population Reference Bureau, 2006 World Population Data Sheet.

south to the Black Sea states of Bulgaria, the Ukraine, and Romania, there is no natural increase. Population loss also results from outmigration.

Europe will experience, to a greater degree than the United States, the problems aging populations present to pension and health schemes. Germany, for example, with a population of 80 million people, has a birth rate of eight per thousand and a death rate of ten per thousand; the population is slowly shrinking in size and aging. We might think with an aging population, and many retirements, work would be plentiful. In fact, over five million Germans are without work and the unemployment rate is near 10 percent.

Migration

Over the last three quarters of a century Europe has experienced large-scale migration of every type: voluntary, forced (involuntary), internal, and international. Before, during, and after World War II there were major forced (involuntary) migrations as the evil empires of Nazi Germany and the Stalinist Soviet Union took territory and rearranged ethnic populations. The term "ethnic cleansing" came into common use with the breakup of Yugoslavia in the early 1990s, but tens of millions of people were forced to move for ethnic reasons in the years 1938 to 1948. (Photo 3.8)

Large-scale displacement of people came after the Soviet Union and Nazi Germany conquered Poland in September 1939. Stalin cattle-carted Poles and peoples from the Baltic states eastward. Germany expelled nearly a million Poles to open up land for Nordic settlement. Hundreds of thousands of Polish airforce, army, and navy personnel found a way to Britain to fight on.

As the war progressed, workers from France, The Netherlands, Czechoslovakia, Belgium, the Ukraine, and Russia were brought into Germany to work in fields and factories to support the war effort. Millions of Jews were transported to death camps in the east.

Photo 3.8 German refugees walking from Poland to Berlin at the end of World War II.

Source: Fred Ramage/Getty Images/NewsCom.

By 1943 it was clear that the allies would win the war and the countries that signed the Atlantic Charter, the founding document of the United Nations, established the United Nations Relief and Rehabilitation Administration (UNRRA). At war's end in Europe the region was full of refugees, displaced persons, forced workers trying to return home, liberated prisoners, and survivors of death camps.

Because of the realignment of borders and ethnic cleansing, the largest group of people to be resettled were displaced persons, people forced from prewar homes. Poles were pushed out of eastern Poland as the Ukraine border moved westward and Germans were expelled from Poland as it received territory at German expense. Then there were the Germans from the Baltics, the Czech lands, Yugoslavia, Hungary, and Romania, who were physically deported or moved because they understood being identified as German would be a deadly liability in a Slavic eastern Europe under Soviet control.

By late 1945 UNRRA, the forerunner of the UN Commission for Refugees, was operating nearly three hundred camps in western Europe, mainly in Germany, for not only Germans were moving west. Millions of Balts, Belarussians, Ukrainians, Hungarians, Romanians, Croats, Russians, and Cossacks had to avoid coming under Soviet and Communist rule. Everyone who had lived in a region occupied by Germany was suspect in Soviet eyes and many in Hungary, Romania, and Croatia, all allies of Germany, had supported the Nazi cause. Ukrainians and Belarussians forced to work for German units were seen as collaborators.

In 1947, the number of UNRRA camps exceeded seven hundred with the organization working to resettle millions of people (Judt, 2005). The onset of the Cold War saw yet another category of migrant: the political refugee seeking asylum.

Gradually the camps emptied as the U.S., Canada, Australia, Britain, and countries in western Europe accepted settlers. All receiving countries favored the entry of young, healthy workers who could contribute to postwar economic growth. Three hundred thousand Jews went to Israel after its establishment in 1948. Most of the Germans forced out of their prewar homes in eastern Europe were settled in east and west Germany.

Economic Recovery and Voluntary Migration

By 1950 European economies were expanding. Labor became scarce. Voluntary migration was reestablished, often with a strong international element. The trains from Sicily and southern Italy filled with young workers moving north to Milan and Turin where industrial production was growing rapidly. In Germany, automobile plants needed labor that came by traditional paths from rural areas and small towns to industrial cities but also across borders. By the 1960s Germany had guest worker agreements with many countries and labor was attracted from southern Europe, Turkey, and North Africa. In the 1970s both France and Germany had over two million foreign workers, making up around 10 percent of the workforce. (Photo 3.9)

The Treaty of Rome (1957) would eventually promote the free movement of manufactured goods, labor, and capital. Workforces in the six original members—France, Germany, Italy, and the Benelux countries—were fully employed at home. There was not much movement of labor within the six, except in the case of Italians moving to Germany, and that migratory stream predated the European Community.

Photo 3.9 Turkish market in Kreuzberg district Berlin held Tuesdays and Fridays.

Source: Alamy Images.

Germany, the fastest growing economy in postwar Europe, drew migrant workers from peripheral areas in southern Europe including Iberia, the Italian south, Yugoslavia, Greece, Turkey, and North Africa. Because of economic growth there is little surplus, mobile labor in southern Europe today and for cultural reasons, migration from North Africa is discouraged. The new source of migrant labor is eastern Europe, particularly the countries admitted to EU in 2004: the Baltics, Poland, the Czech Republic, Slovakia, Slovenia, and Hungary.

All but three of the existing EU countries choose to delay admitting workers from new members joining in 2004. The exceptions are Ireland, Sweden, and the United Kingdom. What attitudes are behind these decisions? In general there is a strong anti-immigration bias, particularly in France. Unemployment is around 10 percent in France, Germany, and Italy so it can be argued new workers are unneeded. However, in spite of unemployment, there are unfilled jobs that unemployed French folk, for example, on generous unemployment benefits, will not do.

Low-paid east Europeans are prepared to be cleaners, laborers, and agricultural workers. Many from eastern Europe provide needed skills as plumbers, carpenters, masons, and metal workers. During the communist era, manufacturing and heavy industry were the basis of eastern European economies. With the collapse of communist regimes in the 1990s many state-run industries closed, downsized, or privatized, shedding labor. East Europe has a surplus of skilled labor.

The United Kingdom, Ireland, and Sweden, all with growing economies and a shortage of plumbers and other traditional crafts, saw the advantage of attracting skilled workers. Since 2004, Britain has admitted approximately a half a million east Europeans who have been rapidly absorbed into the economy. In Dublin there are vibrant Polish areas together with Polish restaurants and pubs. When Polish plumbers began working in France in 2005, there was public pressure to exclude them. In Britain, where a good plumber is hard to find, Poles were welcomed to practice plumbing and other needed crafts.

Sealing Borders?

Eventually free movement of labor will be phased in, across EU, for the new member states of eastern Europe. If the experience of southern Europe is a guide, economies will grow in eastern Europe, and living standards and wages will rise. As economic disparities between west and east decline, so will the migratory streams of workers moving west.

The major migratory problem is not eastern Europe but North Africa and increasingly, sub-Saharan Africa. Every day boatloads of undocumented North Africans try to cross the Mediterranean Sea and make a landfall in an EU country. If vessels reach EU territory the occupants seek asylum and must be processed. Every day European newspapers carry stories of overloaded boats sinking and passengers drowning.

In Morocco, the EU has an African land border, for the Spanish provinces of Ceuta and Melilla are enclosed by Moroccan territory. The provinces are fenced but, armed with ladders and battering rams, migrants force entry.

The Canary Islands, in the Atlantic, lie off the northwest coast of Africa. The archipelago is divided into the Spanish provinces of Santa Cruz de Tenerife and Las Palmas de Gran Canaria. Boatloads of undocumented migrants set off from ports on the

Photo 3.10 A boat with would-be immigrants arrives at the port of Los Cristianos on the Spanish Canary island of Tenerife. In the first wave, migrants are predominantly young men.

Source: DESIREE MARTIN/AFP/Getty Images/.

coast of Africa hoping for landfall in the Canaries. Hundreds of migrants make landfall each day, but vessels flounder or miss their destination and sail out into the Atlantic where rescue is unlikely. (Photo 3.10)

Europe is experiencing a new form of forced migration in which undocumented travelers force their way into the territory of EU members. There is no likelihood that living standards will rise in Africa and reduce the pool of those willing to risk death to find low-paid work in the economically developed world.

▶ FURTHER READING

Bade, K.J. *Migration in European History*. Oxford, UK: Blackwell, 2003.

Livi-Bacci, M. *Population and Nutrition: An Essay on European Demographic History*. New York: Cambridge University Press, 1990.

Livi-Bacci, M. *The Population of Europe: A History*. Oxford, UK: Blackwell, 1999.

Livi-Bacci, M. *A Concise History of World Population*. Oxford, UK: Blackwell, 2001.

Mitchell, B.R. *International Historical Statistics: Europe, 1750–2000*. New York: Stockton Press, 2002.

Population Reference Bureau, *Population Data Sheet 2006*.

Judt, T. *Postwar: A History of Europe since 1945*. New York: Penguin, 2005.

Wrigley, E.A. *Poverty, Progress, and Population*. New York: Cambridge University Press, 2004.

4

SETTLEMENT PATTERNS: FARMS, VILLAGES, TOWNS, AND CITIES

▶ INTRODUCTION

Settlement geography involves the study of the size, spacing, function, and morphology (shape and structure) of urban and rural inhabited places. There is a hierarchy of settlement in which the smallest element is the farmhouse, followed by the hamlet, the village, the market town, the regional city, and major metropolitan areas. As we go up the hierarchy the number of people living in places increases and the number of functions performed by settlements expands. For example, a village would contain a few hundred people and provide some basic services. (Photo 4.1) (Photo 4.2) A market town has a few thousand inhabitants and a greater range of services, including grocery, hardware, and clothing stores together with veterinarians, doctors, perhaps a small hospital, lawyers, schools, and a weekly market. (Photo 4.3) A city would contain a cathedral, university, hospitals, fashionable shopping streets, and specialized financial services.

The farmhouse, the hamlet, and the village are rural settlements; towns and cities are urban settlements. In Europe, the growth of towns has a long history but it is only in the last 150 years that the region has become heavily urbanized. Now a majority of the population lives in urban places. We first review the development of the European network of towns before examining the history of rural settlement, keeping in mind that different parts of Europe have different settlement histories. Eastern Europe, for example, evolved a network of towns much later than western Europe.

▶ THE NETWORK OF ROMAN TOWNS

West of the Rhine and south of the Danube the network of towns is a product of the Roman Empire, although the Romans did not found all the towns on the map today. The Greeks had towns, the Etruscans had urban places in Italy before Rome, the Celts had

Photo 4.1 Hamlets are clusters of houses with few services.

Source: Philippe Colombi/AgeFotostock.

Photo 4.2 Villages provide basic goods and services, including a church.

Source: Miles Ertman/Masterfile.

Photo 4.3   Street market in Brittany, France.

Source: Alamy Images.

towns conquered by the Romans (as at Colchester in southeast England), but the overall pattern of towns in western and central Europe is a product of an empire that consciously created an urban system linked by roads. The roads often followed existing routeways but the Romans paved the routes. Many roads remained in use after the empire collapsed.

Many of the major towns in western and central Europe including London, Paris, Vienna, Lyon, Córdoba, Genoa, Florence, and, of course, Rome are on Roman foundations, although often built over or near to a pre-Roman settlement. At Rome the classical past is still evident in the landscape at the Forum, the Coliseum, and numerous other structures. Genoa and London conceal their Roman past. Chester, in northwest England, still retains Roman walls and the characteristic Roman plan with the two major streets, running north–south and east–west, intersecting at the center of the town. Nîmes, in southern France, retains a Roman coliseum and the huge Pont-du-Gard aqueduct still runs to Arles, although today the town has a modern source of water. Arles was built by the Romans as a port to serve the Rhône valley and it is linked to the sea by a canal to avoid the silty distributaries of the Rhône delta. The huge coliseum and amphitheater remain prominent in the Arles landscape.

▶ THE MEDIEVAL CITY

After the fall of Rome, in the fifth century A.D., Europe is said to have descended into the Dark Ages, but towns were not wholly abandoned and along the routeways there was still some trade. Towns and markets revived in the eleventh and twelfth centuries as well as fairs, that contributed to the growth of trade.

> *Fairs differed from markets in being less frequent, in attracting merchants and goods from a far greater distance and . . . in their specialization in a narrower range of goods . . . the fair at Reims began as a wine sale in the fall . . . cloth fairs emerged at an early date in the southern Low Countries. . . . The Vikings contributed to the growth of fairs in northern Europe, since they used them to dispose of the loot they had gathered on their raids.* (Pounds, 1990, p. 139)

By the thirteenth century it had become profitable to found towns to fill gaps in the network of market towns. Feudal lords would lay out a town with a marketplace and building lots. The lord gained income by land sales, leases, and rents from the new town. In France, many of the new towns grew slowly. In England there were successes like Liverpool (1209), Salisbury (1219), and Hull (1293), as well as failures. An early lesson in what we now term "central place theory" was experienced. If a new market was created close to an existing town, there would not be enough business to attract traders to stalls on market day. The new town would fail. Traders moved from one periodic market to the next. If a town was established in a gap in the network, traders would welcome the new stop and the surrounding population would patronize the more conveniently placed market.

Photo 4.4   Ponte Vecchio over Arno river. Florence.

Source: Age fotostock/SUPERSTOCK.

Central Europe, north of the Danube and east of the Rhine, was short of towns and many new marketplaces were created in the region. Between 1200 A.D. and 1400 A.D., well over a thousand new towns were created in central Europe.

By the end of the thirteenth century, the medieval trading and urban patterns of Europe were well established. Mediterranean cities like Florence, (Photo 4.4) Genoa, and Venice were prosperous on the basis of seaborne trade and trade with northern Europe using the Rhône and Rhine corridors. Towns on the Rhine flourished and in the Low Countries cities like Brugge were important in the woolen trade, much wool coming from England. Genoese and Venetian vessels made journeys out of the Mediterranean to Brugge to bypass the middlemen and the costs of the Rhine routeway. The Hanseatic League, an association of trading cities agreeing to common rules of trade, started in 1241 when Hamburg and Lubeck reached agreement. Other cities, including Bremen, Rostock (f. 1218), Tallinn (f. 1219), and Danzig (f. 1230), joined the League, which was particularly important to Baltic trade and included inland centers like Breslau (Wrocław), (Photo 4.5) and Kraków, both now in modern Poland. In general the thirteenth century was marked by expanding trade, economic growth, colonization of additional agricultural land, and increased agricultural output. Cities grew and prospered in the thirteenth century.

Setbacks in the Fourteenth Century

By contrast, in the fourteenth century, rural and urban populations suffered hardships. Climates were generally mild in the thirteenth century. Crop yields improved. Moorland, marsh, and woodland were brought into cultivation. Early in the fourteenth century

Photo 4.5 Wrocław, Poland.

Source: Mattes/AgeFotostock.

Europe experienced wetter, cooler years, resulting in poor or failed grain harvests and foot rot problems with livestock. Marginal land, brought into cultivation in good conditions, failed to produce crops in wetter, cooler years. There was a Malthusian crisis as harvests shrank and food supply per capita declined. Living standards fell and business decreased at market towns and regional trade centers.

The population problem was about to be solved, unpleasantly. In 1347, ships bringing goods from Asia arrived at Genoa. On board the ships were rats and on the rats were fleas. The fleas carried bubonic plague, the Black Death. Between 1348 and 1352 the Black Death spread across Europe, killing 25–30 percent of the population. The plague produced labor shortages in rural areas. In towns and cities there was plenty of space

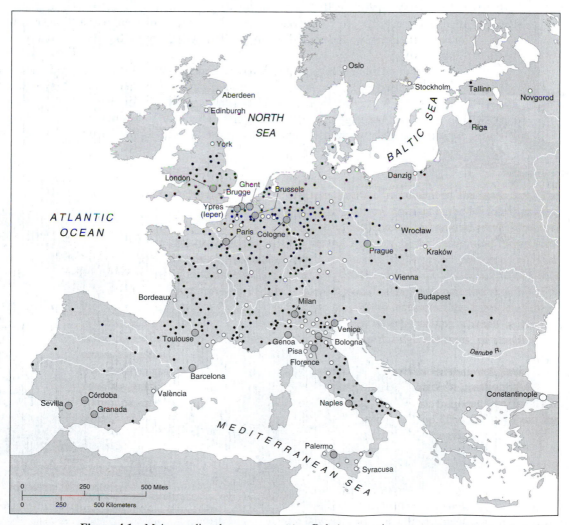

Figure 4.1 Major medieval towns and cities. Relative city sizes shown by proportional circles. (After Pounds, 1990)

for those who wanted to migrate in from the countryside, but in urban places the plague would reoccur and raise death rates.

Urban Recovery in the Fifteenth Century

Urban areas recovered in the fifteenth century and the medieval city reached its full development (Figure 4.1). The larger medieval cities displayed the following characteristics. For trade purposes they were on navigable water and walled for protection. Within the walls high densities of population occurred with poorer inhabitants living in small spaces in crowded alleyways. The range of crafts practiced is indicated by the surviving names of streets: Silver Street, Sadler Street, Leather Street, Broom Street, and Butchers' Row. In addition, open markets operated in central spaces. A small town would have a market one day a week. A larger town would have two market days and cities had several markets specializing in different commodities, for example, a grain market, a cattle market, a hay market, and so on some of which would operate on a daily basis. (Photo 4.6)

Leading citizens occupied fine medieval houses staffed by servants and possessing gardens or courtyards. There was a town hall and the position of Lord Mayor was an honored office. The mayor and the alderman oversaw officials who inspected weights and measures, drainage channels, walls, and gates. Guilds of craftsman and merchants erected guildhalls where they conducted ceremonies and applied the rules that regulated

Photo 4.6 Line engraving, 1617. Leipzig, Germany. Medieval towns were walled and contained high densities of population with many crowded streets.

Source: The Granger Collection, New York.

the practice of crafts and trade. The rules ensured quality and fair practice but were restrictive in that entry to a guild, and the license to practice a trade or craft, was regulated to ensure markets were not oversupplied with goods.

The most pervasive institution in rural and urban areas was the medieval church: the Roman Catholic Church that looked to Rome in western and central Europe, and the Orthodox Church in the east that looked to Constantinople until 1453 when the city fell to the Turks. An important city would have a cathedral and a bishop. The city would be divided into parishes and every parish had a church, perhaps as many as twenty churches within the walls to serve neighborhoods, each with a priest. The Church founded a major urban institution—the university (Figure 4.2). An early European university was

Figure 4.2 Medieval universities.

established at Bologna in 1158 and specialized in reviving and disseminating Roman law. The Bologna model was used to found other universities in southern Europe. Further north the university at Paris, with a collegiate structure, was the model for Oxford and Cambridge, where the colleges dominate the structure of the cities. The Church also created cathedral and parochial schools, which had a large influence on the development of education and were often an important part of the fabric of medieval cities.

On the way to the city were people bringing produce and craft goods and others driving livestock to market. Within a few miles of the city, cultivation was more intense. (Photo 4.7) The fields, within half a day's walk, or cart journey, were a source of milk, meat, fruits, and other foodstuffs. Grain and livestock, brought in by drovers, came from further away but there was intense agricultural activity in the fields around the city and livestock were fattened in nearby fields before going to the slaughterhouse and Butcher's Row. Approaching a medieval city by road or river, the cathedral came into view, then the fortifications, the gates, and the spires and towers of the parish churches. Inside the city walls the streets of medieval towns were often narrow and crowded with adjoining, densely populated alleys. Water came from wells or from the river, which also received the runoff from dirty streets and often served as a sewer. On entering the gate of a medieval city we would

be struck by a barrage of contrasting odors, some of them decidedly unpleasant; we would hear the cries of animals, the rumble of carts, the pounding of horse's hooves, the pealing

Photo 4.7 Medieval city beside a river. It is fall and farmers are sowing the winter grains. Notice the activity on the river.

Source: Fol.10v Month of October: Sowing, from "Breviarium Grimani" by Italian School (16th century) ©Biblioteca Marciana, Venice, Italy/The Bridgeman Art Library.

of the bells, the voices of the people: noises and smells that would vary from one street or piazza to the next, from working days to holidays.

People spent a lot of time out of doors because of the closeness of the houses and the workshops. Everyone knew everyone else, and they liked to chat while resting on benches placed at the doors of houses. (Frugoni, 2005, pp. 63–64)

▶ THE RENAISSANCE CITY

In the fifteenth and sixteenth centuries, new ideas influenced art, architecture, and town planning. New economic opportunities, particularly seaborne trade, had an impact on cities like Sevilla that engaged in Atlantic long-distance trade.

By the mid-fifteenth century citizens had embellished the cityscape of Florence with elegant houses, imposing municipal buildings and new architectural features such as Brunelleschi's dome on the cathedral, for the church and ecclesiastical architecture still shaped major elements in urban morphology. (Photo 4.8) The architectural works of classical authors were reinterpreted and made practical for the times by contemporary writers including Leon Battista Alberti, who was active in Florence and other Italian cities as an architect and town planner in the mid-fifteenth century.

As there was an urban network in western and central Europe, there was not a large demand for new towns, although some were built. Existing towns, however, came under

Photo 4.8 View of Florence featuring Brunelleschi's dome on the Cathedral.

Source: PIXTAL/AgeFotostock.

the influence of new Renaissance ideas in architecture, town planning, and fortification building.

New Walls for Old Towns

The sixteenth century saw the development of efficient artillery, which could knock holes in medieval walls quickly. Medieval walls had been designed for archers, pikemen, and missile hurlers and did not have gun platforms for canon. Military engineers were brought in to build strong curtain walls protected by projecting bastions with artillery positions. The military engineers were the town planners of the sixteenth and seventeenth centuries. When walls were realigned and outworks built, there were opportunities to lay out new streets. The fortification engineers ensured that new urban developments did not compromise defenses. Adequate space had to be left behind fortifications for troops to utilize artillery. Streets had to be straight, and devoid of projecting doorways, to deny any invaders cover. Squares had to be positioned to serve as command stations and mustering points for reserves that might be dispatched to reinforce a section of the fortifications. Many towns did not refortify and fossilized their medieval morphology. Spain is still full of such cities, including Avila in Castile. (Photo 4.9)

New Towns

In Spain the new towns platted in the wake of the *Reconquista* (reconquest from the Islamic moors) in the fifteenth century often display, in basic form, classical ideas on town planning. At Malta, a new capital, Valletta, was laid out in 1566 with a rectilinear

Photo 4.9 Avila, Spain. Medieval fortifications.

Source: Age fotostock/SUPERSTOCK.

Photo 4.10 Starting in 1566, rectilinear Valletta was laid out on the peninsula dominating the mouths of the Grand Harbour left and Marsamxett right. Note the projecting bastions on Fort St. Elmo in the foreground.

Source: RENAULT Philippe/AgeFotostock.

street plan and modern fortifications. (Photo 4.10) In Italy, in the sixteenth century, several new towns including Palmanova (1593) were built using contemporary ideas. (Photo 4.11) Palermo, in Sicily, was expanded and refortified between 1564 and 1570. Other places in the Mediterranean, including Cyprus and Crete, under threat from Islamic expansion, refortified the towns but still fell to the Turks. Danubian cities, like Vienna, were heavily refortified after the Turks pushed up the Danube valley in the sixteenth century.

We can see the style of cities the military engineers wanted to build in the new world. In Spanish America there was a large demand for new towns and those in strategic locations were fortified. In the second half of the sixteenth century, Caribbean cities including Havana, Vera Cruz, and San Juan were fortified. Cartagena, on the Caribbean coast of Colombia, is a well-preserved example and St. Augustine, on the Atlantic coast of Florida, started in 1565, displays all the features that military engineers employed in Europe, including powerful walls defended by artillery capable of engaging an attacker at a distance and a regular street plan allowing the rapid, unimpeded movement of troops and armaments.

In the European town, as defenses improved, architectural styles changed within the walls. In the sixteenth and seventeenth centuries, medieval Gothic gave way to Renaissance and Baroque styles. Sometimes, as at St. Peter's, in Rome, the old church was knocked down (1506) and replaced by a new building, in a new fashion that took over a century to complete. St. Peter's square, started in 1655, is a grand example of the Baroque style and it reflects the trend to create carefully designed squares and plazas.

In the sixteenth century the townhouses of leading families were often rebuilt as small palaces. The poor continued to be crowded into cramped accommodations and as

Photo 4.11   Palmanova northern Italy. A sixteenth century city built to an ideal design of the time.

Source: Alamy Images.

numbers grew, population pressure resulted in the rebuilding of sites with taller structures to house more people.

An Age of Absolutism

Rulers in the seventeenth century, believing they had divine rights, assumed absolute power. Absolutist rulers wanted cities to reflect the grandeur of their presence, for "absolutism demanded perfect settings" (Benevolo, 1993). Space was made for parade grounds, plazas, thoroughfares giving vistas of the royal palace, triumphal archways, fountains, gardens, and statues (Figure 4.3). All these features are on display at Versailles (outside Paris), started in 1671, as a new royal residence for Louis XIV of France. (Photo 4.12) In walled towns absolutist redevelopment often increased overcrowding as buildings were demolished to make room for elegant urban features.

Absolutist styles were widely used in cities across France in the eighteenth century. Bordeaux is an excellent example of eighteenth-century redevelopment. The obsolete fortifications were in disrepair, the gates too narrow to accommodate traffic entering the city, streets were cramped and crooked, often obstructed by buildings projecting into the thoroughfare. Parts of the urban area within the walls were derelict. The city fathers, and the Intendent, had gates and streets widened, filled in the ditch that lay around the walls, and built a ring road. Inside the walls grand open spaces, like Jardin Royal, were created.

German towns were remodeled with Baroque buildings. The outstanding example was Dresden, now restored after being devastingly, and controversially, bombed near the end of World War II. The German lands also had cities with a Versailles-style layout as at Mannheim and Karlsruhe, which are grand, spacious towns laid out as Royal residences. Britain put a stop to Absolutism by beheading Charles I in 1649, then exiling a successor, King James II, in the Glorious Revolution of 1688/9. William and Mary became monarchs and chartered the College of William and Mary in 1693. No great

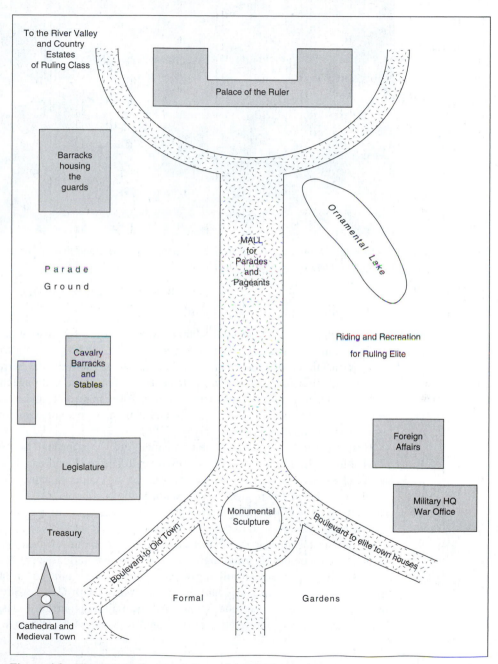

Figure 4.3 Elements in the morphology of the absolutist city.

Photo 4.12 Versailles, France. Absolutist landscape with grand vistas, formal gardens, triumphant monuments, and dominant palace architecture.

Source: Gail Mooney/Masterfile.

royal residential towns were built, although the absolutist urban style is on display when the Queen opens Parliament. The Queen moves by horsedrawn coach from Buckingham Palace, down the Mall through St. James's Park, past a monumental barracks with parade ground, under a triumphal arch to Parliament. The landscape and pageantry, from Palace to Parliament, reflects the eighteenth century. Elsewhere in England resort towns like Bath, with its Crescent, Circus, squares, and classically inspired architecture and grand vistas, display eighteenth-century styles. (Photo 4.13)

The trends in European urbanism from medieval times through the eighteenth century are visible at Edinburgh, the capital of Scotland. The medieval castle was built high on a peak of igneous rock that plugged the vent of an extinct volcano. Beneath the castle a walled medieval town grew with cramped alleys running off the main street, now known as the Royal Mile, that fell steeply down to Cannongate. Beyond the gate was the royal palace, Holyrood, started by the king of Scotland in 1500 with an adjoining park. Today, within the line of the former city walls, few medieval buildings are left for, in the eighteenth century as Edinburgh prospered, old habitations were knocked down and replaced with tall tenement buildings. Overcrowding was relieved in the late eighteenth century when New Town was laid out, to the north, with George Street running from St. Andrew Square to Charlotte Square. All the buildings are in a classical style along broad, straight streets, intersecting at right angles (Figure 4.4). (Photo 4.14) (Photo 4.15)

Urban Population

Although towns and cities grew in the period from 1200 to 1800, most of the population of Europe still lived in rural areas. No European country, in 1800, had an urban majority.

Photo 4.13 The Crescent overlooked and dominated the rural landscape below. These grand townhouses were owned or rented by wealthy visitors to Bath. Jane Austen had a modest house in a less fashionable part of town.

Source: Alamy Images.

In 1500 only 5–10 percent of the population of western Europe was urban. Going to the east, urban populations were smaller and in Russia less than 5 percent of the population lived in towns. By 1600, around 10 percent of people in western Europe lived in towns, but Russia still lagged behind that figure. The seventeenth century suffered a prolonged economic depression. Urban population grew slowly but by 1700, 10–20 percent of people, depending on region, could be described as urban in the west, with the figure falling to the east.

By the end of the eighteenth century in England, Scotland, and The Netherlands there were substantial urban populations but no European country had more than 30 percent of the population in urban places. In the nineteenth century, as industrialization stimulated urbanization, Europe became a region in which rural-to-urban migration created urban majorities. (Figure 4.5)

▶ THE INDUSTRIAL CITY

Preindustrial towns were crowded with the makers of craft goods: silversmiths, cobblers, jewelers, wood carvers, clothiers, milliners, potters, and the makers of furniture, metal goods, and candles, as well as food processors like millers and bakers. Artisans were on fashionable streets, side streets, alleys, and cellars. (Photo 4.16) Noxious trades, animal slaughter and the tanning of leather, were usually pushed to the periphery, but not always.

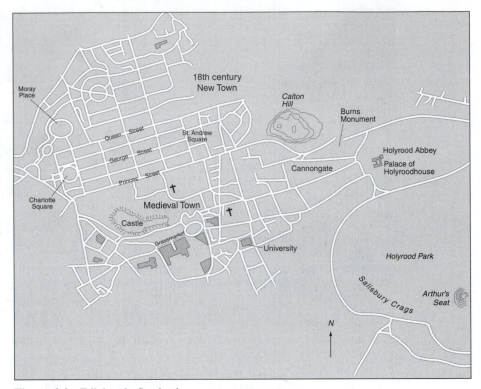

Figure 4.4 Edinburgh, Scotland.

Photo 4.14 View of Edinburgh Castle from Princes Street. Edinburgh, Scotland.

Source: Age fotostock/SUPERSTOCK.

Photo 4.15 Busy Princes Street forms the south edge of Edinburgh's eighteenth-century New Town.

Source: Darius Koehli/AgeFotostock.

There were few large manufacturing operations. Port cities had shipyards, some capable of building many vessels at once as at Venice and Barcelona. The manufacture of armaments, such as the casting of cannon, was undertaken in foundaries.

The surrounding countryside produced products for sale in the city including pottery, wood, reed products, and textiles. Textile making was a cottage industry, often utilizing middlemen who organized spinning, then weaving and the sale of finished products.

Most activities in the city or countryside were handpowered but there were windmills and waterwheels to grind grain and saw timber. When water power was utilized in metal trades, to drive hammers and power bellows, workshops located along streams. In the eighteenth century the textile industry started to use water power to drive spinning and weaving equipment. The use of water power spread manufacturing activity in chains along water courses, as one workshop after another utilized downstream river flow to turn water wheels. By the end of the eighteenth century some factory-type operations were water powered as at the large silk mill in Derby, England. (Photo 4.17)

The use of the steam engine, developed by the Scot James Watt (1736–1819), to power manufacturing brought different locational forces, starting in the late eighteenth century. The steam engine burned coal, so locations on coal fields were favored. The steam engine was uneconomical in a workshop setting. Steam engines were

Figure 4.5 Distribution of European cities c. 1850. (After Pounds, 1990)

economically effective driving multiple machines and that meant factories staffed by many machine minders. The result was, by the nineteenth century, towns on coal fields became manufacturing centers and workers were housed in row upon row of houses built close to factories because workers walked to work.

Industrial cities displayed what we now call environmental injustice with workers living in "toxic wastelands" (Platt, 2005, p. 20) close to factories, while entrepreneurs and professionals inhabited spacious suburbs. Industrial cities degraded air and water quality. Steam engines burned coal, producing smoke and carbon dioxide. Metal smelters threw particulate matter into the air. Warm, polluted waste water flowed from factory to stream. Rivers became industrial drains lacking fish. (Photo 4.18)

Photo 4.16 High Street slum buildings in Glasgow, Scotland.

Source: Mansell/Time Life Pictures/Getty Images.

As seen in Photo 4.16, workers lived in streets of small row houses lacking both running water and sewage disposal. Human excrement was collected and hauled to surrounding fields as fertilizer. Better houses might have a cess pit or a drain that carried wastes to the river. Outbreaks of cholera were common in industrial cities.

Conditions improved in the late nineteenth and early twentieth centuries in parts of western Europe. Public health acts, or city regulations, required the installation of sewers. Water works were created upstream of cities to provide clean drinking water. In central and eastern Europe progress came later, with many parts of Vienna and Berlin imperfectly provided with water and sewage services until the twentieth century.

Urban amenities improved in the late nineteenth century. As towns grew, revenue from property and other taxes increased. Schools, public libraries, recreational space, and, in larger cities, technical colleges and universities were created. Trams—first horse drawn, then electric—allowed the city to spread and create outlying suburbs. (Photo 4.19) The old housing for workers close to the industrial heart of the city remained occupied until post–World War II slum clearance schemes demolished rowhouses and tenements.

▶ RURAL SETTLEMENT

Today, most Europeans live in towns and cities. In the economically advanced countries only a tiny percentage of the workforce is employed in agriculture. The agricultural workforce is larger in eastern Europe. In terms of gross domestic product (GDP), agriculture

Photo 4.17 Silk Mill, Derby. Originally the mill was driven by water power. The smokestacks indicate steam engines have been installed.

Source: Sir Thomas Lombe's Silk Mill, Derby, 18th century (print) by Anonymous. ©Private Collection/The Bridgeman Art Library.

contributes 1 percent in the United Kingdom and Germany and 3 percent in The Netherlands and France. To the east, the figures rise, with Romania drawing 13 percent of GDP from agriculture. In spite of urbanization, the mechanization of farming, and the decline in agricultural workforces, the landscapes of rural Europe are still characterized by farmhouses, cottages, hamlets, villages, and small towns, which provide goods and services to surrounding areas. In the Midwest of the United States, increasing farm size has led to the abandonment of farmhouses and the decline of small settlements in rural areas. This has not generally been the case in Europe. People commute from countryside villages to work in cities. The demand for second homes and vacation homes has meant that old farmhouses and cottages in villages have been renovated by those who want a place in the country. Britons, Germans, and Scandinavians buy or rent farms, villas, and houses in Spain, southern France, and Italy to enjoy an assured warm summer. In the United States many aspire to a condo in Florida. In western Europe the ideal is a Tuscan farmhouse within reach of Florence.

SHEFFIELD SMOKE.
From a Drawing by A. Morrow.

Photo 4.18   Circa 1890. Smoke rising from factory chimneys over Sheffield.

Source: Hulton Archive/Getty Images.

Rural Settlement Types

Nucleated and Dispersed Settlements

Rural settlements are classified as nucleated or dispersed. The most widespread form of nucleated settlement is the village, an agglomeration of residences and buildings providing basic services, including a school, church, village shop, bar, and post office. Dispersed settlements are represented by individual farmhouses and cottages dotted across the landscape with perhaps small clusters of houses at a crossroads. (Photo 4.20) Such small clusters are called hamlets and they provide few services.

It has been recognized that some parts of Europe tended to be characterized by dispersed settlements and other regions by nucleated settlements. For example, it is agreed that the highlands and islands of Scotland, Wales, Ireland, Brittany, western France, northwestern Spain, northwestern Germany, and Sweden have dispersed rural settlements. The midlands of England, northeastern France, western Germany, the Spanish meseta, peninsula Italy, and the Danubian plains have villages and nucleated settlements.

Photo 4.19 1901. A horse-drawn tram. Portsmouth, U.K.

Source: Hulton Archive/Getty Images.

Photo 4.20 Dispersed farmsteads are common in west European landscapes. Note the intensive cultivation at this farm in Verzenay, France.

Source: Age fotostock/SUPERSTOCK.

Origins of Rural Settlements

Explaining the origin and distribution of nucleated and dispersed settlements is not easy. The German settlement geographer August Meitzen, over a century ago, proposed an ethnic explanation. The dispersed settlements of Brittany, the British Isles, and elsewhere were the result of Celtic settlement. The nucleated villages of northern France, Germany, Denmark, and the English Midlands were a product of Germanic settlement after the fall of the Roman Empire. At first view this thesis is attractive. In Brittany the inhabitants speak a Celtic language and live in dispersed farms and cottages. In Cornwall, where a Celtic language was spoken, away from the relatively few towns, the countryside is covered with dispersed farms and cottages.

In the midlands of England, a land of compact, nucleated, rural settlement, the villages all have Germanic names coined by Anglo-Saxon settlers. Common Germanic place-name elements include *ham* and *ton,* as in Birmingham (the settlement of Birm's people). Meitzen's initially attractive thesis does not stand up because there are many areas of dispersed settlement in zones of Germanic colonization, including East Anglia, southern Sweden, Flanders, and Germany.

Other factors that commentators have examined in the search for origins are numerous. In parts of France, for example, the origin of outlying farmhouses may be associated with the distribution of Roman villas, for after the empire contracted, the local labor tended to stay around the villa and perpetuate the settlement site. This explanation cannot be used to explain dispersed settlement in general because some areas of dispersed settlement—for example, central Wales—were not settled by the Romans. In the Po valley the *corti,* villages built around a central courtyard, are said to have started around Roman villas. (Photo 4.21)

Photo 4.21   Hadrian's Villa. Lazio, Italy.

Source: Age fotostock/SUPERSTOCK.

The form of landholding may be of fundamental importance. If a society has individual land ownership many families would choose to live on that holding to be close to crops and livestock. This factor would give rise to a dispersed pattern of farmhouses. This is what happened later on the prairies and plains of North America when governments made land available in blocks that encouraged the development of single-family farms and farmhouses.

However, if a society has communal land ownership, and elements of communal farming, groups would tend to settle in the middle of the communal lands and live in a village, as long as the land was held in common and the villagers farmed the surrounding open fields. Open fields are not enclosed in individual plots. Members of the community have cultivation rights in the fields but not title to land. What were termed "open-field villages" were common from lowland England, through northern France, to the Jutland peninsula and across the lands on the south shore of the Baltic Sea. Open-field villages persisted into the nineteenth century when the fields were often divided into individual farms, and families then moved to farmhouses on the land. A few open-field villages are still preserved as historic sites in England and Denmark. The open-field system worked in the following way. Members of the community lived in the central village surrounded by two or three large fields consisting of many strips of land. Each year lots were drawn to distribute the cultivation strips to households. If you were lucky you drew strips that were near the village and minimized walking. On common grazing land each household was allowed to run a stated number of animals. The origin of the open-field village, with group settlement at the center of the lands the community controlled, may go back to the time of Germanic settlement.

There are other reasons why a group of people might cluster in a community and form a village settlement. If water is available at a prolific and persistent spring, a settlement might form close to an assured supply. People might choose to cluster together for defense. The numerous, large, inconveniently sited hilltop villages on the Italian peninsula are held to be the product of a long history of sea raiding, piracy, and slaving. To be captured meant transport to a North African slave market. This may seem far fetched but remember Thomas Jefferson found it necessary to send frigates to put down Barbary corsairs in the early nineteenth century. Once the seas around the Mediterranean coasts of southern Europe were secure, people moved from hilltop villages to new settlements on the roads running along coastal plains. If you visit the crowded resort coasts of southern Europe you will notice that, apart from ports and other fortified places, most of the building is recent. The old inland, uphill towns and villages are often in decline.

In southeast Europe, particularly on the Danubian plains, there are large villages containing thousands of inhabitants. Some attribute these places to the need for defense but social forces may bring about agglomeration.

The attitudes of large landholders are a factor to consider. A landholder might want to rent, lease, or share crop fields. The landowner may not want the peasants who cultivate the land living on the land because peasants are difficult to dispossess once tied to territory. From the landholders' point of view it may be better to have the peasantry living in the village and walking out to fields each day.

Photo 4.22 Livestock, in front of Benedictine abbey of Saint-Michel de Cuxa, move and graze along a country road. Pyrenees, France.

Source: Sylvain Grandadam/AgeFotostock.

Spacing and Function of Settlements

There are basic factors that influence the spacing of traditional settlements. If a population lives in villages, what is the maximum distance that cultivators are prepared, or able, to walk to fields and back in a day? Whether the acceptable distance to cultivation plots is half a mile or one mile, on an agricultural plain the distance would give regularity to the spacing of villages.

The same type of factor influences the spacing of traditional market towns. How far can you walk or ride to market in the morning, conduct business and return home in the evening? Humans can move 4 miles per hour on foot, faster on horseback, but slower if driving livestock to market. (Photo 4.22) Ten miles to market and ten miles back on foot would represent a long day of travel. In practice, on an agricultural plain, 10–12 miles between markets is the maximum spacing, implying journeys of 5–6 miles in each direction for farmers in the hinterland of the town. Where livestock were involved journey times were slow. "To market to market to buy a fat hog. Home again, home again, jiggety-jog" does not address the difficulty of getting the hog home. (Photo 4.23)

Implicit in the idea of towns drawing customers from the surrounding area is the concept of hinterland. Around any settlement that provides services is a zone from which it draws people to access services. In a village the services are few: a church, post office, café or pub, a village store supplying basic items. People travel short distances to access services.

The town with a market and a greater range of services has a larger hinterland and people travel several miles in both directions to use the services. The inhabitants in the town contribute to the demand for services.

Photo 4.23 A traditional market at Stein in Switzerland. Livestock in the center, stall holders on the periphery.

Source: Age fotostock/SUPERSTOCK.

Cities have many towns and villages in the hinterland and a large resident population. The city provides a greater range of services and draws people longer distances to visit the cathedral, the more specialized markets, and craftsmen. Attempts have been made to construct the geometry of service center distribution. The best known effort is that of the German geographer Walter Christaller. Under the supervision of Professor Robert Gradmann, Christaller noted the regularity of the spacing of villages and towns on the south German plain. (Figure 4.6)

Christaller constructed a service center landscape using a hierarchy of settlement that included hamlets, villages, towns, and cities. There are, in reality and in Christaller's theoretical landscape, more small places than large places, with the city having the largest hinterland. Notice that the hinterlands of all settlements, great and small, are hexagonal. The hexagonal hinterlands (service areas) are a theoretical requirement to achieve efficient nesting. If circular hinterlands were used service areas would either overlap or leave gaps in the interstices between the circles.

In real landscapes there are no neatly shaped service areas. Christaller attempted to isolate the factors of distance and relative location to analyze how those factors influence the spatial distribution of villages, towns, and cities. To focus on relative location and distance Christaller made some basic assumptions. His theoretical landscape of service centers was constructed on an isomorphic plain. There were no hills and valleys, or distinctions between rich cultivated land and poor land used to pasture animals. It was assumed that people living on the plain would utilize the nearest service center.

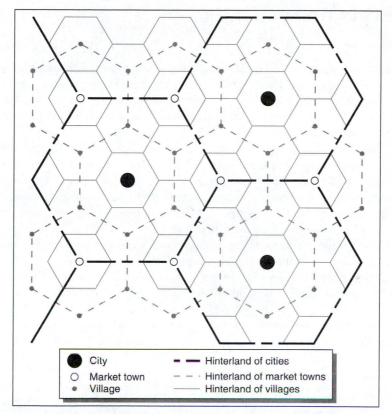

Figure 4.6 A theoretical distribution of settlements, after Christaller.

This was later termed the "principle of least effort." Then Christaller postulated that for every good there was a threshold of demand to be met before it was profitable to offer the good for sale. A village with a small population in the settlement, and surrounding hinterland, would sell lower-order goods and services: in general, everyday subsistence items. A town with more inhabitants and a larger hinterland would reach more thresholds of demand. A town could offer a greater range of goods and services.

In real-world European landscapes, settlements adjust to topography and seek sites that offer advantages. Frequently towns and villages form linear patterns along river valleys or other routeways but the basic factors still operate. Villages are closer together than towns and towns are usually miles apart, and offer more goods and services, than villages or hamlets.

Christaller used his knowledge of settlement distribution in the service of states. In World War II he helped design a network of places to be occupied by Nordic settlers, part of Hitler's plan to occupy *Lebensraum* in the east at the expense of the Slavs. Newly planned settlements were created in Poland and Poles were ejected from the land to make space for new towns. After World War II, Christaller was in East Germany (GDR) planning settlements for the Communist government.

► SETTLEMENTS IN THE MODERN ERA

Rural Settlements

Since the end of World War II farming has become more mechanized, labor requirements have decreased, and farm size has increased. It might be thought that rural areas would display deserted cottages and abandoned farmhouses. Although uninhabited houses are found in marginal zones like western Ireland, and village depopulation is common on the Spanish meseta and the hill regions of southern Italy, in most rural regions the loss of farmers and agricultural laborers has been offset by other factors. (Photo 4.24)

The demand for second homes and holiday houses has meant that farmhouses in France, Spain, and Italy have been purchased and renovated by people who live and work in the city. The purchasers may be nationals wanting to retain links with rural landscapes but often the buyers are Dutch, German, British, or Scandinavians who want a vacation home in southerly, sunnier weather.

Just as small towns in the United States close to major centers became homes for commuters, European towns and villages within reach of larger places have residents who travel to a city to work. The demand for houses in rural places by those with well-paid jobs in urban areas drives up prices, making it difficult for people in lower-paid rural work to afford houses in their home community.

If villages and small towns have survived and often grown, as places to live, the services they provide have diminished. There are several causes of service contraction.

Photo 4.24   Tuscan landscape with dispersed farmhouses.

Source: Simon DesRochers/Masterfile

The population that lives in the village but works in the city accesses services and buys goods in the larger place where prices are more competitive and the range of goods and services greater. In the modern world, services concentrate in larger places. Village schools close and pupils travel to bigger institutions with more facilities and a better range of courses. Hospitals in smaller towns do not have the demand to allow the installation of new technology. Patients are sent to a regional hospital and the local facility closes or assumes a primary care function.

In retailing bigger stores require a higher threshold of demand to be viable. Perhaps this would protect village and small town stores from competition? But those with cars drive to the larger towns with a high (main) street containing supermarkets, chain stores, and banks. The result is village stores and family businesses in small towns lose business and close. As less business is done in smaller places, bank branches, post offices, and other services concentrate in towns with sufficient demand to make money. With wider car ownership the demand for public transport decreases and the frequency of bus services is reduced. Just as in the United States small service stations and garages close as gas is sold in high-volume locations and cars are sold on mega lots. The landscape of Europe is not full of deserted villages and decaying small towns but the range of services offered in rural centers has decreased.

Urban Areas

For nearly a decade after World War II most European urban areas were under stress. Infrastructure had been destroyed or overworked. Public transport was overcrowded, and bomb sites (weedy vacant areas from which the rubble had been cleared) were common. Even in towns and cities not directly involved in the conflict, housing stocks had not increased and accommodation was scarce. The central areas of most German cities had been bombed out. People existed in cellars and makeshift shelters. In Warsaw it was worse. The American ambassador described the city of rubble as a place without lights. In every urban place that had been fought over, in eastern Europe and the Soviet Union, the devastation was terrible because the fighting front had passed through towns several times as the Axis and the Soviets advanced and retreated. Often the damage was worse in retreat as buildings and infrastructure were blown up to deny the attacker shelter and services. (Photo 4.25)

Workforces everywhere were depleted and the task of clearing rubble and reclaiming bricks for reuse was often done by women. Attitudes to reconstruction varied. In places there was a determination to rebuild streets in prewar form, as happened in central Warsaw. Sometimes particular features were restored. At Florence the masonry of a demolished bridge over the river Arno was recovered from the water to form the reconstructed structure. The famous Benedictine monastery at Monte Cassino in southern Italy was the center of a prolonged battle at the end of which the holy place was a pile of rubble. The monastery was rebuilt. At Coventry, in the United Kingdom, the stark, bombed shell of the cathedral was left as a reminder of the cost of war. The restoration of Dresden, brutally bombed near war's end when full of refugees fleeing the Russian advance, was finished recently.

Often as landmarks and prominent streets were restored the rebuilding of surrounding spaces was pragmatic. (Photo 4.26) Proud St. Paul's Cathedral survived the blitz but

Photo 4.25 U.S. soldiers walk through a ruined town in Germany.

Source: Fred Rampage/Keystone/Getty Images/NewsCom.

Photo 4.26 "Standing up gloriously out of the flames and smoke of surrounding buildings, St. Paul's Cathedral is pictured during the great fire raid of Sunday, December 29th." London, 1940.

Source: National Archives and Records Administration 1940. 306-NT-3173V.

proximate buildings were destroyed. The rebuilding was carefully designed but resulted in the Cathedral having office tower blocks as neighbors, leading Prince Charles to the controversial remark that the architects had done more damage than the bombers. Frequently developers took the opportunity to rebuild on sites without reference to what had been in place before or the character of the traditional street. Close to its historic center Warsaw got high-rise buildings reflecting Moscow's Soviet architecture.

In eastern Europe, taken over by communist regimes after the war, gray blocks of housing for workers dominated urban landscapes that centered on squares featuring statues of Marx, Lenin, or Stalin. Planning and control dominated development processes in socialist eastern Europe and the Soviet Union.

Land use planning was common in western Europe. In general, western Europeans wanted to restore cities and conserve traditional landscapes, urban and rural (Ford, 1978). Urban sprawl that had begun to appear before World War II was contained.

In Britain the Town and Country Planning Act of 1947 required that all urban and rural areas produce maps recording and designating the use of land. Land was classified as agricultural, recreational, residential, or developable. Changes in usage could only be made after an extensive, time-consuming process. Most parts of Europe adopted regulations to prevent piecemeal urban growth and in the east the state controlled all development from land acquisition to construction and allocation of accommodation.

New Towns

New housing and new sites for industries had to be provided and countries in eastern and western Europe adopted the concept of New Towns. There was of course nothing new about the idea of new towns. The Romans laid them out as did medieval lords and nineteenth-century industrialists, but in the postwar era they were seen as a way to expand housing stock, providing services and sites for new industries in a planned, logically laid-out urban setting. The creation of towns was common in nineteenth-century America but the concept of laying out complete urban places with residential, manufacturing, and recreational areas, together with shopping services, has not been widely employed in recent decades, although excellent examples do exist at Reston, Virginia, and The Woodlands in Texas.

Many European new towns were built on green field sites. Britain passed a New Towns Act in 1946, and 28 new towns were built including eight around London beyond the Green Belt, which was to contain the sprawl of the capital. Later, more new and enlarged towns were authorized, including Milton Keynes. In this case an existing town with good transportation links and expandable services was enlarged, starting in 1967, until it held around a quarter of a million inhabitants. Many new economic activities came to Milton Keynes, within reach of London, including the Open University. The University allows people to enroll in degree courses, study via distance learning, and attend summer schools on existing university campuses around Britain. The Open University enrolls more students than any other institution of higher education in the United Kingdom.

The Dutch created new towns to serve reclaimed areas around the Zeider Zee. The Randstad city region, which includes Amsterdam, Rotterdam, and The Hague, contains approximately one-third of the population of The Netherlands and densities are high even by the standards of a country that contains, on average, a thousand people for every square mile. The Randstad had a green center and to preserve this planners directed growth outward into new communities and housing areas. The Netherlands—with high population densities and the need to control land use around rivers, coasts, dykes, and canals—is probably the most carefully planned landscape in Europe.

Paris is the largest city in Europe. Since World War II the metropolitan area has increased rapidly in size and population. New towns have been created around Paris to absorb growth. New towns were also built near Lille, Rouen, Lyon, and Marseille in an attempt to deflect growth from Paris. New towns have been developed around Stockholm and the Finns established Tapiola near Helsinki.

Many new communities have grown up along the coasts of Spain, Portugal, and Italy to serve the mass tourist market that developed in the later decades of the twentieth century. Some towns were planned, but most sprawled along the shore, overstressing coastal environments.

The Socialist City

There are many examples of carefully planned new towns in eastern Europe and the former Soviet Union. Magnitogorsk in the Urals (founded 1929) and Nowa Huta (founded 1949) in Poland are examples. The cities had central services, industrial zones, and residential and recreational areas with public transport linking homes, workplaces, and services. The state planned and built the cities. (Photo 4.27)

Most cities in eastern Europe existed before communist regimes took over after World War II. How was the urban fabric altered by socialist governments? Perhaps surprisingly there was much variation from country to country and town to town. War damage was a factor. A badly damaged city was easier to rebuild on a socialist model, but in several cases destroyed central, historic areas were restored. Warsaw is an example. Bucharest, Romania and Sofia, Bulgaria were extensively replanned with major squares, thoroughfares, and buildings to house the state and party apparatus. (Photo 4.28)

Buildings and spaces associated with capitalist-imperialist pasts were not usually destroyed. Rather the regime would co-opt symbolic places, renaming and reinterpreting them. In east Berlin, the Brandenburg gate was reinterpreted as a peace gate, the Lustgarten was renamed Marx-Engels Square, and the Unter den Linden, linking the two sites, was used for parades, as it had been by the Prussians and Nazis. Along the Unter den Linden some historic buildings were restored, while others were demolished to make space for socialist structures (Stangl, 2006). (Photo 4.29)

In the historic core of Prague there are few socialist buildings. (Photo 4.30) At Sofia, capital of Bulgaria, and Bucharest, capital of Romania, city centers were reworked to accommodate ceremonial squares with statues of communist leaders and monolithic blocks to house communist governments and their ministries (Danta, 1993).

Many eastern European cities evolved in the following way in the postwar/Cold War era. Adjacent to the historic core a new city center was laid out. Prague is an exception. In the areas of poorer housing that had surrounded historic cores, demolition was common as streets were widened, infrastructure improved, and offices and housing blocks were erected. Further out, beginning in the 1950s, five-story blocks of worker housing were built, along with welfare and consumer services. New industrial zones were built on the outskirts (Hamilton, 1979).

Photo 4.27 Boulevard Unirii and Parliament. Bucharest, Romania.

Source: SUPERSTOCK.

Photo 4.28 Council of Ministers Building, Sofia, Bulgaria. Notice the prominent images of Communist leaders on the front and side of the building in the Soviet era.

Source: SUPERSTOCK.

Since the end of the Cold War, more effort has been made to see the parallels between postwar urban development in eastern and western Europe. Prague, with a metro system linking the now privatized housing projects to both the historic core and the industrial zones, employs a similar concept to Stockholm, where postwar development concentrated on linking dwellings, workplaces, and the central city. There is a rich town planning literature going back to the garden city themes of the late nineteenth century. Planners worldwide are aware of the theories and, in the postwar world when housing was short and economies growing, it is not surprising that similar concepts were used in eastern and western Europe (Figure 4.7).

Many socialist towns are unsuited to the rise of consumerism and the use of cars. Public transport overloaded as consumers made more shopping trips. Public housing projects in the 1950s and 1960s, in both eastern and western Europe, made little provision for car ownership. As public projects were privatized, in east and west, the new owners turned small yards and sidewalks into parking spaces as car ownership became the norm.

Shopping, Traffic, Commuting, and Suburbs

After World War II, the percentage of women in the workforce was still low. Many housewives shopped every day, visiting the local butcher, baker, and grocer. No private

Photo 4.29   Unter den Linden, Berlin. The Brandenburg gate is top right.

Source: Walter Bibikow/SUPERSTOCK.

cars were made during World War II and they were slow to reappear on roads after the war as gasoline was scarce and expensive. Most people used public transport to visit the city center.

By the late 1950s and early 1960s, as living standards improved in western Europe, Fiat Toppolinos, Volkswagen Beetles, Renaults, Morris Minors, and small, cheap Citroens were crowding city streets. Congestion and parking became problems. Traffic engineers tried to improve traffic flows by widening and straightening streets, but problems got worse as bomb sites that had been parking lots were redeveloped, often with high-rise buildings that generated more traffic. Traffic engineers kept trying to improve flows within city centers, even when it meant degrading historic urban landscapes.

One classic proposal of destruction concerned Christ Church Meadow in Oxford. The Meadow is two blocks south of High Street in an area of wetlands and hayfields where the rivers Thames and Cherwell come to a confluence. Christ Church College owns the Meadow but opens it daily to visitors, locals, and members of other colleges rowing on the river. In the early 1960s, the planners proposed compulsory purchase of land through the Meadow and the construction of a sunken highway to relieve traffic congestion in central Oxford. Against opposition the proposal made progress and Prime Minister Harold Macmillan, an Oxford man, had to call in the Minister of Transport and tell him to stop the nonsense. Christ Church Meadow was saved. Other historic towns in Europe, lacking powerful friends, had historic streets altered to improve traffic flow and surrounding beauty spots run over by ring roads.

Photo 4.30   Prague Castle and Vltava river in Prague, Czech Republic.

Source: Age fotostock/SUPERSTOCK.

By the 1970s and 1980s efforts were made to reduce traffic flows in cities by improving public transport and park and ride systems. Pedestrian precincts, (Photo 4.31) peripheral parking, one-way systems, and traffic wardens tried to reduce the flow of cars into town centers. Only those with access to private parking or who could afford to pay for public parking could come to the city to conduct business without using public transport or walking a long distance.

Until the 1980s most western European countries strived to maintain planning regulations that kept cities compact and out of the countryside. With growing affluence and more women in the workforce, the demand for convenient shopping grew. In the 1980s France built hypermarkets: large, outlying shopping centers with motor car access. Other countries relaxed their compact city planning regulations to allow well-designed shopping centers on the urban edge.

Because western European cities had tight planning regulations the movement to the suburbs was less abrupt than in the United States. New suburban areas were built, but at a slower pace. European city centers were not decayed by the flight to the suburbs in the same way that U.S. CBDs were. In most large- and medium-sized European cities the central streets remained vibrant with a strong retail presence in the downtown. (Photo 4.32) Small towns, as in North America, have suffered as family businesses cannot compete with national and international chains that offer a greater range of goods in the larger places.

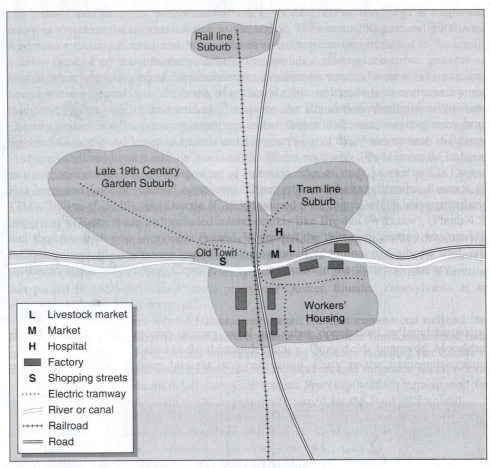

Figure 4.7 Elements in the morphology of European cities c. 1900.

Photo 4.31 Pedestrianized street. Germany.

Source: Waldkirch/AgeFotostock.

European cities have not been without problems close to the city center. In the nineteenth century rowhouses were built for workers within walking distance of the center or close to factories. The housing was often crowded into alleys and courts. Rowhouses were built back to back with only one entrance and no inside sanitation. Even late nineteenth-century housing for workers tended to be two rooms up, two rooms down, and a small yard at the back to house a privy and give access to a narrow lane between streets.

After World War II countries adopted slum clearance programs. But before slum demolition could begin new accommodation had to be provided for the slum dwellers. Municipalities built new housing projects that were often unsatisfactory. People were put into high-rise units unsuited to families with children. Other blocks were not so high but had the same dull, drab appearance of housing for workers in communist eastern Europe. An unwanted side effect was that slum clearance destroyed the communities that had existed in the areas of poor housing. Some took the view that while slum clearance was necessary to improve health and living standards, visually, one eyesore was replaced by another.

Changing City Structure

Pierre Laborde (1994) examined the manner in which the structure of the European city had evolved in the postwar years. He provided detailed study of French towns but the trends he identified are common to most towns and cities in northern and western Europe (Figure 4.8) (Figure 4.9).

▶ ETHNIC COMMUNITIES IN EUROPEAN CITIES

Port cities have always had ethnic neighborhoods as traders and sailors from foreign lands settled and did business. In the second half of the twentieth century, as European economies grew, there was frequently a demand for less skilled labor to perform tasks native-born Europeans did not want to do. Frequently, migrants arrived from former colonial territories to work and live. As migration chains developed from colony to a community in a European city, new arrivals tended to settle together. Cheap accommodation was frequently found in working-class streets near city centers and close to jobs. Sometimes the new arrivals were absorbed into the mainstream of city life; frequently ethnic communities stayed apart to reinforce and protect cultural traits and religious practices. France, Britain, The Netherlands and other countries found it difficult to assimilate Muslims. Major French cities have significant Muslim communities. Efforts to reduce the sense of separateness, including the banning of religious head dress in state schools, are controversial. Hopes that integration would come as second and third generations were born in Europe have not been fulfilled. The July 2005 London tube bombers were born in Britain. They blew themselves up and killed over 50 passengers. Pakistani and Saudi Arabian officials suggested that Britain, in the name of freedom of speech, had allowed fanatics to preach terror.

The origins of ethnic communities are widespread. Amsterdam has populations derived from the former Dutch East Indies (Indonesia), Surinam (many of whom are Muslims), and The Netherlands Antilles. France has Muslim communities that came from Tunisia and Algeria. Former French West African colonies are another source of

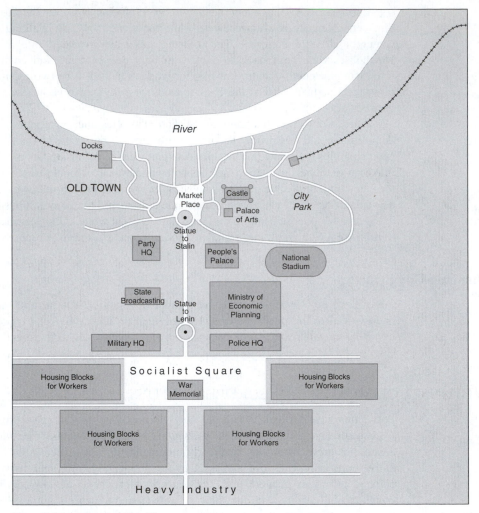

Figure 4.8 The Socialist City.

migrants. Portugal has drawn people from Angola, Mozambique, and the former Far East colony of Macao. Britain has urban communities derived from the West Indies, particularly Barbados and Jamaica, and the Indian subcontinent. The Muslim communities from Pakistan have not assimilated. People of Indian heritage are much more likely to be absorbed into national life in the way many migrants are assimilated in the United States.

In addition to former colonial territories, Europe itself provides migrants who move from one country to another looking for work. When there were large wage disparities between southern and northern Europe, Italians, Greeks, and Spaniards moved into France and Germany to take work under EU free movement of labor rules.

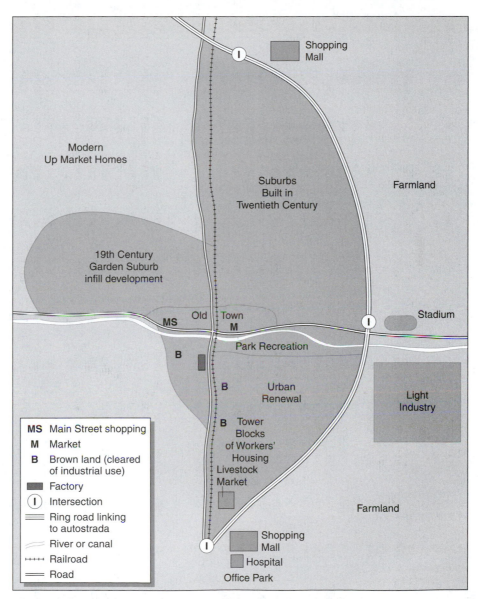

Figure 4.9 Elements in the morphology of the contemporary European city.

Turkish workers moved into jobs in German factories, often living in ghetto-like conditions and finding it difficult to become German citizens. Vienna, too, houses Turks and people from the former Yugoslavia. Vienna, the former capital of the multinational Austria-Hungary Empire, has long had many ethnic communities. More recently, as eastern Europe has joined the EU, communities of Poles have formed in Britain, Ireland, and Sweden.

TABLE 4.1 The Growth of European Cities (in Thousands)

	1750	1800	1850	1900	1910	1920	1930	1940	1950	1960	1970	1980	1990	Latest
Moscow	130	250	365	989	1,533	1,050	2,029	4,137	NA	5,046	7,061	8,203	8,747	10,400
Paris*	576	581	1,053	2,714	2,888	2,907	2,891	2,830	4,823	7,369	8,197	9,707		9,645
Istanbul														9,120
London	675	1,117	2,685	6,586	7,256	7,488	8,216	8,700	8,348	8,172	7,452	7,678	6,803	7,172
St. Petersburg	150	336	485	1,267	1,962	722	1,690	3,191	NA	3,321	3,950	4,676	4,437	4,660
Berlin	90	172	419	1,889	2,071	3,801	4,243	4,332	3,337	3,261	3,208	3,057	3,438	3,382
Madrid	109	160	281	540	600	751	834	1,089	1,618	2,260	3,146	3,188	2,991	2,882
Kiev	23	23	50	247	505	366	514	846	NA	1,104	1,632	2,248	2,643	2,663
Rome	156	163	175	463	542	692	1,008	1,156	1,652	2,188	2,800	2,831	2,828	2,656
Bucharest	NA	32	120	276	341	309	631	648	886	1,226	1,475	1,929	1,934	1,921
Budapest	NA	54	178	732	880	1,185	1,006	1,163	1,571	1,805	1,945	2,059	2,017	1,852
Barcelona	50	115	175	533	587	710	783	1,081	1,280	1,558	1,745	1,755	1,668	1,787
Minsk	8	11	24	91	101	104	132	239	NA	509	907	1,333	1,613	1,729
Hamburg	75	130	132	706	931	986	1,129	1,682	1,606	1,832	1,794	1,649	1,661	1,702
Warsaw	23	100	160	638	872	931	1,179	1,266	601	1,136	1,308	1,596	1,654	1,618
Vienna	175	247	444	1,675	2,031	1,866	1,874	1,918	1,616	1,628	1,620	1,531	1,540	1,608
Belgrade	NA	25	15	69	91	112	267	NA	368	585	746	1,088	1,136	1,594
Kharkiv	NA	10	25	175	236	220	417	833	NA	934	1,223	1,485	1,622	1,494
Novosibirsk	Town founded 1893, name Novosibirsk in 1925													1,390
Gorky (Nizhniy Novgorod)	9	14	31	90	109	106	222	644	NA	942	1,170	1,367	1,441	1,350
Tbilisi	NA	30	35	161	188	327	294	519	NA	695	889	1,095	1,300	1,310
Milan	124	135	242	493	579	836	992	1,116	1,260	1,583	1,724	1,635	1,549	1,302
Munich	32	40	110	500	596	631	735	828	832	1,085	1,294	1,299	1,237	1,194
Sofia	NA	NA	NA	68	103	154	287	401	435	687	877	1,057	1,220	1,192
Prague	59	75	118	202	224	677	849	928	922	1,005	1,080	1,193	1,212	1,178
Zagreb	NA	13	14	61	79	108	186	NA	321	438	574	650	704	1,047
Perm (Molotov)	6	3	13	45	50	68	120	255	NA	629	850	1,018	1,099	1,018
Rostov-na-Donu	NA	4	13	120	121	233	308	510	NA	600	789	957	1,027	1,006
Odesa	2	6	90	405	506	428	421	604	NA	667	892	1,072	1,096	1,002
Naples	305	427	449	564	723	722	839	866	1,011	1,183	1,233	1,211	1,208	1,000
Volgograd (Stalingrad)	NA	4	NA	56	78	90	15	445	NA	592	818	948	1,006	996
Birmingham	24	74	233	523	840	922	1,003	1,053	1,113	1,107	1,015	1,007	944	977
Cologne	43	50	97	373	517	634	757	768	595	809	848	977	956	968
Brussels	60	66	251	599	720	685	840	913	953	1,020	1,075	1,000	954	965

*1950, 1960, 1970, 1980, and latest numbers reflect population of greater Paris

Source: Mitchell, *International Historical Statistics Europe, 1750—2000*. New York: Stockton Press, 2002.

Photo 4.32 Below the castle, and across the Charles bridge is Old Town Square, Prague.

Source: Age fotostock/SUPERSTOCK.

▶ SUMMARY

The cities of Europe (Table 4.1) have ghettos, traffic problems, overcrowding, and much substandard accommodation. In general, cities, as a result of planning and attempts to curb sprawl into the countryside, have maintained viable downtowns, with high (main) streets full of shops and shoppers. Many complain that because of chain stores central shopping streets look too much alike from town to town. Frequently the historic buildings have been preserved and restored. European city centers are vibrant with shoppers, visitors, promenaders, and sidewalk cafés. Few U.S. cities achieve this. The greater availability of land, the growth of suburbs and shopping malls, has meant that the majority of U.S. cities no longer have a CBD with a large retail function, even if financial services are still important downtown.

▶ FURTHER READING

Benevolo, L. *The European City*. Oxford, UK: Blackwell, 1993.

Blacksell, M., and A.M. Williams. *The European Challenge: Geography and Development in the European Community*. New York: Oxford University Press, 1994.

Clark, P. (ed.). *The European City and Green Space: London, Stockholm, Helsinki, and St. Petersburg, 1850–2000*. Burlington, VT: Ashgate Publishing, 2006.

Crouch, D.P. *Geology and Settlement: Greco-Roman Patterns*. New York: Oxford University Press, 2003.

Danta, D. "Ceauseseu's Bucharest." *Geographical Review* 83(2) (1993): 170–182.

De La Pradelle, Michèle. *Market Day in Provence*. Translated by Amy Jacobs. Chicago: University of Chicago Press, 2006.

Dickinson, R.E. *The West European City: A Geographical Interpretation*. 2nd edition. London: Routledge, 1961. A classic study of the site, situation, and urban morphology of many European cities.

Ford, L. "Continuity and Change in Historic Cities: Bath, Chester, Norwich." *Geographical Review* 68(3) (1978): 127–144.

Frugoni, C. *A Day in a Medieval City*. Chicago: University of Chicago Press, 2005.

Hall, P. *Cities in Civilization: Culture, Innovation and Urban Order*. New York: Pantheon Books, 1998.

Hall, P. *Urban and Regional Planning*. 4th edition. New York: Routledge, 2002.

Hamilton, F.E.I. "Spatial Structure in East European Cities," in R.A. French and F.E. Ian Hamilton (eds.), *The Socialist City:*

Spatial Structure and Urban Policy. New York: John Wiley, 1979: 195–261.

Hill, M. *Rural Settlement and Urban Impact on Countryside*. London: Hodder Murray, 2003.

Hohenberg, P.M., and L.H. Lees. *The Making of Urban Europe, 1000–1994*. Cambridge, MA: Harvard University Press, 1995.

Kaplan, D.H., James O. Wheeler, Steven R. Holloway, and Thomas A. Hodler. *Urban Geography*. New York: John Wiley, 2004.

Laborde, P. "The Spatial Evolution of West European Cities 1950–1992," in M. Blacksell and A.M. Williams, *The European Challenge*. New York: Oxford University Press, 1994.

Labrianidis, L. (ed.) *The Future of Europe's Rural Peripheries*. Aldershot, UK: Ashgate, 2004.

McCarthy, L., and D. Danta, "Cities of Europe," in Stanley D. Brunn, Jack F. Williams, and Donald J. Zeigler, *Cities of the World*. 3rd edition. Lanham, MD: Rowman and Littlefield, 2003.

Pacione, M. *Urban Geography: A Global Perspective*. 2nd edition. New York: Routledge, 2005.

Platt, Harold L. *Shock Cities: The Environmental Transformation and Reform of Manchester and Chicago*. Chicago: University of Chicago Press, 2005.

Pounds, N.J.G. *An Historical Geography of Europe*. New York: Cambridge University Press, 1990.

Stangl, P. "Restoring Berlins Unter den Linden: Ideology, World View, Place and Space." *Journal of Historical Geography* 32 (2006): 352–376.

Woods, M. *Rural Geography: Processes, Responses and Experiences in Rural Restructuring.* Thousand Oaks, CA: Sage, 2005.

5

ECONOMIC GEOGRAPHY: AGRICULTURE, INDUSTRY, AND SERVICES

▶ INTRODUCTION

Europe has a diverse resource base segmented by political boundaries. In the south is a zone of subtropical Mediterranean climate. In the west is a mild temperate maritime climate that becomes progressively more continental, with colder winters, to the east. In the northern tundra and taiga regions, little agriculture is possible but the tundra supports reindeer herds, which in turn support reindeer herders. The coniferous forests (taiga) lying south of the tundra are a rich source of timber and wood pulp.

In terms of early industrial revolutions, parts of Europe had excellent resources with plenty of coal and sufficient iron ore and nonferrous metals to support the first phases of industrialization. Britain, Belgium, northern France, Germany, Poland, the Czech lands, and the Ukraine possessed iron, coal, and nonferrous metals such as lead, copper, tin, zinc, and manganese. Southern Europe, Spain, southern France, and Italy lacked widespread coal deposits. Industrialization came later with hydroelectric power.

At the start of the twentieth century with the beginning of the automobile age, and the adoption of diesel power by industry, Europe was not well supplied with home-produced oil. The oil and gas fields in and around the North Sea were unknown. Quickly, private European capital invested in early Caspian fields, off Baku, in northern Iraq around Mosul, in Persia (Iran), and further afield in Mexico. Europe had acquired the economic momentum to import raw materials from many parts of the world.

Resources within Europe were not equally distributed. France envied Germany's abundant coal, and Germany coveted the Lorraine iron ore field in France. Wars were fought to secure resources, both agricultural and mineral, the most spectacular conflict being Nazi Germany's attempt to take over resources from the Soviet Union in World War II, including the grain fields and coal and steel capacity of the Ukraine (Figure 5.1).

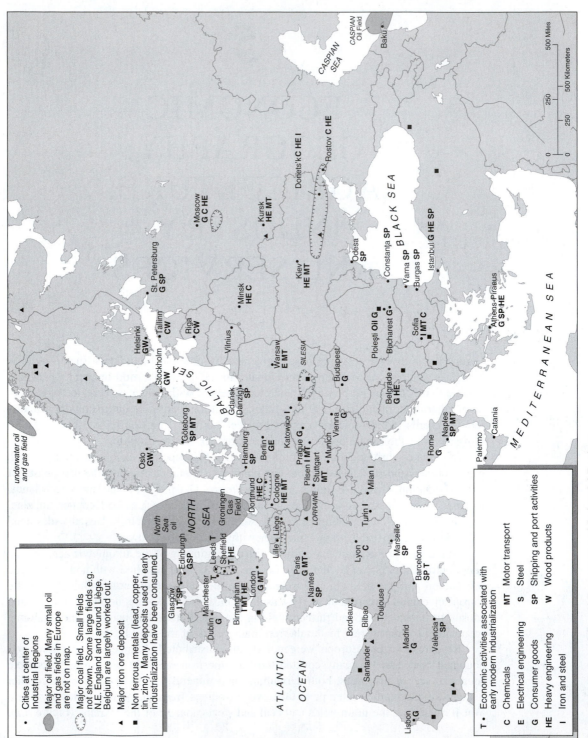

Figure 5.1 Europe: Mineral Resources and Industrial Regions.

One of the forces underlying the creation of EU has been the attempt to make resources of the region equally available as far as transport costs allow. By removing tariffs and other restriction on trade, agricultural products, mineral resources, and manufactured goods enjoy large markets and the old games of making German coal expensive in France and French ore overpriced in Germany have been eliminated by EU rules. In one area the opening of resources has been a failure. The Common Fisheries Policy, which opened traditional fishing grounds to all members, is based on an overoptimistic assessment of fish resources. Among the reasons that Iceland and Norway are not members of EU is the need to protect their fishing grounds.

► SECTORS OF THE ECONOMY

Traditionally economies are divided into sectors: (1) a primary sector consisting of farming, fishing, forestry, mining, and quarrying; (2) a secondary sector of manufacturing and construction; and (3) a tertiary sector including government, education, tourism, health care, wholesale, retail, transportation, and financial services. More recently a quaternary, or information, sector has been recognized.

Analysis of the structure of employment, by sector, is one indicator of level of economic development. In preindustrial, traditional societies the greatest part of the workforce is in the primary sector of the economy as farmers, foresters, fisherman, and mineral extractors. In modern developed countries, as in Europe and North America, most workers are employed in the tertiary, or service, sector.

There are large differences in the percentage of gross domestic product derived from farming, manufacturing, and the structure of employment among the countries of Europe, as Table 5.1 shows. Eastern European countries often earn relatively high percentages of GDP from agriculture compared with countries in the west.

► AGRICULTURE

If we examine the agriculture practiced around the Atlantic in 1700 we would encounter contrasts. With the exception of The Netherlands, the only areas of intensive land use were on sugar plantations of the Caribbean and the Atlantic shore of Brazil. The tobacco fields of Maryland and Virginia were intensively cultivated but after a few years the fields had to be fallowed to restore fertility. The same was true for most farmland in colonial America, but fallowing was not a problem because there was plenty of land. Europe did not have the same abundance.

All over Europe, except in The Netherlands, continuous cultivation of grain fields depleted soil fertility and reduced yields. (Photo 5.1) Land had to be fallowed and left uncultivated, even if it did provide some grazing for livestock. Farming communities coped with the need to fallow land in different ways.

In regions including lowland England, the north European plain from the Jutland Peninsula to Lithuania, northern and southern France, and Andalucía, the basic technique was to cultivate grains for a year or two and then leave land fallow to restore fertility. The best-known variant of this technique was the three-field system, used in a belt stretching from England, through France, Denmark, and north Germany to Poland and Lithuania. Village communities had three large, open fields divided into unfenced cultivation strips among villagers. Each year two fields were cultivated and one field

TABLE 5.1   Percentage of GDP Derived from Farming, Manufacturing, and Services

	Avg. annual % growth rate, 2000–2004	Approximate % of GDP by sector			GNI PPP* per capita, 2005
		Agriculture	Industry	Service	
Austria	1.2	2	32	66	33,140
Belgium	1.2	1	26	72	32,640
Bulgaria	4.7	10	27	63	8,630
Cyprus		No data			
Czech Republic	2.9	3	39	57	20,140
Denmark	1.2	2	26	71	33,570
Estonia	−0.5	Partial data			15,420
Finland	2.2	3	31	66	31,170
France	1.4	3	24	73	30,540
Germany	0.5	1	29	69	29,210
Greece	4.1	7	24	69	23,620
Hungary	3.5	4	31	65	16,940
Ireland	5.4	3	42	55	34,720
Italy	0.8	3	28	70	28,840
Latvia	7.5	4	25	71	13,480
Lithuania	7.5	7	33	60	14,220
Luxembourg		Partial data			65,340
Malta		Partial data			18,690
Netherlands	0.3	3	26	72	32,480
Poland	2.83	3	31	66	13,490
Portugal	0.3	4	29	68	35,514
Romania	5.5	13	40	47	8,940
Slovakia	4.6	3	29	68	15,760
Slovenia	3.2	3	36	61	22,160
Spain	2.5	3	30	67	25,820
Sweden	2.0	2	28	70	31,420
United Kingdom	2.2	1	27	72	32,690

*GNI Gross National Income, PPP Purchasing Power Party
Source: Human Development Report, New York: UN, 2004 Population Reference Bureau, 2006.

fallowed. As some plots were more fertile than others, lots were drawn to distribute the cultivation strips among villagers who usually ploughed grain fields communally. There were many variants. Some areas employed a two-field system in which half the arable land was rested every year, as opposed to the third fallowed in the three-field system. The infield–outfield system was used in Scotland and parts of Belgium. The intensely cultivated infield, closest to the settlement, received more inputs of labor and manure than the outfield, which was used for grazing and cultivated intermittently. Another technique was called convertible husbandry, in which pasture land was grazed and fertilized by animals and then ploughed and sown to produce good crops of grain until yields declined. The land was put back to grass and grazing, the farmers then ploughing up another parcel of pasture land. (Photo 5.2)

Photo 5.1 Innovative Dutch agriculture required large labor inputs. "Harvesting Scene," Brueghel the Younger.

Source: National Museum of Art, Bucharest, Romania/SUPERSTOCK.

In eastern Europe, land use was less intensive. (Photo 5.3) In the Slav lands it was not uncommon to encounter shifting cultivation in which plots were cultivated for a few years before reverting to grassland or woodland for six to twenty years. Shifting cultivation does not require intensive labor and, if land is plentiful, yields a steady subsistence.

In areas using the three-field system, if population grew and the demand for food increased, a way to improve agricultural output was to cultivate the land continuously, which could be achieved if fields were fertilized with manure. More manure meant more livestock. But how were additional animals to be fed? Here we come up against a fundamental problem of farmers who grow grain for subsistence all over the world. Livestock need fodder but the arable land is fully employed, growing grain for the community to eat. It is not possible to turn grain fields into grazing, or hayfields, without reducing the food supply of humans. The solution was to grow fodder crops on the fallow land in the three-field system. Fodder crops allowed more livestock to be fed, more manure to be produced to fertilize fields, and human diets to include more meat and dairy products. Additionally, fodder crops like clover, lucerne, and alfalfa improve soil fertility by fixing nitrogen in the soil.

By 1700, Dutch farmers had developed mixed farming systems that combined the production of arable crops with livestock raising. In winter, when the grass stopped growing and grazing was scarce, animals were kept in barns and fed clover hay. The manure

Photo 5.2   After the harvest, gleaners recovered grain that had fallen to the field.

Source: The Gleaners, 1857 (oil on canvas) by Jean-Francois Millet (1814–75) ©Musee d'Orsay, Paris, France/Giraudon/The Bridgeman Art Library.

accumulated in the barn in the winter months. In spring the dung was spread onto fields prior to the sowing season. Cattle produced milk, meat, and hides, allowing the production of butter, cheese, and leather goods. Agricultural productivity rose and better yields of grain were obtained from fertilized fields, but labor requirements went up. Now all fields were cultivated each year, little land rested in fallow, the dairy cattle had to be milked, the butter churned, and the cheese made. There were products to take to local markets, which required additional time and effort but produced income.

The new mixed farming did not spread quickly for a variety of reasons. Traditional farming communities resist change because it involves risk and subsistence farmers cannot afford risk, as failure means famine. Not all communities wanted to supply the additional labor the mixed arable and livestock system required. Systems of land tenure retarded development. In France, for example, *metayage* (share cropping) was widely employed, with the landowner taking a considerable portion of the harvest. In crude terms the *metayage* system meant that if peasants worked harder, half the extra produce went to the landlord.

The mixed farming system developed by the Dutch did spread to Britain in the eighteenth century where estate owners assumed the costs of adapting the Dutch system. In East Anglia the Dutch were particularly influential, frequently building the waterways

Photo 5.3 Particularly in eastern Europe, traditional equipment is primitive.

Source: J.D. Dallet/AgeFotostock.

that drained marshes and fens. In Norfolk landowners used a four-course rotation of crops and, on poorer land, turnips were grown as fodder for sheep that were pastured on the fields. The animal droppings improved soil quality, after which grain could be grown for a few years before the turnips and sheep returned to restore fertility.

If in 1800 an observer had gone east or west from the improved farming areas of the Low Countries and eastern England, the intensity of land use decreased. In eastern Europe much land was still in long fallows. In Wales, northern England, and the Scottish uplands the land was better suited to raising sheep and cattle than arable crops. The animals were driven to distant markets for fattening and then slaughter. Across the Atlantic farmers in New England, Pennsylvania, Maryland, and Virginia used the land unintensively because land was plentiful and labor short. George Washington, a large landowner operating several farms, declared it was cheaper to clear new land than to fertilize old. In the Carolinas the demand for raw cotton from new, mass-production textile mills in western Europe helped expand the acreage in the crop.

Crops Grown

Much of Europe is a temperate land well suited to the growing of grains, provided there is sunshine in July, August, and early September to ripen the crop before harvest. Wheat is an important crop but barley is widely grown for fodder, malting, and the brewing of beer. On the north German plain, Scandinavia, and Scotland with shorter growing seasons and moister climates, rye and oats are better suited to the environment.

New World crops, including the potato and maize (corn), were not instantly adopted by European farmers but by the late eighteenth century the potato was widely grown in

Photo 5.4 Vineyard, Nuits-St-Georges, Burgundy, France. Traditional methods for a high value crop.

Source: Mauritus/SUPERSTOCK.

the British Isles and on the northern European plain. The damp, clayey soils of glaciated regions suited the potato and when lifted the crop left a weed-free field with a good tilth.

Maize (corn) was not grown in northern Europe as a food crop because the growing season was too short to allow ripening. In southern Europe, where temperature conditions are suited to the crop, corn was less widely grown than might have been anticipated because foodways alter slowly. Today, corn is grown in temperate areas of Europe but the crop is often cut green and used as animal fodder.

Northern, western, central, and eastern Europe have a temperate climate except in the far north where temperate gives way to tundra. It is difficult to grow crops in the short tundra summer. In the long winter the ground is frozen and in the short summer thawing results in water-logged soils. There is some summer grazing and herds of reindeer do survive and provide food and hides for Lapps in northern Norway, Sweden, and Finland.

Southern Europe beside the sea has a Mediterranean climate. On coastal plains frosts are rare and temperate crops, grains, potatoes, and other roots can be raised in the winter months for harvest in spring. Citrus fruits are produced in the winter, the fruit being picked in January and February. Mediterranean summers are hot and dry. Vineyards are productive and the new world crops, including tomatoes, zucchini, melons, pumpkins, and beans, yield abundantly in the intense sunshine if irrigation water is provided. Like California, which has a Mediterranean-type climate except in the northern part of the state, southern Europe produces fruit and vegetables year round for urban markets in the north. Although the more northerly French vineyards of Burgundy, Champagne, and Bordeaux have leading reputations, southern Europe is a major wine-producing region. (Photo 5.4) Spain and Portugal are well known for fortified wines. Sherry, produced around Jerez de la Frontera, is shipped from the Gulf of Cádiz. The port wines of the Douro valley are shipped from Oporto. In addition, there are numerous wine-producing areas in Spain, southern France, Italy, the Dalmatia coast, and Greece. Northern Europe brews much beer but produces little wine, importing large quantities from vineyards further south.

Agriculture and Industrialization

Towns are markets for agricultural output of surrounding areas. Industrialization resulted in the growth of towns and expanding markets for agricultural products. Adam Smith, in *The Wealth of Nations* (1776), commented on the impact of towns on agriculture and how the town increased the value of land lying close by:

> *The corn which grows within a mile of the town, sells for the same price with that which comes from twenty miles distance. But the latter must ... pay the expense ... of bringing it to market.*

Thus farmers close to town avoided the costs of carting produce long distances to market. Adam Smith went on ...

> *Compare the cultivation of the lands in the neighbourhood of any considerable town, with that of those which lie at some distance from it, and you will easily satisfy yourself how much the country is benefited by the commerce of the town. (quoted by Wrigley, 2004, p. 270)*

This idea was taken up by the German landowner Heinrich von Thünen, who developed Smith's comment into a model. In *The Isolated State,* von Thünen constructed a model of land use on an isolated plain at the center of which was a town providing the market for all agricultural output. So that we would concentrate on location, and distance to market, Von Thünen made his plain isomorphic, with the same topography, soil fertility, and climate in every place. The only variable was distance to market; farmers further away had greater transport costs than those close to town.

Von Thünen's diagram displays the land use arrangement that he thought likely (Figure 5.2). The land close to town was used to produce fruit, vegetables, and dairy products that were higher-value, perishable products. Moving away from the city, in any direction, the intensity of land use decreased until ground was used for the raising of livestock, which would be driven on the hoof to market, probably being fattened up by a nearby farmer before becoming part of the urban meat supply.

We are all surprised to find, in Von Thünen's theoretical arrangement, that woodland occupies relatively high-value land close to market and here we are back to transport costs. Timber, for fuel and construction, is bulky and costly to move in preindustrial societies but it is essential to frame houses, heat homes, and fire bakeries. Timber is a crop with a market in urban areas and commentators have suggested that a constraint on growth in preindustrial cities was poor fuel supplies. In Von Thünen's view, timber commanded a sufficient price to make it worthwhile to grow trees close to town.

In the modern world Von Thünen's model of towns surrounded by agricultural land that produces milk, and in season, fruits and vegetables has been nearly eclipsed by the technology of transport and refrigeration. Today our lettuces, cucumbers, and tomatoes

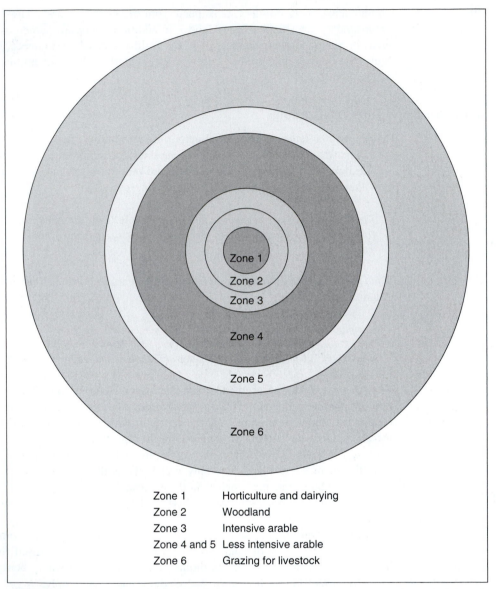

Zone 1	Horticulture and dairying
Zone 2	Woodland
Zone 3	Intensive arable
Zone 4 and 5	Less intensive arable
Zone 6	Grazing for livestock

Figure 5.2   Von Thünen: Land Use Zones Around a Central City.

no longer arrive from local market gardens but are trucked in, year round, from Florida, Texas, California, or Mexico. Salads for Europeans are raised in hot houses in Holland or under plastic in Spain. Cheap transportation even allows Dutch cucumbers to be sold in U.S. East Coast supermarkets.

The process of specialized agricultural regions supplying distant markets started in the second half of the nineteenth century when national railroad networks allowed produce to be moved to major markets. By the 1850s the Great Western Railway, starting at

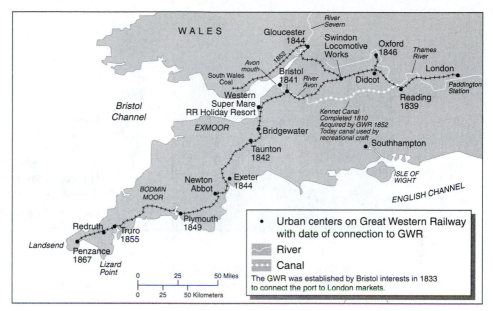

Figure 5.3 Great Western Railroad: Paddington to Penzance.

Paddington station in London, had extended to Bath, Bristol, Exeter, and Plymouth in the southwest peninsula of England (Figure 5.3). Soon west country milk was transported by train, overnight, to supply the London market. The southwest peninsula, sticking out into the Atlantic and the Gulf Stream, has a mild climate and a comparatively long growing season. Early vegetables were transported to London for sale. With the development of railroads, fishing ports supplied the capital with fresh fish.

Similar trends developed as railroads were built on the European mainland. The important line from Paris to Lyons and Marseille linked southern France, with a Mediterranean climate, to the Parisian metropolitan market, allowing flowers and fruits to be sold in the capital when the temperate lands of the Paris basin produced little in the winter months.

As Europe industrialized and urbanized, demand for foodstuffs stimulated agricultural production. The Danes and the Dutch, in the nineteenth century, produced meat and dairy products for sale in adjoining industrial countries like Britain and Germany. In the Mediterranean lands of southern Europe, flowers, fruits, and vegetables were produced for sale in northerly markets. Overall the industrialization of western Europe did not lead to the rapid development of efficient agriculture. At the outbreak of World War II, agriculture in France, Germany, Italy, and, of course, eastern Europe was largely unmechanized. The Germans yearned for agricultural land and *lebensraum* in the east, but German farms were inefficient with low inputs of fertilizer, high hand-labor requirements, and limited mechanization. Europe was a large net importer of food and grains. During World War II there were widespread food scarcities because labor was drafted from the land and not replaced, except in Britain, by mechanization.

Europe began to produce food surpluses under the Common Agricultural Policy (CAP) developed by the European Economic Community (now EU) after 1957. Farmers

Photo 5.5 "Fancy a Farmer?" A group of lonely dairy farmers have put their photographs on milk bottles in a bid to find dates. The isolation of farming means social outlets are rare, so the farmers took direct action.

Source: Neil Jones/©AP/Wide World Photos.

were paid prices well above market rates to produce surpluses of grains, wine, dairy products, vegetable oils, and meat. The EU produces food surpluses, but many products, including grains, could be imported more cheaply. Further, CAP policies produce large surpluses that are stored at community expense. For a time the stocks were so large that newspapers referred to the "butter mountain" and the "wine lake." Deploying surpluses as food aid is controversial. Farmers in receiving countries have their markets disrupted.

In recent years payments to farmers under the CAP have altered and so have EU attitudes to the countryside. Previously farmers were paid above-market prices to grow crops and raise livestock. (Photo 5.5) This led to overproduction and food surpluses and the cultivation of marginal land that would have been better unploughed. Now farmers receive an annual payment that is not linked to specific crops. Further there are EU grants to restore wetlands and conserve countryside.

The old, expensive CAP policy produced surpluses, but it also stimulated mechanization and other inputs to farming that improved efficiency and allowed much labor to move to better-paying work in cities. The new policy is certainly needed in western Europe, but how will CAP policy impact farming in the newly admitted EU countries of eastern Europe? Paying farmers a direct subsidy because they are farmers may perpetuate old and inefficient methods on small farms when what is needed is the mechanization and modernization of eastern European farming, still struggling to emerge from the collective agricultural practices of the communist era.

▶ INDUSTRIALIZATION

Western Europe started industrial revolutions in the late eighteenth and early nineteenth centuries. In an industrial revolution goods are mass produced for sale in national and international markets. The earliest industrial revolutions developed on coalfields in Britain in the second half of the eighteenth century (Figure 5.4). The British technologies did not spread rapidly across the Channel to mainland Europe until the end of the Napoleonic wars (1815) but were then adopted to improve the output of products like iron and textiles, which were already made using craft methods.

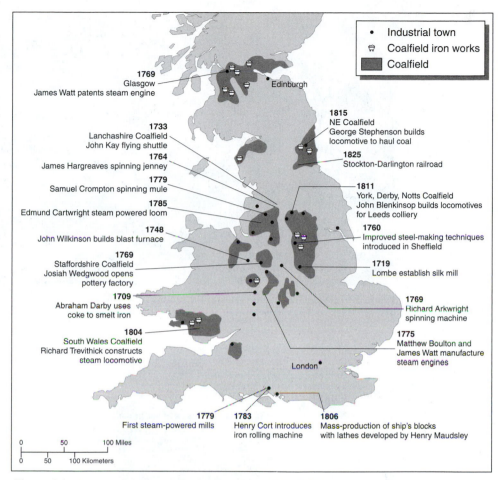

1769
Glasgow
James Watt patents steam engine

Edinburgh

1815
NE Coalfield
George Stephenson builds
locomotive to haul coal

1733
Lanchashire Coalfield
John Kay flying shuttle

1825
Stockton-Darlington railroad

1764
James Hargreaves spinning jenney

1779
Samuel Crompton spinning mule

1811
York, Derby, Notts Coalfield
John Blenkinsop builds locomotives
for Leeds colliery

1785
Edmund Cartwright steam powered loom

1760
Improved steel-making techniques
introduced in Sheffield

1748
John Wilkinson builds blast furnace

1769
Staffordshire Coalfield
Josiah Wedgwood opens
pottery factory

1719
Lombe establish silk mill

1709
Abraham Darby uses
coke to smelt iron

1769
Richard Arkwright
spinning machine

1804
South Wales Coalfield
Richard Trevithick constructs
steam locomotive

1775
Matthew Boulton and
James Watt manufacture
steam engines

London

1779
First steam-powered mills

1783
Henry Cort introduces
iron rolling machine

1806
Mass-production of ship's blocks
with lathes developed by Henry Maudsley

Legend:
• Industrial town
⚒ Coalfield iron works
▪ Coalfield

0   50   100 Miles
0 50 100 Kilometers

Figure 5.4 Industrialization on British Coalfields 1700–1825.

Most of the early products of industrialization had been made for centuries using traditional methods. There was no shortage of processing activities in the traditional countryside. Windmills and watermills for grinding grain and sawing timber were widespread. (Photo 5.6) (Photo 5.7) Ore was smelted into metal using charcoal-fired furnaces. Armament makers cast cannon to arm the sixteenth-century vessels that sailed on trading voyages. In specialized regions, like Toledo and Milan, steel was forged into weapons and armor. Households spun and wove woolen textiles. Local markets contained locally produced pots and craft goods. In the cities, those with wealth purchased jewelry, fashionable clothing, and furnishings.

Industrialization moved the focus of manufacturing from the workshop, where goods were made one at a time, to the factory where products were made in large quantities using repetitive techniques on early production lines, with workers performing a part of the process and not making the whole product as craftsmen had usually done.

Photo 5.6 Windmills and a castle on the Spanish Meseta.

Source: Markus Bassler/AgeFotostock.

Photo 5.7 Bayeux, Normandy. The dam creates the mill pond from which water is diverted to the waterwheel, which provides power to drive machinery.

Source: David Newham/Alamy Images.

The movement from workshop to factory, from small-scale output to mass production, was not easily achieved, as illustrated by the industrialization of the cotton industry. A series of related technological problems had to be overcome before large-scale production was possible. For example, in the traditional Lancashire, cottage cotton industry, imported cotton was hand-spun into thread, which was then hand-woven into cloth. A series of spinning machines, including the Spinning Jenny (1764), the roller spinning frame (1769), and Compton's Mule (1779), allowed a huge increase in the production of cotton thread. Producing machines to weave the thread into cloth was more difficult but achieved, in early form, by 1790. The output of raw cotton improved with the invention of the cotton gin (1793) in the United States by Eli Whitney.

In addition to inventing machines to spin and weave cotton, new sources of power replaced human hand power. By the 1770s water power was driving machines and in the 1780s the first steam engines entered the Lancashire cotton industry. Water power continued to be important for many years but, as spinning and weaving machines were improved and became larger and more efficient, the steam-powered factory became the industry norm. A series of interlocking inventions industrialized the cotton industry and rapidly reduced the price of cotton cloth. With falling prices, home and overseas markets expanded.

Coal, Railroads, Iron, and Steel

By the sixteenth century coal was being used to heat houses in London. On cold, still days the capital was covered in smog. London's coal was shipped by sea from the Northeast coalfield via ports on the rivers Tyne and Wear. Growing demand for coal spurred inventiveness. The industry evolved from holes in the ground (bell pits), to tunnels driven into hillsides, to vertical shafts that reached below surface strata, to deeper seams of coal. Shaft mines collected water from underground sources. If the water could not be drained via a tunnel to a hillside, it had to be pumped out. By the mid-eighteenth century the Newcomen Atmospheric engine powered pumps. After 1780, the James Watt steam engine, was widely used in mines.

Railroads evolved from mining. Planked wagon ways were used to move coal carts that ran from the mine to shipping staithes on the river Tyne, in the northeast of England. In the eighteenth century iron rails were laid on wagon ways and the carts fitted with metal, flanged wheels. Now several carts would be moved in a horse-drawn wagon train.

Early in the nineteenth century mine engineers, the men who maintained the steam engine in the pump house at the pit head, began to develop locomotives. Steam engines were used to power a small locomotive that could pull a train of coal carts from mine to shipping point on metal rails. (Photo 5.8) (Photo 5.9) Several mining engineers built locomotives in different parts of Britain, including Cornwall (Richard Trevithick), Leeds in west Yorkshire, Shropshire, and at Killingworth mine near Newcastle-upon-Tyne. When Quaker bankers put up the capital for the Stockton and Darlington Railway Company, they asked the engineer from Killingworth, George Stephenson, to build a railroad from the inland South Durham coalfield to the river port of Stockton-on-Tees.

The Stockton to Darlington Railway, the first public railroad, opened in 1825. (Photo 5.10) The rails were 4 feet, $8\frac{1}{2}$ inches apart and that distance remains the standard gauge for railroads to the present. By 1830 a railroad was open from the port of Liverpool

Photo 5.8 Factory with a blast furnace.

Source: Factory with blast furnace, from a series of books on 'L'Univers et l'Humanité' by H. Kraemer, published in Paris at the end of the 19th century (colour engraving) by August Dressel (b.1862) ©Private Collection/Archives Charmet/The Bridgeman Art Library.

to the city of Manchester, the commercial center of the Lancashire cotton industry. From this time major lines were built in Britain to create a partial network of railroads by 1850. Railway mania and the proliferation of competing lines came later.

The major coalfields of Europe are shown in Figure 5.1. Most large coalfields became manufacturing regions. Coal is heavy, dirty, bulky, and difficult to move. If you carry coal on a horse-drawn cart or a pack animal, the cost of transport is high. Railroads reduced the cost of moving coal and allowed industry to develop away from coalfields. Heavy industry, which used large quantities of coal, found coalfields an attractive location. From Glasgow on Clydeside in western Scotland, through northern France, the Liège region of Belgium, the Ruhr, Silesia, Bohemia, and eventually in the Ukraine, coalfields were the basis of iron and steel, chemical, and heavy metal manufacturing.

Countries like Spain and Italy, which had few coalfields, industrialized later when coal, exported from Britain and Germany, became available or hydroelectric power plants were built in the Pyrenees and northern Italy. Not all countries that lacked coal deposits were late industrializers. Switzerland, on the trade routes between northern and southern Europe, rapidly applied innovative techniques to the manufacture of existing products like clocks, watches, and instruments. Some major manufacturing regions were not on coalfields. London and Paris, large markets for goods, were industrial centers in addition to their governmental and cultural functions. Most ports, centers of shipbuilding and the

Photo 5.9   Old Hetton Colliery, Newcastle. Primitive coal mine.

Source: Old Hetton Colliery, Newcastle by Anonymous ©Beamish, North of England Open Air Museum, Durham, UK/The Bridgeman Art Library.

processing of imported raw materials, were not on coalfields, as at Liverpool, Hamburg, Marseille, and Genoa. At these locations coal was brought in from the hinterland or imported by boat. By the second half of the late nineteenth century every major port in the world had coal bunkers to refuel the steam-powered vessels that altered maritime commerce. Coal for distant ports was shipped from Europe in vessels called colliers, just as oil today is moved around the world in tankers.

The importance of coal as a locational force altered with time and technology. When coal, rather than charcoal, was used to smelt iron and steel, the iron industry was pulled to coalfields, a tendency reinforced by the existences of thin beds of iron ore at the top of coal deposits. The pull of coalfields, for the iron and steel industry, decreased over time. Blast furnaces became more efficient and burned less coal to smelt a ton of metal. The thin bands of iron ore in the coal measures were insufficient to sustain iron making on a large scale. When the invention of the Gilchrist–Thomas steel-making process allowed the use of phosphatic iron ores, French steel makers started to use the ores found in a band of territory running from southern Luxembourg through Lorraine in western France. Now there was a decision to make: Do you move the ore to the coal field or the coal to a blast furnace on the ore field? The economics of transportation provided the answer.

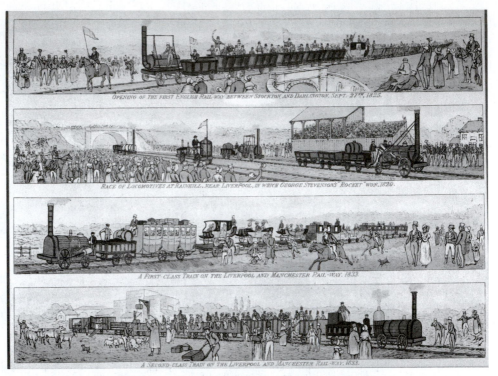

Photo 5.10   The opening of Stockton, Darlington in 1825 and other railways.

Source: The opening of the Stockton and Darlington railroad, 1825; Locomotive race at Rainhill, near Liverpool, won by George Stephenson's 'Rocket', 1829; a first class train on the Liverpool and Manchester Railway, 1833; a second class train on the Liverpool and Manchester Railway, 1833 (colour litho) by English School (19th century) ©Private Collection/The Bridgeman Art Library.

It took a ton of coal to smelt a ton of metal. Four tons of Lorraine ore, with around 25 percent iron content, were needed to smelt a ton of iron. Iron makers, using Lorraine ores, tended to locate on the ore fields.

The German Alfred Weber, in the *Theory of Location of Industries* (1909), constructed a model, particularly applicable to heavy industries like iron and steel making, in which transport costs were analyzed as a factor in the location of industry. Weber did not neglect the cost of getting products to market and the cost of labor. He also noted that industries obtained advantages when they *agglomerated,* providing goods and services to each other. Once established in a location it was difficult for an industry to move, for the move broke established linkages.

Other locational solutions were possible. Steel makers on the Ruhr coalfield did not want to abandon capital equipment and the market linkages they had developed with steel users in the Ruhr region. Ruhr steel makers imported high-grade Swedish iron ore with a 60 percent iron content.

In the modern world, when iron ore and coal are just commodities trading on world markets, and not strategic resources to be denied competing countries, iron and steel works often seek coastal locations and cheap raw material imports. A large, integrated iron and steel plant was built, postwar, near Genoa using imported raw materials, as Italy had little iron and coal.

Industrialization in Germany did not accelerate rapidly until the second half of the nineteenth century and the creation of a unified German empire. By the time German industrialization occured, rail lines were in place. Manufacturers found that coal was widely available away from coalfields. German cities not on coalfields became manufacturing centers in the late nineteenth century, including Stuttgart, Cologne, and Munich.

The Ruhr is the richest coalfield in western and central Europe. Exposed part of the Ruhr coalfield had been exploited on a small scale for centuries. Exploitation of the concealed coal field (coal seams lying beneath other sedimentary rocks) took off in the second half of the nineteenth century. In 1850 the Ruhr region produced less than two million tons of coal. By the early years of the twentieth century the Ruhr was mining 80 million tons of coal per annum.

Demand for Ruhr coal was generated by the iron, steel, engineering, and armaments industries in cities like Dortmund, Essen, and Bochum, often associated with industrial combines like Krupps and Thyssen. By 1900 Germany was the largest steel producer in Europe, second in the world to the United States. Not all the steel was produced in the Ruhr, as Germany had other major industrial regions and coalfields in the Saar and Silesia, for example. (Photo 5.11)

Bohemia, in the Czech lands of Austria-Hungary, was a major producer of coal, iron, steel, and engineering equipment. The Russian Empire was a late industrializer.

Photo 5.11 Power station at Croix-Wasquehal, near Lille, France.

Source: Power Station at Croix-Wasquehal (oil on canvas) by Hippolyte Lety (1878–1925) ©Musee des Beaux-Arts, Tourcoing, France/Giraudon/The Bridgeman Art Library.

Exploitation of coal, iron ore, and manganese (a steel hardener) in the Ukraine did not start until the 1870s. The region possesses the best resource base for iron and steel making in Europe. By the beginning of the twentieth century industrialization in the Russian Empire proceeded rapidly but from a low starting point. World War I disrupted industrialization but heavy industry was a major priority in Soviet Five Year Plans, the first launched by Stalin in 1929, but backdated to 1928, the year in which the economy began to rebound.

Diffusion of Industrial Technology

After the Napoleonic Wars the new industrial technology diffused from Britain to continental Europe. Among the innovations to diffuse were steam engines, railroads, and cotton textile machinery. The new textile technology was adopted rapidly because there were already regions producing woolens, cottons, silks, and linens on the basis of cottage industries. The new spinning and weaving machines were employed as far to the east as Russia by the 1830s.

The process of diffusion on the continent was complex. Boulton and Watt had patented the steam engine. The technology had to be bought from Birmingham. Sometimes manufacturers from Britain would set up on the continent, a famous case being the factories in the Liège area of Belgium, where English workers helped to diffuse soccer to the region. Technology was pirated but most manufacturers at first bought machines from Britain, although it was not long before local artisans saw ways to improve the imported equipment. By the 1830s manufacturers in Switzerland were making and exporting textile machinery to other European countries.

The British lead did not last long. By the second half of the nineteenth century and the arrival of new industries, other European countries became the leaders: France and Germany in the chemical industry and Germany, France, and Italy in the auto industry.

Industrialization and Urbanization

The changes initiated by industrialization were lasting. The adoption of the steam engine concentrated manufacturing into factories. (Photo 5.12) A steam engine produced more power than could be economically used in a workshop. Manufacturing activity was organized on a larger scale with products being mass produced for sale in regional, national, and international markets. Production had to be at nodes in transportation systems to allow for the intake of raw materials and the marketing of products. All of these factors—steam engines, factories, larger workforces, and railroads—helped to increase the pace of urbanization, the growth of industrial cities, and rural-to-urban migration. In 1800 western Europe was rural. More people lived in farms, villages, and hamlets than lived in towns. Just over a century later, most western European countries had urban majorities. In eastern Europe, increased urbanization started later but the trend was the same.

Industrialization diversified economies. Relatively, the importance of the primary sector—farming, fishing, forestry, and mining—decreased with the growth of manufacturing and the service sector. As machines replaced human physical effort they became more efficient and complex. Companies making machinery invested in research and development. Companies using machines had to employ maintenance engineers. By the

Photo 5.12 "The Dinner Hour".

Source: The Dinner Hour, Wigan, 1874 (oil on canvas) by Eyre Crowe (1824–1910) ©Manchester Art Gallery, UK/The Bridgeman Art Library.

end of the nineteenth century technological colleges and universities were developed all over Europe, with Germany providing the lead. With the use of machines productivity per capita rose, prices dropped, and living standards eventually improved, although for most factory workers, living conditions were bad, nutrition poor, health care limited, life expectancy low, and infant mortality high.

The Changing Character of Industry

The first phase of European industrialization was based on coal mining, iron and steel, shipbuilding, textiles, the making of machinery to equip factories, railroading and basic consumer products, pottery (china for the elite), clothing, footware, and food processing. Ports grew as the import of raw materials and foodstuffs, to feed the new urban populations, increased. After 1870 new industries and products emerged, including synthetic chemicals, coal-burning gas works to provide gas lighting and heating, coke and by-product gases for the chemical industry. In the 1880s, electric trams and bicycles were widely produced and by the end of the century automobiles were manufactured. The movement of labor from farm to factory helped stimulate the production of agricultural equipment. Consumer goods industries advertised and sold into national markets and many famous-brand products have marketing origins in the late nineteenth century, including Bass beer and Palmolive soap.

In the synthetic chemical industry German scientists were the leaders. In 1880, von Baeyer synthesized the blue dye used to color blue jeans. Formerly the natural dye was extracted by rotting the cut indigo plant and you needed to rot four hundred pounds of plant to extract a pound of dye. German science produced many technological innovations. In higher education Germany was the European leader. German universities were a model frequently followed by state universities in the United States, particularly in the Midwest. The technical training tradition remains important in Germany and when, in the late twentieth century, BMW built a plant in Spartanburg, South Carolina, a major locational force was the existence of a state system of technical education.

In the twentieth century aeronautical, automobile, and electronic industries developed rapidly, often given impetus by defense contracts for aeroplanes, trucks, tanks, and radio and electronic equipment. Consumer goods became increasingly important in the late nineteenth through the twentieth century to the present day. As living standards rose so did the demand for clothing, footwear, fashionable attire, furniture, cars, public transport, newspapers, books, household equipment, restaurants, seaside resorts, spas, and innumerable brands of tobacco, preserves, pickles, mustard, and beer.

In the later decades of the twentieth century, Europe experienced the trends common to the economically developed lands. In western Europe the production of iron and steel declined and many plants closed. Railroads lost freight traffic to trucks, and private cars became the preferred form of transportation, although passenger rail traffic was maintained by government subsidies on the grounds that public transport slowed the growth of cars on overcrowded roads. Service industries became the mainstay of economies. Manufacturing and farming employed a declining percentage of the workforce. Many easily assembled consumer goods were imported from countries where labor costs were low.

In eastern Europe and Russia, the heavy industries were fully employed until the 1990s when the centrally planned economies of the region, associated with communist ideas on economic development, were opened to market forces. Much heavy manufacturing closed but many products could compete, after modernization, including cars made in the Czech Republic and buses made in Poland. The 2004 European Union expansion into eastern Europe introduced competitive economic forces to the region. There were plant closures, but with lower labor costs, eastern Europe makes a range of goods for sale in more prosperous western Europe.

▶ THE STAGES OF EUROPEAN ECONOMIC GROWTH

An American commentator, W.W. Rostow, looking at the economic history of the western world and particularly Europe, suggested five phases of economic development: (1) *traditional societies* that (2) established the *preconditions* for industrialization leading to (3) *a take-off* into economic growth followed by (4) a *drive to maturity* and (5) an *age of high mass consumption.*

Traditional Societies
Traditional societies are agrarian. The greatest part of the population depends directly on the land for a living, craft industries supply the limited demand for goods, the middle

class is small and consists largely of doctors, lawyers, and priests. Wealth and power is in the hands of a land-owning aristocracy and institutions like the crown and the church.

European traditional societies altered from the sixteenth century on as the countries of the region created an Atlantic world, sending soldiers, missionaries, migrants, and traders out to exploit new sources of fish, timber, precious materials, and agricultural commodities like sugar and tobacco. The Atlantic world enhanced the importance of European ports, stimulated ship building and trade, and brought into being an extensive merchant and entrepreneurial class who financed enterprises, and if successful, accumulated capital.

Preconditions

Overseas trade impacted economies. Imports had to be moved to inland markets, which stimulated the development of turnpike roads, improvements in river navigation, and the building of canals. Overseas trade created demand for products that could be sold or bartered abroad. Profits on trade created capital and successful merchants became merchant bankers who financed the enterprises of others. Most West Indian plantations relied upon a merchant in London, Bristol, Liverpool, Bordeaux, Antwerp, or Amsterdam to sell the sugar and provide credit to the planter to buy supplies, including salted fish from North America and slaves from West Africa.

By the seventeenth century banks operated in Antwerp, Amsterdam, Hamburg, Frankfurt, London, and Edinburgh. Improvements in trade, transport, and finance helped establish the preconditions for economic growth. By the late eighteenth century small improvements in living standards and increased demand for goods created conditions, in parts of Europe, in which manufacturing activities could grow. In England, Scotland, The Netherlands, and other prosperous parts, demand improved for fuel, foodstuffs, clothing, and some household goods. The size of the middle class was enlarged by merchants and others engaged in shipping, finance, and commercial enterprises. These people, along with landowners and aristocrats, began to buy more goods including fabrics, fashionable dress, cutlery, pottery, porcelain, wines, spirits, carriages, clothing for servants, furniture, paintings, jewelry, and silverware.

By the end of the eighteenth century, major cities were developing shopping streets in which the well-to-do could buy fashionable goods of the time. Producers profited from the demand. The output of porcelain increased as makers like Josiah Wedgwood, in the English "Potteries" of Staffordshire, produced high-quality china in volumes that undercut the price of china from the Orient. Wedgwood used early forms of mass production and cut costs by using transfers to decorate products, rather than handpainting.

The Take-Off into Economic Growth

Take-off is an industrial revolution tied to new methods of production that results in a modernizing society. Remember how the history texts spoke of the eighteenth century enlightenment, the age of reason, and the development of the scientific method. We can see these intellectual developments as a part of the preconditions for modern economic growth.

Rostow assigned time periods for take-off into modern economic growth (see Table 5.2).

Britain was the first country to generate a take-off. France and Belgium followed decades later. In Rostow's view the United States started to take off in 1843. We can

TABLE 5.2 Rostow's Approximate Dates for Take-Off into Economic Growth

Country	Take-Off
United Kingdom	1783–1802
France	1830–60
Belgium	1833–60
United States	1843–60
Germany	1850–79
Sweden	1868–90
Japan	1878–1900
Russia	1890–1914
Canada	1896–1914

Rostow (1990).

argue with the assigned dates and, if you are from the Lancashire cotton manufacturing region in northwest England, you would probably say that by 1783 many of the crucial inventions and breakthroughs had already been made in the cotton industry.

The Drive to Maturity

By the 1880s economies in western Europe and North America were producing more complex products. In the take off phase the major industries were iron and steel, railroading, heavy engineering, and textiles. Those basic industries continued but in the Drive to Maturity the new growth industries were synthetic chemicals, electrical equipment, machine-tools, and, later, cars.

By the 1880s Britain was no longer the leader in industrial development. French and German manufacturers led in the synthetic chemical industries, the manufacture of electrical equipment and, by the end of the century, the production of automobiles.

Living in an automobile society in North America, it is hard to understand that the motor car was not invented on this side of the Atlantic. German inventors developed the technology, French manufacturers were the first to successfully market cars, and the production techniques that made cars widely affordable were first developed in North America.

The Automobile Industry

In 1886 Karl Benz patented the first practical vehicle powered by an internal combustion engine burning gasoline. Gottlieb Daimler developed a car about the same time and set up the Daimler Motor Company in 1890. The company became Daimler-Benz in 1926 and remains a leader in the automobile world.

However, it was French industrialists during the 1890s who made and sold cars profitably using the technology the German inventors had patented. The major British heavy engineering corporations showed little interest in the new product. The UK imported cars from France and it was not until 1903 that the United States produced more vehicles than the French manufacturers. French companies had interests in early British car companies and in the first years of the twentieth century there were few British manufacturers with the notable exception of Rolls Royce, established at Derby in 1907. Ford entered the British market before World War I with a plant producing low-cost vehicles at Trafford Park near Manchester.

In Italy between 1899 and 1907, approximately forty automobile companies manufactured cars, including Fiat (*Fabbrica Italiana Automobile Torino*), Alfa Romeo, and Bugatti in Milan and Lancia in Turin. The cars were well designed and innovative but there was no mass market for automobiles in Italy and sales were small. Cars were also made in Austria by Porsche, and in Catalonia, Spain, prior to World War I.

A low-cost British car industry emerged prior to World War I as Austin at Longbridge in Birmingham and Morris in Oxford produced cheap cars along with Ford. Morris was in the bicycle trade and he used the model for making bicycles to make cars. In bicycle production outsource suppliers delivered the components to the factory where the bike was assembled. Morris used this model to build cars at Cowley, a suburb of Oxford. World War I helped production and the plant doubled in size. By the 1920s both Morris and Austin were making small, cheap family cars, with Morris quickly taking a third of the British market. In 1925, having failed to take over Austin, General Motors bought Vauxhall, a car maker in Luton. Then in 1932 Ford opened a large plant at Dagenham, east of London, with its own steel-making, body-pressing capacity, and an engine plant. Almost unnoticed, a motorcycle side car manufacturer started to make Jaguar cars on the Italian *Gran Turissmo* model. Jaguar is now in the Ford group.

In World War II motor car manufacturers enjoyed high demand for trucks, tanks, jeeps, troop carriers, and half tracks. Ford made trucks at Dagenham for Britain and trucks at Cologne for the *Wehrmacht*. Renault built tanks for the German army. Fiat and Alfa Romeo got capital injections from Mussolini's fascist government and Nazi Germany set up Volkswagen, with Hitler opening the factory in 1938. The factory survived the war and became a major player in the world automobile industry.

Since World War II motor car manufacturing in Europe has rationalized with a few large corporations now dominating the market. In Europe and North America, in the early phases of car production, there were large numbers of small producers in many towns. Most plants lacked the capital to expand and serve large markets. In Britain many small producers, in addition to the big three—Austin, Morris, and Ford—survived to World War II when they made money on war work. After the war, small plants, on cramped sites, returned to civilian car production and received government allocations of scarce steel that allowed them to make vehicles.

When rationalization of the industry came, it happened quickly. Austin took over Morris. Ford, which had been a free-standing British company, was taken over by the Detroit parent. Leyland, a manufacturer of trucks and buses, bought up car companies including Triumph and then Austin Morris to create the British Motor Corporation (BMC). The BMC conglomerate was unwieldy and when, after joining the EEC in 1973, the British market opened to French, Italian, and German cars, the British products rapidly lost market share. BMC was taken over by the government and sold to BMW, who tried unsuccessfully to build cars profitably at Longbridge, Birmingham. In 2000 BMW gave Longbridge away, continued to make the Mini in Oxford, and sold Land Rover to Ford. GM still assembles cars at Luton. Ford assembles cars at Dagenham and builds engines at the plant. BMW bought Rolls Royce and built a new plant, having learned from the

Longbridge experience it was cheaper to build a new factory than try to modernize an old plant with entrenched work practices. Bentley is in the Volkswagen group. The Morgan sports car company is one of the last independent British car manufacturers.

After World War II the main French producers—Citroën, Peugot, and Renault—with considerable government help, rapidly built up production. Renault had taken contracts for German war work, as had many other French corporations, and the factory near Paris was confiscated by the government. France subsidized Renault and used it to start joint ventures with other motor car manufacturers.

In Germany, the Ford plant in Cologne was heavily damaged and workers and materials were scarce. At Stuttgart the Mercedes plants were rehabilitated. The Volkswagen factory, at Wolfsburg, had only received light damage. It was restarted by the British army and became a private corporation in 1954. BMW lost a factory in East Germany and did not resume making cars at Munich until the mid-1950s.

In Italy, Fiat concentrated on producing affordable cars. The *Nuova Cinquecento* was launched in 1957 with a two-cylinder engine and four seats. It sold millions. Alfa Romeo was used as a tool for regional development by the government and opened a plant in Naples, in addition to production in Milan.

The motor car industry is illustrative of the drive to maturity as the product requires a large number of components and accessories and creates numerous industrial linkages.

The Age of High Mass Consumption

This phase of economic development, in which we currently live, is based on the production of consumer goods to supply mass markets. In the age of high mass consumption Europeans, like North Americans, import an increasing share of manufactured goods and make fewer goods at home. This trend is particularly marked in Britain. After World War II the government took over industries including coal mining and iron and steel. As industries like automobile manufacturing got into trouble, the government felt pressure to take over failing producers. State support stopped in the 1980s with the government of Margaret Thatcher, who believed in the virtues of competition and privatized (sold) the state corporations the government owned. The older industries, including coal mining, steel, shipbuilding, textiles, and car manufacturing, were denied subsidies and opened to external competition. Only profitable operations survived. Many products are now imported from low-cost producers. By letting older industries fail, Britain has enjoyed higher rates of economic growth than France, Germany, and Italy. But growth in service industries and neglect of old industries has created gaps in industrial structure. The liner QEII was built on the river Clyde at Glasgow, in the heart of a major shipbuilding region. Now, when the vessel needs a refit, it goes to a German shipyard. The new Queen Mary liner was built in France.

A reason for the decline of Britain's older industries was the inflexibility of labor and the militancy of unions. It became easier to allow the mining industry, for example, to decline and use North Sea oil and gas than to invest in mine modernization with unions insisting that no jobs be lost.

All countries in Europe have seen older industries decline and historic sites are being preserved, some as part of the European Route of Industrial Heritage. Sites include

Ironbridge (UK), the Volkslinger iron and steel works in Germany's Saarland, together with steel and coal plants in the Ruhr.

The age of high mass consumption is reflected in the landscapes of Europe. All the major cities of western Europe have high streets full of chain stores selling imported goods at competitive prices. On the margins of cities are technology parks, warehousing zones adjacent to modern highways, and shopping malls. In general the urban sprawl around European cities is more contained than in the United States. City centers in Europe remain vibrant retail locations. In the United States, with few exceptions, retailing has left the downtown and moved to peripheral malls. Sleek, high-speed trains on modern, electrified track link the major cities of western Europe. Concern for the environment is increasingly displayed in the wind farms composed of tall towers, which dominate landscapes but produce power without burning fossil fuels. Many object to the skyline visual impact of the wind towers.

► THE IMPACT OF GLOBALIZATION

Across Europe in the decades between World War I and World War II, partly as a result of the Great Depression, a view emerged that governments had to manage economies to provide steady employment, a welfare safety net for those out of work or in ill health, and attempt to create stability in prices and wages. Some countries, like the Soviet Union, had a government that managed every aspect of economic life: housing, work, wages, health care, and education. Fascist Italy had a corporatist approach, with the government acquiring stakes in corporations and providing funds to develop key industries. A part of Hitler's appeal was that he abolished persistent unemployment, while regulating prices and wages. In the years leading to World War II, living standards improved for ordinary Germans who had suffered in the hyperinflation of 1923–24 and economic collapse in the Great Depression of the early 1930s.

France had a statist approach, with the government helping to shape investment decisions and providing subsidies and protections for important industries. In many countries, including Sweden and Switzerland, there was in the 1930s a confrontational relationship between labor unions and employers. In both the neutral countries a set of understandings developed between labor, employers, and the state. Governments would provide infrastructure and a social safety net. Unions and industrialists would cooperate to address wages and employment issues. Norway and Denmark created a similar system.

The attitudes described above survived World War II and strengthened in the postwar era, as generally Social Democrat governments managed economies. The attitudes were a fundamental part of the thinking on European economic cooperation. The establishment of the coal and steel community meant old coal mines and blast furnaces would shut down. There were no rapid closures resulting in heavy unemployment in affected communities. Workers were kept employed and retrained for other work. By agreement within the European community, governments were allowed to make subsidy payments, encourage regional development, and practice a degree of protection as long as each country abided by common guidelines.

The arrangements worked well in the period of economic growth, which lasted from the end of World War II until 1974, when high oil prices created stagflation—economies stagnated but prices and wages inflated.

Prosperity returned in the 1980s but in the 1990s globalization arrived, Europe had high wages, security of employment, high taxes, and strong environmental regulation. Workers enjoyed generous maternity leave and, by U.S. standards, vacations were long and the work week short. Labor costs were high, although most successful European manufacturers had invested in efficient plant and modern technologies. In the 1990s many basic goods could be imported more cheaply than they could be manufactured in Europe. European corporations closed plants or moved production to lands with cheap labor. European unemployment has risen and, in EU core countries like France and Germany, is around 10 percent of the workforce. The benefits system shields workers from the worst economic impacts but new forms of employment are not appearing rapidly enough. High unemployment rates in the new EU of eastern Europe have a political impact. Polls consistently show that many recall "the good old days" under communism, when everyone had work. The trend has been for the parties that promoted free market economics, and EU entry, to lose power and for more cautious parties to make advances.

France wants the EU to establish a "Globalization Fund" aimed at helping workers cope with Asian competition. Countries like Sweden and The Netherlands, which do compete successfully in global markets, are against the fund, seeing it as a costly mechanism to perpetuate the present conditions of slow EU economic growth. The President of the European Commission, José Manuel Barroso of Portugal, warned that demands for protectionism should be resisted. Europe should embrace globalization, with its opportunities and impacts, for the region has the capacity to be successful in a competitive global environment.

In enlarging economies creating stability and managing the environment, the EU was broadly successful in the second half of the twentieth century. In the twenty-first century, with emphasis on opening markets to competition from low-cost producers, the statist assumptions that countries like France have taken for granted are being eroded. Many in France see open markets as a threat to their economic security and this was a factor in the French electorate voting against the proposed EU constitution in 2005.

In eastern Europe the new members of EU have been forced to open up their economies to competition from established members. Many older eastern European industries have failed but cheaper labor has attracted investment and new, efficient plants are replacing older, obsolete facilities. In general the economies of Poland, the Czech Republic, Slovakia, Slovenia, Hungary, and the Baltics are growing but many people have changed jobs and unemployment is high. In Poland a sixth of the workforce is unemployed and people migrate to find work in western Europe.

The Impact of EU on Agriculture, Industry, and Services

This chapter began by recognizing primary, secondary, and tertiary sectors of economies. What impact has the EU had on each of the sectors?

The Common Agricultural Policy (CAP)

When the EEC was founded in 1957 slightly over 20 percent of the workforce in the original six members still worked the land. In the 1950s memories of near starvation, during World War II, and postwar scarcities were vivid. The CAP policy promising food self-sufficiency in temperate crops, better living standards for workers in rural areas, and the modernization of agriculture was easy to promote politically.

By the 1980s production had increased dramatically and there were huge surpluses; storing and disposing of excess production was costly. Sixty percent of the EU budget was spent on agricultural subsidies, yet food prices remained high. Further, the environmental impact of increased production was not benign. Encouraged to produce, farmers applied more inputs of artificial fertilizer and herbicides. Marginal land was ploughed to grow crops because the price paid for crops was well above world prices. Poor land, better left in pasture, was exposed to erosion. Increasing field size, as a result of mechanization, caused the loss of hedgerow habitat, the clearing of woodland, and the draining of wetland.

On the benefit side, apart from ensuring food supply, CAP delivered improved rural living standards, better farm income, incentives for older farmers to retire, and a slowly increasing farm size.

The Treaty of Amsterdam, coming into force in 1999, contained a provision that environmental considerations had to be a part of all EU policies. Now in European agriculture there is more emphasis on setting land aside, and EU grants are available to restore or protect wetlands and other habitats.

CAP Payments

Payments made under CAP policies are unequally divided. France receives 21 percent of the total, Spain 14 percent, Germany 13.5 percent, Italy 11 percent, and the United Kingdom 9 percent. France is the largest agricultural producer in Europe and was before the payments began. At the farm size level allocations are profoundly unequal, with 70 percent of all monies going to 20 percent of the farms, and the small farmers, who account for nearly half the farms in the EU, receiving less than 10 percent of the payments! Big grain farmers with large acreages do well out of the EU. Small family farms that struggle to make a living are not well looked after.

Widespread fraud is practiced in several countries. The claimed area in olive groves in Italy quickly doubled when subsidies were paid for olive oil, although it takes several years for an olive tree to mature. When citrus production was in surplus, and payments were made to transfer to other crops, nonexistent citrus groves in Sicily received money. To reduce overproduction of milk and dairy products, CAP fees were paid to cull herds. The application had to be accompanied with an ear of the culled cow. Italian pastures filled with one-eared cows (Rothacer, 2005, p. 75)!

The CAP has moved toward payments that do not encourage overproduction and are more environmentally friendly. The whole system still results in high food prices, payment inequalities, and, via export subsidies, distortions in global agricultural trade.

Common Fisheries Policy

The policy was drawn up in the early 1970s, prior to the entry of Ireland, Denmark, and Britain, which possessed large fishing fleets, to present them with an established policy. The policy was a major factor in Norwegians voting no in the referendum on joining. The Norwegians were right to be wary.

Outside a twelve-mile limit the policy pooled the fishing resources of the community. Further, funds were made available to modernize fishing fleets and their fishing gear. The more efficient vessels have overfished the resources and fish stocks have been depleted.

Manufacturing

The creation of EEC theoretically created an extensive European home market and encouraged larger-scale production. It was not until the Single European Act (1986) that administrative obstacles to moving manufactured products across national borders within the community were removed. Previously trucks stopped at borders, with drivers producing documentation on the origin of goods being carried. Now the movement of goods is freer but governments still find ways to favor national manufacturers.

The EEC/EC/EU has been successful in attracting investment into Europe by multinational corporations wanting to operate within the territory of the world's largest market for manufactured goods. European corporations have not been so successful in using Europe to become major players in the world economy. Oil companies like BP and Royal Dutch Shell are global corporations, so are companies like Siemans, Philips, Ericksen, Volkswagen, Daimler, and, in the luxury car market, BMW. Volvo is a major truck and bus producer but the car division is now part of Ford. Several car producers including Fiat, Peugeot, Citroen, Alfa Romeo, and MG (Morris Garages) have ceased to compete in U.S. markets or have gone out of business, which was the case with the British sports car manufacturers.

Services

In the service sector there is not a European Union open market for many products including telecommunications and financial services. Nor is there a Europe-wide open skies policy allowing competition between airlines in the region.

▶ SUMMARY

In overall social and economic terms how do European countries compare with the rest of the world? The UN Human Development Index (HDI) compares countries on the basis of life expectancy, literacy, education, and gross domestic product per capita. In the world league the countries with the highest HDI scores are Norway and Sweden, followed by Australia, Canada, The Netherlands, Belgium, and Iceland (Table 5.3). The U.S. is in eighth place, Japan at ninth, and the positions between tenth and seventeenth are all European: Ireland, Switzerland, the United Kingdom, Finland, Austria, Luxembourg, France, and Denmark. Notice that three non-EU, European countries—Norway, Iceland, and Switzerland—are highly ranked. In general Scandinavia and western Europe score highly. Southern Europe—Spain, Italy, Greece, and Portugal—do less well but are still in the top thirty as are Cyprus and Malta. The eastern European countries,

admitted to the EU in 2004, are all in the top fifty, with Slovenia and the Czech Republic being the highest ranked. Bulgaria, Romania, and Turkey are not in the top fifty countries, along with Russia, Belarus, and the Ukraine. The Human Development Index suggests that it will be more difficult for the EU to expand eastward in the future.

The Human Development Index compares country with country. We can change the scale and look at living standards within countries. No countries avoid having disparities between rich and poor regions. West Germany has much higher living standards than East Germany for political reasons. Southern Italy is poorer than northern Italy, for historical reasons. Scotland and the north of England, with early, wealth-producing industrial revolutions, have lower living standards than the southeast where London, the UK Treasury, and policymaking are concentrated.

TABLE 5.3 Human Development Index (Countries in HDI Rank Order)

	Life expectancy (2006)	GNI per capita (2005) in US$	Infant mortality rate (2006)	% of population over age 65 (2006)
Norway	80	40,420	3.1	15
Sweden	81	31,420	2.4	17
Australia	81	30,610	4.9	13
Canada	80	32,220	5.3	13
Netherlands	79	32,480	4.9	14
Belgium	79	32,640	4.8	17
Iceland	81	34,760	2.5	12
United States	78	41,950	6.7	12
Japan	82	31,410	2.8	20
Ireland	78	34,720	4.7	11
Switzerland	81	37,080	4.3	16
United Kingdom	78	32,690	5.1	16
Finland	79	31,170	3.0	16
Austria	79	33,140	4.1	16
Luxembourg	78	65,340	3.9	14
France	80	30,540	3.6	16
Denmark	78	33,570	4.4	15
New Zealand	79	23,030	5.1	12
Germany	79	29,210	3.9	19
Spain	81	25,820	4.0	17
Italy	80	28,840	4.1	19

Source: Human Development Report, New York: UN, 2004; Population Reference Bureau, 2006.

► FURTHER READING

Butlin, R.A., and R.A. Dodshon (eds.). *An Historical Geography of Europe*. Oxford, UK: Clarendon Press, 1998.

Church, R. *The Rise and Decline of the British Motor Industry*. Cambridge, UK: Cambridge University Press, 1995.

Dinan, D. *Ever Closer to Union: An Introduction to European Integration*. Boulder, CO: Lynne Rienner, 2005.

Hugill, P.J. *World Trade since 1431: Geography, Technology, and Capitalism*. Baltimore: Johns Hopkins University Press, 1993.

Landers, J. *The Field and the Forge: Population, Production, and Power in the Pre-industrial West*. New York: Oxford University Press, 2003.

Pounds, N.J.G. *The Ruhr*. Bloomington, IN: University of Indiana Press, 1962.

Rostow, W.W. *The Stages of Economic Growth: A Non-Communist Manifesto*. 3rd edition. New York: Cambridge University Press, 1990.

Rothacer, A. *Uniting Europe*. London: Imperial College Press, 2005.

Wrigley, E.A. *Poverty, Progress, and Population*. New York: Cambridge University Press, 2004.

6

POLITICAL
GEOGRAPHY

Since the French Revolution, starting in 1789, there have been attempts to dominate Europe ideologically and territorially. The EU can be seen as a mechanism to engage the major states of the region, allowing them to compete, and cooperate, under a common set of rules. World War I and World War II resulted when the international system failed to contain the ambitions of major states. After World War II most countries in western Europe were prepared to seek a solution that put cooperation before national aggrandizement. To understand how the countries of Europe came to believe that a European Union was needed, let us see how they interacted over the last two centuries.

At the end of the Napoleonic wars (1815) the Congress of Vienna produced a lasting peace that reintegrated the defeated power, France, into Europe. For decades the boundaries of Europe were relatively stable. The Greeks, by revolt against Turkish rule, established an independent state in 1829. Belgium and The Netherlands separated in 1831 and Britain, France, Prussia, Russia, and Austria guaranteed Belgium neutrality in 1839. In the Crimean War (1854–1856), Britain and France supported Turkey to prevent the Russian Empire from expanding to the Bosporus (the channel leading from the Black Sea to the Mediterranean) and becoming a Mediterranean power.

On the world scene the 1860s were a period of unification that often involved war. Japan displaced regional rulers with a central government by the Meiji restoration of 1868. The U.S. fought a civil war to maintain the Union. In Europe, the Kingdom of Italy, after a long war, unified in 1862. Venetia joined the Kingdom in 1866, and Rome in 1870, becoming the capital in the following year. The Dual Monarchy of Austria-Hungary was created in 1867. Both countries had parliaments and constitutions but shared foreign, defense, and economic policies, while recognizing one emperor. After taking territory from Denmark and Alsace-Lorraine from France, the German Empire was proclaimed in 1871 at the end of the Franco-Prussian war.

In southeast Europe, Ottoman power declined. Romania emerged from the Ottoman Empire in 1862. Britain leased Cyprus from the Turks in 1878 and Greece enlarged at Turkish expense in 1863 and 1881. By the 1890s the world was in a phase of imperial competition. The United States annexed Hawaii and went to war with Spain (1898) in the Philippines and the Caribbean, prior to starting the Panama Canal in 1903. Japan was expanding imperially, taking Taiwan from China in 1895 and preparing for a

Figure 6.1 Boundaries of Europe before WWI.

successful Far Eastern war with the Russian Empire in 1904. But the crucible of competition was Europe where the imperial, territorial, and security interests of Germany, Austria-Hungary, France, Britain, Italy, Russia, and the Ottoman Empire rubbed against one other (Figure 6.1).

► FRIEDRICH RATZEL

The territorial pressures were reflected in the writings of the German geographer Freidrich Ratzel (1844–1904). Ratzel suggested that the inhabitants of states bonded with the territory they occupied. Successful states enlarged territory and expanded their *Lebensraum* (living space). States did not have permanent borders; powerful organisms grew at the expense of weaker states. The logic of Ratzel's notion of *Lebensraum* was that larger states would emerge and continue to compete for territory. The Swedish political scientist Rudolf Kjellén coined the term "geopolitics" and in 1905 published *Stormakterna,* the Great Powers. The book advocated the development of a northern European superstate under German leadership. No one took much notice of Kjellén's argument in Sweden, but his book went to twenty editions in Germany!

▶ MACKINDER AND BOWMAN

The British geographer Halford Mackinder (1861–1947) suggested that great power competition for territory in Eurasia would lead to domination by one power or an alliance of powers, possibly Germany and/or Russia. It took two world wars and the Cold War to prevent Mackinder's prophesy in "The Geographical Pivot of History" (1904) coming true. Mackinder was correct in seeing that the powers of Europe were in competition for territory, resources, productive capacity, and routeways. Looking back on World War I the American geographer Isaiah Bowman commented:

> It was not the death of a Grand Duke at Sarajevo or the invasion of Belgium, nor was it the ambitions of the German ruler or the Pan-German dreams of the German Junkers—it was not any one of these things that produced the war. It was a combination of all of them, colored by a desire to control the seats of production and the channels of transportation of all those products, like coal and oil and hemp and cotton and iron and steel and manganese, that are foundations of the modern industrial world. These commodities may become sources of unrivaled power, and in the hands of unscrupulous men or nations they may be used for the devastation of the entire world. And by this we do not mean merely the destruction of its material wealth, but also the destruction of its political and religious liberty. (1921, pp. 8–10).

▶ WORLD WAR I

In 1914 war was started by client states: states supposedly controlled by greater powers. The assassination of Austrian Archduke Franz Ferdinand, heir to the Dual Monarchy of Austria-Hungary, by a terrorist with Serb connections in Sarajevo, Bosnia, part of the Austro-Hungarian Empire at the time, set up conflict. (Photo 6.1) The Austrians, backed by Germany, made demands on Serbia. Serbia, backed by Russia, agreed to most but not all conditions. The system of alliances hastened the onset of war as countries mobilized and others followed, fearing they would be unable to resist invasion.

In August 1914, Germany precipitated Britain's entry into the war by attacking Belgium to get at France. As the Foreign Secretary told the House of Commons, if Britain stood back, did nothing, and allowed Germany to enlarge itself by taking over smaller states, the United Kingdom would be powerless to reverse the result.

German War Aims

At first the war went well on the eastern and western fronts for the Central Powers: Germany, Austria-Hungary, and the Ottoman Empire. In September 1914 the German

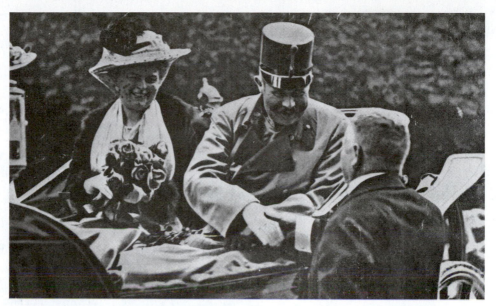

Photo 6.1   Archduke Franz Ferdinand and his wife Sophie riding in an open carriage at Sarajevo shortly before their assassination.

Source: Henry Guttmann/Getty Images/NewsCom.

Chancellor drafted the terms he would present to the Allied Powers: France, Belgium, Britain, and the Russian Empire. In eastern Europe, the Russian hold on non-Russian people would be broken, resulting in the creation of new states like Poland, Belarus, Ukraine, and Georgia. In western Europe, France would give up territory in the strategic Vosges Mountains, on the west bank of the Rhine, and more of the Lorraine iron ore resources. Belgium would be partially absorbed into Germany, along with Luxembourg. Antwerp would become a German port. Much of the Belgium coast would be under German control, as would the French ports of Calais and Dunkirk. Germany was proposing to control the south shore of the Strait of Dover (Figure 6.2)! Overseas, Germany wanted an enlarged African empire that would link her colonies in east and west Africa.

The German Chancellor was not solely interested in territory. He wanted a European Economic Association (EEA) that would include at least France, The Netherlands, Belgium, and Austria-Hungary. This would enable German goods to be traded more easily in Europe and protective tariffs, around EEA, would disadvantage goods of nonmembers. Britain was to be specifically excluded from the EEA, as Napolean had tried in his Continental System a century earlier. The EEA was to be controlled by Germany but here was the basic idea for a European Common market. After World War I the concept of a European Union emerged from the remnants of Austria-Hungary and was taken up by France in the late 1920s.

In World War I, Germany won the war in the east and by the Treaty of Brest-Litovsk (1918) broke Russia's hold on the non-Russian people of eastern Europe. Unfortunately for Germany, the war in the west still had to be won and rather than accepting Woodrow

Figure 6.2   A projected Europe shaped by German victory in WWI.

Wilson's fourteen points as the basis of peace talks, Germany held to a hard line, refusing to withdraw from Belgium. Huge, unsuccessful German attacks on the western front in the spring of 1918 depleted manpower. In September, realizing they could not win, the German high command resigned. A civilian government was left to seek peace, helping to build the myth that the German army had not lost the war but had suffered a stab in the back from politicians at home.

Postwar Settlement

The terms that Germany got, first in the armistice of November 11, 1918, and then at Versailles in 1919, were harsh and from a German perspective did not reflect the fourteen points they had rejected in the early months of 1918. Germany was forced to give up all the eastern gains, withdraw from France and Belgium, accept allied occupation of the strategic Rhineland, relinquish large quantities of armaments, and agree not to rearm.

The Treaty of Versailles started with the covenants of the League of Nations, largely drafted by President Woodrow Wilson. If Germany hoped that the founding of the League was the start of an era of international cooperation, it was wrong. Germany was not to be a founding member of the League.

The body of the Versailles Treaty took territory from Germany and required the payment of reparations. A member of the British delegation, the economist John Maynard Keynes, resigned over reparations, predicting in *Economic Consequences of the Peace* (1919) that if Germany were crippled economically the economies of all Europe would suffer. Keynes was correct. Other treaties were signed with Austria (St. Germain), Hungary (Trianon), Bulgaria (Neuilly), and Turkey (Sèvres). Sèvres did not hold, fighting broke out between Greece and Turkey, and the issues were not settled until 1923, by the Treaty of Lausanne.

New Boundaries and New States

Figure 6.1 shows the boundaries of states prior to World War I. Figure 6.3 depicts the postwar boundaries. In 1914 Central Europe was dominated by Austria-Hungary and the German Empire. The Russian Empire controlled eastern Europe. The Ottoman Empire had extensive territory in the Middle East. In World War I the Empire of Austria-Hungary dissolved. Austria and Hungary were no longer linked after the war and lost territory as the result of the creation of new states. The imperial capital of Vienna became, in function, a provincial city, losing population and suffering unemployment and poverty. In Figure 6.3, due to the Russian Revolution, the Russian Empire has been replaced by the Soviet Union, with boundaries far to the east of the prewar empire.

The German Empire was replaced by the Weimar Republic. The new Germany had lost territory in the east to Poland and in the west to France, primarily Alsace and Lorraine, and Belgium (Eupen and Malmédy). The Polish Corridor gave Poland access to the Baltic Sea via the ports of Danzig (Gdańsk) and Gdynia. There were more Poles than Germans in the Corridor, by a small margin, but the Corridor divided Germany from its eastern territory of East Prussia. The Free City of Danzig and surrounding territory was under the protection of the League, but in practice the indisputably German city was used by Poland for purposes of trade.

Comparing Figure 6.1 with Figure 6.3, the most striking features are the number of new states and new boundaries on the map in central and eastern Europe. Running from the Baltic Sea to the Adriatic and the Black Sea is a tier of new states, or states with significantly altered boundaries. Finland, Estonia, Latvia, Lithuania, Poland, Czechoslovakia and Yugoslavia are new states not on the map in 1914. Austria, Hungary, and Bulgaria, allies of Germany in World War I, have been downsized. Romania, a western ally, has enlarged. Serbia and Montenegro are part of Yugoslavia. Some states briefly independent

Figure 6.3 European Boundaries, post WWI.

at the end of the war, including Belarus, Ukraine, and Georgia, were absorbed into the Soviet Union.

Central and eastern Europe had apparently been reorganized on the nationality principle, what President Wilson referred to as self-determination. A problem with the principle was that language, religious, and national groups merged into each other (Smith, 2003). There were few clean lines to be drawn on the boundary maps. Czechoslovakia, for example, contained Germans, Poles, Hungarians, and Ruthenians in addition to the Czechs and Slovaks in an uncertain union. In Yugoslavia, the Serbs, Croats, Slovenes, Bosnians, and Montenegrans were frequently hostile to each other until the country broke up in the 1990s. Self-determination was not applied to Germans. Many Germans were outside what became the Third Reich, and Hitler would use the "nationality principle" to rally domestic support and extort territory from neighbors in the runup to World War II.

▶ WHO RULES EAST EUROPE COMMANDS THE HEARTLAND

The empire of Austria-Hungary had gone but Germany, with clipped wings, and the Soviet Union, successor to the Russian Empire, were still threats to stability. Austria-Hungary, which had protected its interests in Central Europe against Germany and in southeast Europe against Russia, was lost as a balancing power. Now central and eastern Europe were divided among several weak, ill-organized states, all engaged in boundary disputes with each other.

Isaiah Bowman, the American geographer who accompanied President Wilson to Paris and helped construct the new boundaries of Europe, was pessimistic. Bowman, in *The New World* (1921), commented that every mile of new boundary had increased the friction between unlike and unfriendly people. War would come again in a few years.

The British geographer Halford Mackinder identified the dangers in *Democratic Ideals and Reality* (1919). The strategic structure of eastern Europe was weak. The many new, ill-prepared states made it easier for a major power to control the region as a stepping stone to control of Eurasia. Mackinder, in 1919, enlarged his *Geographical Pivot of History* (1904) and renamed it the Heartland. He gave his view of the strategic situation in the lines:

Who rules East Europe commands the Heartland:

Who rules the Heartland commands the World-Island:

Who rules the World-Island commands the World.

There were only two powers that could command eastern Europe—Germany and the Soviet Union—and just before World War II, in August 1939, they signed the infamous Nazi–Soviet Non-Aggression Pact, which divided the region between them. In June 1941, Germany attacked the Red Army, starting a war to establish the paramount power in Europe.

▶ IDEAS FOR A UNITED EUROPE

World War I was a disaster for all the European countries involved but it was a catalyst for union. In the early years after the war, people in Austria and Hungary rapidly understood what they had lost economically by the disintegration of their empire. The empire had been a customs union with a common currency and a common external tariff on goods entering from other countries. Now goods moving between newly independent Czechoslovakia, Austria, and Hungary had to pass through customs and incur tariffs. Payment for goods was more complicated as Austria (schillings), Hungary (the pengö), and Czechoslovakia (crowns) had different currencies, replacing the common currency issued by the state bank of Austria-Hungary.

The economic experience of Austria-Hungary can be seen as a forerunner of the European Union. An influential group of political scientists and economists in Vienna and Budapest examined the idea of European economic and political union, including Count Richard Coudenhove-Kalergi, who had been brought up in the Czech lands of Bohemia and educated in Vienna. In *Pan Europe* (1923), Count Richard argued for a united Europe that ran from Poland to Portugal but excluded the Soviet Union.

The French geographer Albert Demangeon described the *Decline of Europe* (1920) and later developed the idea of a united Europe, which interested Aristide Briand, the

French foreign minister. Briand took the idea to the League of Nations where the German Foreign Minister Gustav Stresemann spoke in favor of European integration. Many countries saw the idea as another scheme to advance French interests. But notice the convergence of ideas. Germany wanted a European Economic Association in World War I. Coudenhove-Kalergi, from Austria-Hungary, argued for European Union in 1923 and in the late 1920s France and Germany raised the concept at the League of Nations. However, the idea of creating a cooperative Europe was overwhelmed by the Great Depression and the emergence of totalitarian regimes striving for hegemony over the region.

▶ TOTALITARIAN REGIMES AND ECONOMIC AUTARCKY

After World War I Europe was unstable. The Russian Empire was replaced by the USSR, a totalitarian, communist dictatorship. Germany had numerous outbreaks of lawlessness. In 1919 Spartacists (radical socialists) were involved in a week of street fighting in Berlin. The Munich Bierkeller Putsch (1923) led to the imprisonment of Nazi leaders (including Hitler) and an economic breakdown started in the same year as French and Belgium troops occupied the Ruhr. Mussolini established a Fascist regime in Italy (1928); Hitler came to power in Germany in 1933. By 1936 all the new states of Europe, except Czechoslovakia, had dictatorial governments.

Totalitarian regimes established one-party rule, enforced loyalty to the state, and tried to establish autarkic (self-sufficient) economic systems. A country aiming for autarky will grow its own food, process its own raw materials, and manufacture its own goods. Autarky involves reducing imports to a minimum and producing as much as possible on home territory. The concept of comparative advantage, the idea that countries will produce the products to which they are best suited and exchange those goods in international trade to the benefit of all, has little place in an autarkic system where paying for imports is seen as a loss. Autarky is the antithesis of free trade.

Economic autarky is ultimate protectionism. Raw materials and goods will be produced on home territory even if some products could be imported more cheaply. Fascist Italy, Nazi Germany, and the Communist Soviet Union all strove for autarky. In the case of Italy, it was impossible to achieve and for Germany difficult to achieve. After World War I there was a trend toward nationalistic, autarkic economic policies and, when world trade declined in the Depression, protectionist policies deepened the world recession.

The Soviet Union

In theory the USSR had a resource base capable of providing nearly all raw materials needed by a modern, industrial state. The agricultural resources were sufficient to produce all the temperate crops and, around the Black Sea, some subtropical crops could be grown. Stalin, having secured power and established a repressive regime, started the first five-year development plan, dated 1928, with the aim of making the Soviet Union economically self-sufficient. The economy was centrally planned, with the state deciding which industries to develop and what infrastructure to construct. In general, the aim was to develop heavy industries producing iron and steel, trucks and tractors to support the construction of hydroelectric power plants, railroads, pipelines, and new industrial cities.

Agriculture came under central control with the creation of state farms and collective farms.

Fascist Italy

It was difficult for Italy, under Mussolini, to become self-sufficient as it did not have the necessary resources of coal, iron ore, and oil. The country had to import wheat. Incentives to grow more wheat came at the expense of crops Italy was well suited to grow. The draining of marshlands increased the area of agricultural land, reducing the incidence of malaria, a disease still widespread in Italy until after World War II.

To overcome raw material and product shortages, Italy became a corporate state, which is to say the government created corporations, or entered into agreements with existing businesses, to produce raw materials and goods that were in short supply. A state-controlled corporation, AGIP, was created to develop oil and gas resources. Motor vehicle producer Alfa Romeo was taken over by the state and Fiat organized production to meet the needs of the fascist state.

Economic autarky for Italy was impossible unless, and this was Mussolini's aim, the Mediterranean could be turned into an Italian sea (*mare nostrum*). Expansion could only come at the expense of Yugoslavia, Greece, Albania, France (which controlled Corsica, Tunisia, and Algeria), and Britain with bases at Gibraltar, Malta, Egypt, and the Suez Canal.

Nazi Germany

Germany made an interesting attempt to create self-sufficiency in the Nazi era. Germany had resources, including the Ruhr and other coalfields, but lacked iron ore, oil, manganese, and other nonferrous metals.

When the Nazi party came to power it did not have a well-defined economic policy. Hitler knew he wanted to work Germans in the service of the Fatherland and relieve unemployment. State-financed projects, like the building of *autobahns,* provided work and improved infrastructure. With public works and rearmament, the economy picked up rapidly. To avoid inflation, a real fear in Germany after the experience of 1924, prices and wages were controlled. Imports were kept out because Germany had little hard currency to pay for foreign goods. Bilateral trade agreements with Middle Eastern and Latin American countries provided some raw materials. Under the agreements Germany made payment in *marks* that could only be used to buy German goods and services. German trade with European countries was reduced as Germany strove for economic self-sufficiency.

Under the Nazi economic regime, the big corporations like Krupp and Thyssen were not taken over by the state but produced to the needs of the state. In 1936, a central economic planning authority and a four-year plan were started. Synthetics were developed to reduce imports. German chemists, before World War I, had produced nitrogen from the atmosphere and in the 1920s made synthetic rubber (*buna*). Petroleum was distilled from coal and numerous *ersatz* (substitute) products were developed, including coffee, to reduce imports.

The state funded production projects, such as the *Volkswagen,* the people's car. The basic car, designed by Porsche, was assembled in a large, new factory opened by Hitler at

Wolfsburg in 1938. German families made regular advance payments until the purchase price was accumulated. Few German civilians got cars. As war approached the factory made vehicles for the *Wehrmacht.*

However hard Germany worked to increase production and develop synthetics, the country could not become self-sufficient within the boundaries of the state. Some overseas sources of supply, like iron ore from neutral Sweden, were well established but the lack of oil fields was a fundamental problem.

German agriculture was inefficient and not much Nazi effort went into improvement. There was little mechanization, fertilizer use was limited, and livestock, dairy products, and animal fats were imported when productivity could have been greatly increased. But in Hitler's mind self-sufficiency, agricultural surpluses, and raw materials would be achieved by gaining *Lebensraum* (living space) in the east. Hitler did not want overseas colonies, which he considered expensive and difficult to defend. Hitler wanted to settle Germans, and other Nordics, on land taken in the east from Slavic peoples. As Hitler proclaimed at the 1936 Nürnberg rally, if he could acquire the ores of the Urals, the forests of Siberia, and the unending grain fields of Ukraine, Germany would swim in plenty. *Lebensraum* was about space, race, and creating a new geography in the Slav lands. The Poles, Ukrainians, and Russians were to be killed, starved, or deported.

> A German governing class would rule the region, supported by a network of garrison cities . . . around which would cluster settlements of German farmers and traders. Plans were drawn up for a web of high-speed motorways to link the regional centers with Berlin and a wide-gauge double-decked railway. . . . The Slav population would be displaced from the land, employed as menial labor or transported beyond the Urals into Siberia. (Overy, 1998, p. 132)

German Territorial Expansion

Before starting territorial expansion, Germany remilitarized the Rhineland (1936), disregarding Versailles and the Treaty of Locarno (1925). At first territorial expansion involved bringing German speakers into the Reich, a policy advocated from the 1920s by the Munich geopolitician General Karl Haushofer who wanted the creation of a greater Deutschland inhabited by people of German speech and culture. Expansion started in March 1938 when German forces took over Austria in contravention of the Treaties of Versailles and St. Germain, without opposition. *Anschluss* (union) was proclaimed on March 13. Hitler arrived in Vienna in an open car, greeted by approving Austrians as church bells rang and Swastikas flew. (Photo 6.2) Austria became a province of Germany until 1945.

In the summer of 1938, Hitler demanded that the Sudentenland of Czechoslovakia, containing many German speakers, should be joined to Germany or there would be war. Representatives of Britain (Chamberlain), Italy (Mussolini), France (Daladier), and Germany (Hitler) met at Munich in September 1938. Czechoslovakia was not in the conference and at the end was told to hand over the Sudentenland, and other German-speaking areas, to Germany. In March 1939, Germany took the remaining Czech lands. Slovakia became a German protectorate.

The takeover of the Czech lands brought Slavs into Deutschland. Haushofer protested. Hitler dismissed him as a silly old man who did not understand the grand design. In March 1939, President Franklin Roosevelt suggested that Hitler's aims had gone beyond bringing contiguous German populations into the Reich. There were now no limits to German expansion.

Photo 6.2 Adolf Hitler arrives in a motorcade at the "Adolf Hitler palace" in front of Vienna's town hall. Notice the large welcoming crowds.

Source: Hulton Archive/Getty Images/NewsCom.

With the Czech lands absorbed, Hitler prepared to attack Poland. Britain and France gave Poland a territorial guarantee. Hitler assumed that Britain and France would no more fight for Poland than they had for Czechoslovakia. The Soviet Union was a different matter. The USSR was unlikely to allow German occupation of eastern Poland, where many Ukrainians lived, without resistance. A deal was made. By the Non-Aggression Pact between Germany and the Soviet Union of August 23, 1939, the countries agreed not to attack each other and, in secret understandings, divided up Poland and eastern Europe.

The public part of the Non-Aggression Pact produced strategic shock in the West, for it meant the suspension of hopes that the Soviets could be used as a counterweight to Germany. Stalin viewed the Munich agreement as an attempt, by Britain and France, to turn Germany east and countered with the signing of the Non-Aggression Pact that lasted for a short time.

Hitler saw the Non-Aggression Pact as a temporary accommodation with the Communist Soviet Union. This was not so with his foreign minister Ribbentrop who, along with many German strategists, saw the pact as a way to avoid war on two fronts. General Haushofer welcomed the pact. Haushofer had studied Mackinder's "Geographical Pivot

Photo 6.3 Reproduction of Cartoon of 19th May 1941 by Dr. Seuss. University of California, San Diego, Mandeville Special Collection (Dr. Seuss Collection).

of History" (1904) and understood that an alliance between Germany and Russia was the way for Germany to become a dominating world power.

On September 1, 1939, Germany attacked Poland. On September 3, France and Britain, honoring the commitment to Poland, declared war on Germany. On September 17, the Red Army invaded Poland from the east. By the end of the month Poland was gone from the map. In western Poland the Gestapo, using prepared lists, arrested Polish politicians, nationalists, and left wingers. In the east the Soviet Secret Police, using prepared lists, removed those who represented Polish national identity. Later, the Soviets murdered many Polish reserve officers and buried them in the forest at Katyn. The reserve officers were educated, professional men and Stalin did not want any educated Poles unless they were Moscow oriented.

When Britain and France declared war on Germany, a surprised Hitler exclaimed, "Now what?!" The British and French response was muted. The Royal Air Force flew leaflet raids over Germany. The French army sat in the Maginot Line.

In the spring of 1940, Germany attacked neutral Denmark, Norway, The Netherlands, and Belgium and invaded France. (Photo 6.3) Italy entered the war against France and Britain on June 10. France sought an armistice on June 22. The Royal Air Force contested control of the air over the Channel in the Battle of Britain and prevented a German invasion of the United Kingdom.

By the fall of 1940 Germany was in control of Europe from the Atlantic to the Vistula, and as a result of deals with Hungary and Romania, much of the Danube valley flowing to the Black Sea. The USSR expanded too in 1939–40, taking territory by force from Finland, absorbing the Baltics, and pressuring Romania to give up Moldova.

Photo 6.4   Reproduction of Cartoon of 25th June 1941 by Dr. Seuss. University of California, San Diego, Mandeville Special Collection (Dr. Seuss Collection).

In December 1940 Hitler, having failed to take Britain, ordered his generals to plan the invasion of the Soviet Union. (Photo 6.4) There was to be war on two fronts. The *Wehrmacht* attacked the Red Army on the night of June 21–22nd, 1941. Soviet losses were horrendous. German forces reached Smolensk two hundred miles west of Moscow by September. But Hitler had also ordered thrusts toward Leningrad, through the Baltics, and to the Ukraine, along the north shore of the Black Sea. When the cold continental winter arrived as it did every year, objectives had not been reached. Stalin, told by a spy that Japan would strike into southeast Asia, moved forces from the Soviet far east to Moscow. In early December these fresh troops stopped the German advance, which had already lost impetus on long supply lines. (Photo 6.5) On December 7, 1941, Japan attacked Hong Kong, Pearl Harbor, and southeast Asia. On December 11, Germany declared war on the United States.

▶ THE NEW ORDER IN EUROPE

At the beginning of 1942 Germany controlled Europe from France to western Russia (Figure 6.4). Neutral Spain and Portugal were fascist. Neutral Sweden and Switzerland were integrated with the economy of Germany. In the west, only the offshore islands of Britain and Malta were resisting German hegemony. In the east, the Red Army remained a powerful fighting force.

Germany started to create the New Order politically and economically. Under the armaments minister Albert Speer, western European industries were organized to support the German war effort. Contracts were given to non-German companies to supply aircraft components, tanks, trucks, electronics, and guns to the *Wehrmacht.* France became a major supplier of raw materials, agricultural commodities, and armaments to Germany. Prior to the war France exported 1.8 million francs worth of goods to Germany; by 1943 the figure was 29 million francs! France did not make money from trade, for the export earnings went to pay German army occupation costs.

Photo 6.5   A German supply column during the German invasion of the Soviet Union. German forces often lacked motorized transport.

Source: Ullstein bild/The Granger Collection, New York.

The greatest degree of industrial cooperation was between the coal, iron, and steel industries of France, Belgium, Luxembourg, and Germany. The first major postwar economic cooperation venture was the coal and steel community, involving the countries listed plus Italy and The Netherlands.

European agriculture declined during the war. Countries with efficient agriculture, like Denmark and The Netherlands, relied upon imported fertilizers and animal feed. Imports stopped with the war and output declined. Inefficient farming in France, Germany, and Italy depended on high labor inputs. The labor was drafted into war work elsewhere. Every country in Europe suffered food shortages in World War II and some, like The Netherlands, where tulip bulbs were fried for food, starved (de Blij, 2006). After the war the experience of shortages helped to sell the Common Agricultural Policy, with its emphasis on full production and subsidized surpluses.

▶ NEW EUROPEAN BOUNDARIES AFTER WORLD WAR II

The basic pattern of postwar Europe was dictated by the position of the Soviet armies at the end of hostilities. There were no new states in the aftermath of war but boundaries were moved (Figure 6.5).

Figure 6.4 European Boundaries, 1942 (height of Hitler's power).

Figure 6.5 European Boundaries, post WWII.

Finland ceded territory to the Soviet Union close to Leningrad, and on the Arctic Sea, the port of Petsamo (Pechenga). This latter transfer gave the USSR a land boundary with Norway. The Baltics did not reemerge from the Soviet Union. Poland was moved westward at the expense of Germany, eastern Poland being incorporated into the Ukraine. East Prussia was divided between Poland and Russia. Königsberg, the former capital of East Prussia, with its hinterland, became Kaliningrad, a detached area (exclave) of the Russian S.S.R. Czechoslovakia gave up territory in the east occupied by Ruthenians to Ukraine.

The results of the boundary changes are shown in Figure 6.5. Notice that the changes gave the Soviet Union land frontiers with Norway, Czechoslovakia, and Hungary where none existed prewar.

Communist governments took control in all eastern European countries; the last to be taken over was Czechoslovakia in 1948, the same year that Yugoslavia broke with Moscow but remained communist under Marshall Tito. Many Poles and Czechs had to leave their countries twice: once at the beginning of World War II to avoid the Gestapo and again after the war, to evade the secret police of Communist governments. The Czech diplomat Josef Korbel got to Britain as the Nazis took over. He returned to serve his country after the war but had to slip away to the United States to avoid the Communists. He taught political science in Colorado and his Czech-born daughter, Madeleine Albright, became the U.S. Secretary of State.

Winston Churchill's metaphorical iron curtain of 1946 rapidly became reality as Moscow-orientated communist regimes took over all of the countries of eastern Europe. Secret and political police enforced party rule. Travel between states, or to western Europe, was difficult for eastern Europeans. The frontier between the NATO countries and the Communist bloc was marked by fences, floodlights, guard towers, dogs, and armed patrols.

The Soviet takeover of eastern Europe did not suppress opposition. There were uprisings in East Berlin (1953), Hungary (1956), and Prague (1968), and constant rumblings in Poland. Protests by Polish shipyard workers led to the creation of the Solidarity movement in 1980 led by Lech Walesa, which did get the Communist rulers to relax many restrictions.

Eastern Europe struggled to rebuild after World War II under repressive regimes, but western Europe quickly attained prewar levels of production and, in the 1950s and 1960s, experienced rapid economic expansion. Economic growth was epitomized by West Germany where expansion was so rapid that labor shortages resulted in a "guest worker" influx. If East Germans could get to West Germany there was a market for their labor. Large numbers of predominately young workers left the GDR, helping to perpetuate economic stagnation and an aging population in the east.

The Cold War

The Soviet takeover of eastern Europe made the Cold War inevitable. As Winston Churchill declared in a speech at Westminster College, in Fulton, Missouri (1946), "an iron curtain has descended across the continent" from Stettin on the Baltic to Trieste on the Adriatic. The following year the Truman Doctrine announced that the U.S. would help countries that wanted to remain free of Communist regimes. Later, in 1947, the Marshall Plan provided assistance to speed European economic recovery and to ensure that the U.S. had access to the profitable export markets. The Soviet Union declined to take part in the Marshall Plan and told the countries of eastern Europe to stay out.

NATO (1949) was designed to deter further Soviet takeovers of European countries. When the western occupation zone of Germany became an independent state and joined NATO (1955) the Soviet response was to create the Warsaw Pact (1955), with headquarters in Moscow. The members were Poland, Czechoslovakia, Hungary, Romania, Bulgaria, Albania (withdrew 1968), and the German Democratic Republic—East Germany.

Mackinder (1904) had suggested that threats from the east had encouraged European nationhood and made the region look to the Atlantic. The Cold War forced the countries of western Europe to cooperate politically, economically, and militarily. The Council of Europe was established in 1949, and the Treaty of Brussels (1948) created a defensive framework, which was enlarged by the signing of the North Atlantic Treaty (1949) by Belgium, Britain, Denmark, France, Iceland, Italy, The Netherlands, Norway, Portugal, the United States, and Canada.

▶ ECONOMIC COOPERATION IN WESTERN EUROPE

Belgium and Luxembourg had been in a customs union since 1921 and in 1948 The Netherlands joined to create Benelux. In allowing the free movement of goods, labor, and capital between the member states, Benelux embodied the basic concept that was to be utilized by the European Economic Community. The Organization for European Economic Cooperation (OEEC), established in 1948, oversaw the European recovery developed by the Marshall Plan (1947).

The Coal and Steel Community

Before the creation of EEC the six core countries—France, Germany, Italy, and the Benelux states—created a Coal and Steel Community. The linkages between the coal, iron, and steel industries of France, Belgium, Luxembourg, and Germany were long standing. France imported coal, Belgium imported ore, and Luxembourg had to sell its steel into larger economies. Germany, which imported ore from Sweden and France, exported coal to France, Belgium, and Luxembourg. By manipulating subsidies, tariffs, and freight rates countries tried to advantage their own iron and steel industries and disadvantage competitors. The competition was intense between France and Germany.

The Schuman Plan, devised by Jean Monnet, for a European Coal and Steel Community (ECSC) was joined by France, Germany, Italy, and the Benelux countries (Treaty of Paris, 1951) and began functioning in 1952 (Figure 6.6). The ECSC created a common market in coal, iron, and steel, although transport costs still reflected the distance resources and products traveled. The removal of tariffs and subsidies meant that old metal works and coal mines closed. ECSC was phased in and helped retrain displaced workers in Belgium and elsewhere when old coal mines closed.

There was a politics to the plan, for it placed Germany in a European framework. France benefited from access to cheaper coal. Britain was not consulted on the structure of the ECSC. However, politically it was impossible for the UK to join. In July 1945 a Labour government was elected. Churchill was no longer prime minister. The new government implemented radical policies including nationalizing the railroads, coal mines, and iron and steel industries. The Labour government could not hand over decision making to the supranational ECSC. The Labour Party drew powerful support from unions that would have resisted outsiders telling Britons how to run their industries. Britain became a member of ECSC when it joined the European Economic Community in 1973.

Figure 6.6 Europe in the 1950s.

The European Economic Community

France, Germany, Italy, and the Benelux countries moved rapidly toward greater economic cooperation, signing the Treaty of Rome (1957) to establish the European Economic Community (EEC). The treaty provided for the eventual free movement of goods, labor, and capital between the member states. The members created a customs union with the removal of tariffs on goods moving between states, where goods originated in the territory of members. A Common External Tariff was placed on goods entering the territory of the six and protected industries in the enlarged home market from external competition.

The EEC, often referred to as the Common Market, reflected the idea that members were trying to establish common business conditions across the territory of the six. In order to create and administer the rules of economic cooperation a central authority was created at Brussels. The six agreed to some common economic policies, the best known being the Common Agricultural Policy (CAP), under which farmers were paid above world market prices to grow crops such as wheat. Cheaper wheat from Canada and the U.S. paid a duty, on entering EEC, to bring it up to the community price. Again, Brussels administered the complicated subsidies for crops, livestock, dairy products, wine, and olive oil.

Characteristics of EEC

The EEC represented an effort to create a partnership between France and Germany in place of the conflict that had resulted in the wars of 1870, 1914, and 1939. Once France and Germany agreed to create an economic community, the Benelux countries were likely to join. The Benelux countries already had a customs union and a huge proportion of their European trade was with France and Germany. The Rhine flowed to the North Sea through The Netherlands and past Dutch ports like Rotterdam. Antwerp, in Belgium on the river Schelde, was a port that served northern France and parts of Germany (Figure 6.6).

The character of the EEC was strongly influenced by the economic histories of France and Germany. From the seventeenth century on, France had developed mercantilist (protectionist) policies. Its largest trade partner in the 1930s was not a European country but the North African colony of Algeria. In the nineteenth century, German states created the *Zollverein,* a customs union with protective tariffs. Both countries, in the interwar years, had strived to reduce imports and to promote self-sufficiency. France, Germany, and Italy had used statist policies to promote economic development. The state had invested directly in economic activity, often establishing state-controlled corporations in important sectors of the economy. All of these features were reflected in the EEC with its protective common external tariff and the CAP designed to promote self-sufficiency in food production.

Nonjoiners of the European Economic Community

The European Economic Community that developed out of the Treaty of Rome had protectionist and statist characteristics. There would be free trade among the members, enlarging the home market for manufacturers in the six, and the manufacturers would receive protection through the Common External Tariff.

Under present World Trade Organization (WTO) rules, tariffs follow international agreements, but at the outset the EEC had many protectionist characteristics. This was not attractive to the United Kingdom, a traditional free trader, which wanted a European free trade area with no common external tariff to discriminate against trade partners like the United States, Canada, and Australia. Britain did not want the Brussels bureaucracy, nor the CAP, which would have the effect of cutting the UK off from cheaper food imported from the United States and Commonwealth countries.

By the creation of a supranational administration at Brussels, the EEC started a process of transferring elements of sovereignty to a central authority. Technically this infringed upon the neutral status of a number of countries and those states were not anxious, at the time, to join the EEC.

European Free Trade Association (EFTA)

By the Stockholm Convention of 1959, Austria, Denmark, Norway, Portugal, Sweden, Switzerland, and the United Kingdom created a free trade association allowing for the free movement of industrial products between the members. Iceland, Finland, and Liecht-enstein jointed EFTA later but several EFTA countries, starting in the 1970s, joined the European Economic Community.

EFTA and the European Community signed an agreement in 1992, creating a Euro-pean Economic Area (EEA) allowing free movement for goods, services, capital, and labor between EFTA and EU countries. The EFTA countries agreed to accept certain EU rules relating to the environment, social policy, consumer protection, and company law. Iceland, Liechtenstein, Norway, and Switzerland remain as members of EFTA. All enjoy the advantages of free movement of capital, people, and goods within the Euro-pean Economic Area without accepting regulation from Brussels and avoiding common policies on agriculture, fishing, and energy. In the case of Iceland and Norway, fishing is important and they have retained control over their fishing waters. When ten countries joined EU in 2004, all had signed the EEA agreement and EFTA countries benefited from the enlarged trade area. EFTA countries, particularly oil-rich Norway, agreed to provide funds to help economic transition in east Europe (EFTA Bulletin, 2004).

Enlarging the European Economic Community

By 1960 there was an EEC of six states joined in a common market, with a common external tariff and an EFTA. The EEC six were a contiguous group of countries. The EFTA seven were in the north (Norway, Sweden, the United Kingdom), the center (Aus-tria and Switzerland), and in the west (Portugal) (Figure 6.6). The six contained several large economies; the seven only possessed one large, diversified economy—the United Kingdom.

The United Kingdom, along with Denmark, Norway, and Ireland, made application to join the EEC in 1962. A number of large manufacturing corporations in Britain felt a competitive disadvantage as their goods had to pay tariffs on entering the EEC markets. Major financial units in London wanted to maintain their role in Europe as London was, and remains, a global financial center. British political orientation had changed. Harold Macmillan became prime minister (1957–63) of a Conservative government, succeeding Anthony Eden in the aftermath of the 1956 Suez debacle in which France and Britain had invaded the Suez Canal zone and had been forced to back down by the United States. Eden had been pro-Commonwealth and overseas development. Macmillan was a committed European who sped up the process of decolonization and was prepared to reduce mutually favorable trade with Canada, South Africa, Australia, and New Zealand to join the EEC. The United Kingdom application for membership was vetoed by French President de Gaulle in 1963, and the General said 'Non!' again in 1967. De Gaulle thought Britain was influenced by the United States. At Suez (1956), Britain had caved in to U.S. pressure. Britain was seen as a potential reviser of many EEC protectionist policies that the six had created. Above all, if Britain joined, she would be a power in EEC councils, at least the equal of France, and France was establishing herself as the leader of Europe.

The reestablishment of French prestige was a major focus of Gaullist policies. In 1940 French forces collapsed in a few weeks under German attack. De Gaulle had not

Photo 6.6 General Charles de Gaulle inspects a unit of "Free France" soldiers in London.

Source: Agence France Presse/Getty Images/NewsCom.

surrendered and led Free French forces from Britain, where he resented his subordinate status. (Photo 6.6) Postwar, with Germany divided, humbled by defeat, and anxious to be accepted into the European community, de Gaulle was establishing the authority of France. (Photo 6.7) He spent heavily on armaments, developed an independent French nuclear force and, in 1966, removed French forces from NATO command. (Figure 6.7) De Gaulle's spending on armaments came at the expense of education and social services. In May and June 1968 student demonstrations in Paris, backed by a sustained general strike, resulted in de Gaulle losing power in 1969. EEC enlargement became possible.

Early in the 1970s Denmark, Norway, Ireland, and the United Kingdom applied for membership. Britain, Denmark, and Ireland joined on January 1, 1973. The government of Norway, having negotiated entry, referred the issue of EEC membership to voters. The government lost the referendum by a narrow margin and resigned. Norway did not join EEC and remains outside the EU.

Criteria for Membership

Until the Copenhagen criteria (1993), three unwritten tests were applied to applicants for membership of the European club. Are you a democracy? Can you compete economically? Are you "a good European?" It was this last test that de Gaulle used to exclude Britain in 1963 and 1967. Other countries were not considered for membership because they lacked democratic governments. However, once democracy was achieved the European community was eager to embrace a country and confirm democratic institutions.

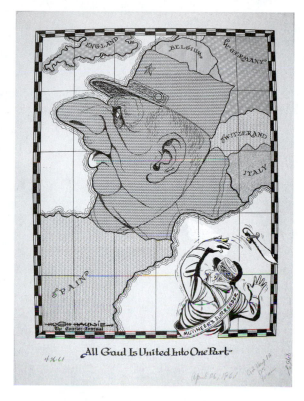

All Gaul Is United Into One Part

Photo 6.7

Source: Hugh Haynie Cartoon Collection, Special Collections Research Center, Earl Gregg Swem Library, College of William and Mary.

In 1993, at a meeting in Copenhagen, the conditions of EU membership were set out:

> Stability of institutions guaranteeing democracy, the rule of law, human rights, and respect for and protection of minorities ... a functioning market economy, as well as the capacity to cope with competitive pressure and market forces within the Union; (Bainbridge, 2002, p. 96)

Greece, Spain, and Portugal

After World War II, Greece was in a civil war for several years and lacked political stability. In 1967 a military *coup d'etat* resulted in the government of Greek Colonels, which held power until 1974. Greece acceded to the European community in 1981 but economically has had difficulty competing. Spain and Portugal became Fascist in the 1930s under General Franco and António Salazar, respectively. Portugal only started to establish a modern democracy with the election of 1976. General Franco took power as a result of the Spanish Civil War (1936–39) and, by staying neutral in World War II, was able to remain in office after the defeat of Fascist dictators like Mussolini and Hitler. Toward the end of his life (he died in 1975) Franco made arrangements to transfer power back to the Spanish monarchy. Democratic government was then established.

Figure 6.7 Europe in 1973.

Spain and Portugal joined the European community in 1986. Both countries have prospered economically and their economies have diversified. In sum, the 1980s resulted in the addition of three southern European countries—Greece, Portugal, and Spain (Figure 6.8). Each country still had a large agricultural sector placing additional demands on the budget of the Common Agricultural Policy, which provides subsidies to producers of many crops.

Creating a Single European Market

The European Community moved slowly from 1957 to 1986 to create a single, integrated market. It took twelve years to remove tariffs on goods moving between member

Figure 6.8 Europe in the 1980s.

states and when the tariffs were gone there were still many nontariff barriers to the free movement of goods. Crossing borders was cumbersome. Trucks had to stop, line up, and produce documentation showing that the goods being carried had originated in the territory of member states. Member countries still imposed their own health and safety rules. Not all countries were committed to the free movement of capital. In short, state policies and bureaucratic rules could, and did, prevent the creation of a single, integrated European market.

In 1986 the member states of the European Community signed the Single European Act, which removed many, but not all, problems and promoted some common policies on taxes, health, safety, the environment, employment, and finance. Tax policies of EU members vary greatly, producing large anomolies in prices. Cars tend to be cheaper in Belgium than elsewhere. Up-market cars purchased in The Netherlands avoid the country's luxury tax if the car is bought by a resident from another EU country. The British government levies heavy taxes on alcohol and tobacco, and many Brits take a day trip to France to stock up. (Photo 6.8) The Swedes go by ferry to Denmark or by bus over the bridge to Copenhagen to avoid buying high-priced liquor in the state monopoly Systembolaget. As in Britain, Sweden has an informal network of travelers who haul cheap drink home for friends and neighbors (Ekman and Fuller, 2005). There is a long way to go before a common market in goods is achieved, and the service sector is not open to cross-border competition.

The Maastricht Treaty on European Union

The European Union came into being on November 1, 1993, after negotiations at the Dutch town of Maastricht in December 1991 led to the formal signing of the Maastricht treaty in February 1992. The treaty provided for the creation of a European currency and common policies on defense, foreign affairs, and social issues. The treaty was ratified by member states in 1993 but the Danes required two referenda to gain approval of voters

Photo 6.8 A British daytripper fills his trolley with cheap wine and beer at Eastenders in Calais, France, to avoid U.K. taxes.

Source: GARETH FULLER/PA Photos/Landov LLC.

and some countries, notably Britain, opted out of the common currency and social policy. Presently Denmark, Sweden, and the United Kingdom retain their currencies, although stores and corporations accept payment in Euros.

The Fall of the Wall

On August 13, 1961, with Soviet approval, East German forces sealed off most of the crossing points between East and West Berlin and built a wall. The U.S. moved a tank division into West Berlin and the high concrete wall became an image of the Cold War. Later, as the Cold War crested, President Ronald Reagan stood in West Berlin and demanded that Mr. Gorbachev "tear down that wall." Gorbachev was too busy with political problems in the Soviet Union but East and West Berliners, armed with hammers, chisels, and picks, started the job in November 1989, when free access between the sections of the city was restored. In 1990, West Germany took over East Germany. As the Communist hold on eastern Europe fell away, the Soviet Union started to fall apart, and the Warsaw Pact was dissolved on March 31, 1991. During the course of the same year the Baltics, the Ukraine, Belarus, Moldova, Georgia, Azerbaijan, and Armenia, independent for a time after World War I, became sovereign states, as did the central Asian states of Kazakhstan, Kyrgyzstan, Tajikistan, Turkmenistan, and Uzbekistan. By the end of 1991, all that remained of the Soviet Union was a Commonwealth of Independent States (CIS) and George Kennan's prediction, that the Soviet system carried within it the seeds of its own decay, was fulfilled. (Photo 6.9)

Breakup was not confined to the Soviet Union. The Czechs and Slovaks, who had never been content together, formed separate states in 1993. Yugoslavia lost Slovenia and Macedonia relatively peacefully in 1991. In a series of civil wars and "ethnic cleansings" the multiethnic regions of Croatia and Bosnia-Herzegovina had long and bitter wars in and around their territory before emerging as independent states. Montenegro separated from Serbia in 2006. Kosovo is under UN administration.

► EU ENLARGEMENT IN 1995

On June 1, 1995, the formerly neutral countries of Austria, Finland, and Sweden joined the EU (Figure 6.9). This was linked to the collapse of the Soviet Union. After World War II Austria, part of Germany as a result of the Anschluss of 1938, was divided into occupation zones. The Austrian State Treaty of 1955 gave the country independence again on condition of strict neutrality. The Soviet Union, one of the occupying powers and a party to the Austrian State Treaty, would have regarded any Austrian attempt to join the European Community as a violation of strict neutrality and by joining a European organization in which Germany was the largest member, a dilution of the provision, in the Austrian State Treaty, that there was to be no Anschluss with Germany. With the disappearance of the Soviet Union there was no powerful voice to contest these issues.

Finland was invaded by the Soviet Union in World War II and lost territory to the powerful neighbor. At the end of the war, Soviet bases were established in the country and Finland adopted a policy of neutrality and nonalignment to avoid interference by the USSR. The Soviet Union objected to Finland even joining EFTA, although it did tolerate associate status, from 1961, and free trade in industrial goods with EEC from 1974. With the fall of the Soviet Union Finland was free to join EU and quickly did so.

Photo 6.9 East Berliners going west after the fall of the Berlin Wall, 1989.

Source: Sipa Press/NewsCom.

Sweden had a long tradition of neutrality. In the Cold War it had NATO neighbors in Norway, Denmark, and Germany but had to contend with Soviet power in the Baltic. Although well armed, Sweden had a policy of strict neutrality and did not apply to join EU, although it was a member of the European Free Trade Association. EFTA involved no dilution of neutrality because a member simply agreed to free trade with other member states. The EU involves the merging of aspects of sovereignty and the transfer of some powers to European institutions. An EU member is no longer strictly neutral in a technical sense. With the end of the USSR, Sweden, which was tightly linked with the economies of Western Europe, felt able to join EU, while retaining a policy of neutrality in military affairs.

NATO Expansion

The dissolving of the Warsaw Pact with the collapse of the Soviet Union left a strategic vacuum in eastern Europe. Several countries were powerfully armed and in some cases—for example, Hungary and Romania—there was the potential for conflict as borders were not always aligned with ethnic distributions of population. Fortunately, Germany had given an undertaking not to reopen the controversial issue of the border with Poland.

Figure 6.9 Europe in the 1990s. The European Economic Area (EEA) promotes free trade between EU and EFTA members.

It was easy to see that arms buildups and conflicts could be avoided if the forces of eastern European countries were brought within NATO command. On the other hand, by enlarging to the east, NATO extended chains of command, supply, and reinforcement. The Czech-born U.S. Secretary of State Madelaine Albright (née Korbel), understood the security concerns of east European countries, newly free of Soviet control but possessing vivid memories of 1938–1948, when many countries were occupied and repressed by Germany and then the Soviet Union. Now was the time to enlarge NATO, an organization that had achieved the task of "containing" both the Soviet Union and Germany. In

1999 Poland, the Czech Republic, and Hungary became NATO members, as belatedly, geopolitical debts were repaid. The Czechs had been abandoned in the 1938 Munich crisis, Poland was allowed to fall into Stalin's grasp after World War II, and the Hungarians, in spite of what the Voice of America radio station had told them, had received no help in the 1956 uprising against Soviet rule. Bulgaria, Romania, and Slovenia joined NATO in 2004.

East European Economic Reform and Entry into the EU

The countries of East Europe had never enjoyed democratic government and economic prosperity. As they emerged as independent states, after World War I, they were engulfed in the Great Depression and a collapse of world trade, which ruined export markets. All countries, except Czechoslovakia, came under the rule of dictatorial leaders. In World War II, the region was fought over and exploited by Germany and the Soviet Union. After the war communist regimes established centrally planned economies in which all economic activity was state controlled. Economic goals were often determined by Moscow, and inevitably, state-run industries became bureaucratic and inefficient. Wages and living standards were low, compared to western Europe. Following the Soviet model, economic growth concentrated on the exploitation of raw materials, heavy industries, and infrastructure. The production of consumer goods was given a low priority. There was some specialization reflecting advantages and established activities. Poland, with an ice-free Baltic seaboard, developed large shipyards at Gdánsk, along with plants producing tractors and trucks. Romania, rich in natural resources, produced oil, iron ore, bauxite, and nonferrous metals. The Czech lands, an important manufacturing region in the Austria-Hungary empire, made motor vehicles and armaments. When the Soviet bloc did an arms deal, the source of many weapons was Czechoslovakia. Hungary had electronic industries.

Eastern Europe was not cut off from trade with the west but, as the region had little convertible currency, exports were emphasized and it was difficult to import goods. Most countries of eastern Europe exported agricultural products, with Hungary, Romania, and Bulgaria selling wine. The Czechs exported some consumer goods including glassware and the Hungarians had exports of electronic equipment. Few western goods were allowed in, but technology was purchased. Fiat was commissioned to build car factories in Poland and the Soviet Union, resulting in cars like the Lada, based on Fiat designs. Czechoslovakia had a large auto industry before World War II and the Skoda car was sold into western markets.

Not all economic activity was state owned. Small businesses were allowed and Hungary evolved a mixed economy with numerous small enterprises employing a few people outside the state industry system. Eastern Europeans were not completely unprepared for a free market economy but the big, state-owned, heavy industries were overstaffed, inefficient, technologically backward, and environmentally unfriendly.

The problems of absorbing eastern European economies into western Europe were apparent when West Germany took over East Germany. The industries in the East could not compete with those in West Germany. Unemployment, in what was East Germany, remains high and living standards low. The costs of integrating East Germany into West Germany were huge. The German central bank, the Bundesbank, had to print money to replace each east mark with one Deutsch mark. On the black market, seven east marks

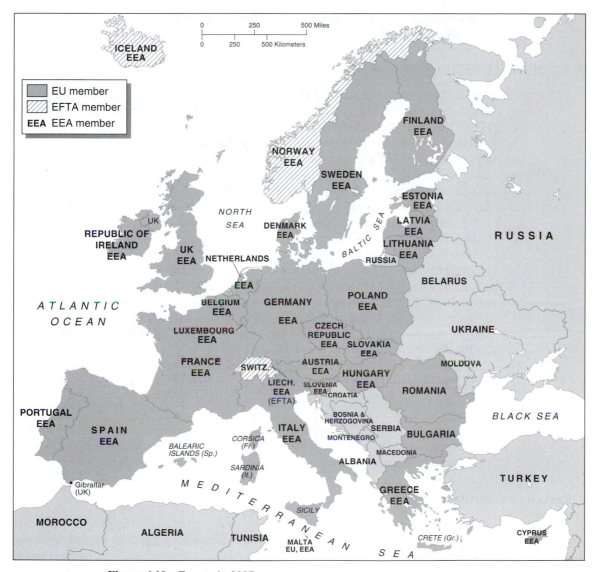

Figure 6.10 Europe in 2007.

had bought one Deutsch mark. To avoid inflation the Bundesbank pushed up interest rates, which had the effect of raising interest rates across Europe and deepening the recession of 1990–91.

It is remarkable that the countries of eastern Europe have progressed so rapidly in recent years. In the 1990s they painfully restructured to meet the demands of competitive free market forces. Yet in little more than a decade after the collapse of the Soviet Union, most countries of eastern Europe were accepted into the EU. Although eastern Europe had an outdated economic structure, countries had advantages. Some had good raw material endowments. Relatively cheap labor was trained or trainable, and investment funds were available from the West.

▶ WIDENING VERSUS DEEPENING

German unification, the dissolution of the Warsaw Pact, and the collapse of the Soviet Union took place as the EU countries negotiated and signed the Maastricht Treaty in 1993. The treaty deepened the relationship between the members. There were members that wished to take the deepening process further before widening EU. In practice, deepening and widening took place together as the Euro currency was introduced in 2002, and ten new members were admitted in 2004: Poland, the Baltics (Estonia, Latvia, and Lithuania), the Czech Republic, Slovakia, Slovenia, Hungary, Cyprus, and Malta. Bulgaria and Romania joined in 2007. (Figure 6.10)

Britain and Germany favored bringing in the countries of eastern Europe. Germany historically has seen eastern Europe as part of *Mitteleuropa,* a central Europe in which Germany exercises economic and political influence. Britain, a traditional free trader, was skeptical of deepening the power of European institutions and eager to expand the economic reach of EU. France, as the author of many European institutions, was an advocate of deepening European ties. Several countries with large agricultural sectors, dependent on CAP subsidies, had reservations about admitting the countries of eastern Europe because all had large farming industries that would place additional demands on CAP funds.

Eventually the decision to bring in eastern Europe echoed Coudenhove-Kalergi's idea of a European Union that ran from Portugal to Poland with the intention of creating better living standards for all. To leave Europe divided into a prosperous west and a relatively backward east could have created political instability with the possibility of the resurgence of dictatorial governments. The eastern European countries now joined to EU have accepted EU standards of human rights, and this is the best available safeguard of democracy in eastern Europe.

A European Constitution?

The admission of ten new members in 2004 was a triumph for those who wanted to enlarge and widen Europe. The next year was a failure for those who wanted to deepen EU institutions.

In the summer of 2005, the twenty-five members of EU were asked to ratify a European constitution. Some countries, like Spain, ratified by a vote in parliament. Other countries offered citizens a referendum. First the French voted *non,* followed by the Dutch voting *nee,* at which point Britain, and other governments planning referenda, said there was no point in seeking the opinion of citizens on the constitution. In Germany, where the legislature had approved the European constitution, the President said he would await a ruling from Germany's constitutional court on whether or not the European constitution was compatible with the constitution of the Federal Republic.

The 2005 version of a European constitution was dead. The cause of death was debated. Poor economic performance within the EU, unpopular politicians, fear of immigrants and foreigners, together with French distrust of neoliberal policies and "Anglo-Saxon economics," were all blamed.

The proposed constitution had over 450 clauses, many loosely worded. A well-informed citizen could not predict what powers were being created and how the document would be interpreted in the future. A substantial proportion of the French and Dutch, who voted against the constitution, did so because they distrusted the document. What was needed was a constitution that described EU, set out decision-making processes, and defined the rights of citizens. Hundreds of clauses are not needed to achieve that aim and French and Dutch voters feared that much baggage was being hauled to Brussels under constitutional cover.

The next 2005 setback was the failure to agree on a budget. Britain, since the 1980s, had received a rebate on its annual EU financial contribution to compensate for the low level of farm subsidy payments received in comparison with France and Germany. French President Chirac, looking around for someone to blame for his failure to get a Yes vote out of his electorate, targeted Britain and the UK rebate.

Britain said it was willing to give up the rebate as part of a package that would revise EU farm funding and the budget in general. Chirac did not see the offer as helpful. France receives huge farm subsidies from EU and, at the suggestion of any reduction in funding, French farmers block roads with tractors, wagon loads of dumped produce, and trucks full of burning sheep.

The June 2005 EU Summit was acrimonious, with Britain and other countries making large net budget contributions, rejecting the proposed budget.

Different countries have different visions of Europe. Britain, Denmark, The Netherlands, and Sweden see the Brussels bureaucracy as a mechanism to harmonize regulations across Europe so that it is easier for a company based in one country to do business in the territory of all EU members. Other countries want more common economic policies developed and controlled by Brussels. In addition, some powerful members want common foreign and defense policies.

France wants the EU to be a "pole of power" comparable with the United States (*The Times,* June 29, 2005, p. 35). Further, France claims that the Common Agricultural Policy is a success. Like so many EU issues, CAP evaluation can be approached from several perspectives. The CAP has helped modernize farming by the use of subsidies. It has made Europe produce crop surpluses, replacing previous shortages, and it has helped to transform, in an orderly way, traditional farming regions employing too much labor into more productive areas, as people have left the land and moved to better-waged work in town.

On the debit side, the CAP comes at an enormous cost, still taking 40 percent of the EU budget. Crop patterns are distorted by subsidies. For example, large acreages of good land grow subsidized sugar beet, when the sugar could be imported from Brazil or Mozambique at one-third the cost. EU consumers have high food costs and this is a major criticism of the CAP. It is a policy that pays big farm subsidies and delivers high food prices.

The Euro

The Euro, launched in 2002, initially fell against the dollar, but now the value of the Euro has risen as more central banks hold reserves in Euros and reduce dollar holdings. As the Euro rose, the dollar fell, making visits from North America to Europe more

Photo 6.10 ▶ Much of Europe now has a common currency, although countries are permitted to include national symbols.

Source: Age fotostock/SUPERSTOCK.

expensive, and a higher valued Euro made European exports less competitive. (Photo 6.10)

The original argument for accepting the Euro was clear: there cannot be one European market unless all European countries have the same currency, otherwise the costs of converting currency and fluctuations in the value of currencies will distort competition. Most multinational corporations operating in Europe held this view.

The major argument against the Euro is of a different nature. In adopting a European currency, countries lost control of aspects of economic policy, including interest rates, to a European central bank. Recently, differing economic performance within the EU showed the disadvantages of a European central bank setting interest rates. The German economy was relatively stagnant and needed cheaper money to stimulate growth. Italy, which had frequently devalued the lira to take care of rising production costs, found that a high-valued Euro overpriced Italian goods in world export markets, but Italy could not devalue the Euro unilaterally. In both Germany and Italy there is strong criticism of the Euro.

There are advantages to a single currency and there are disadvantages when countries require differing interest rates to promote growth or curtail inflation. Britain, Denmark, and Sweden did not adopt the Euro, and national central banks set interest rates. If there is sustained economic growth across the EU, the Euro and the European central bank policies draw little attention. If economies are stagnant and unemployment high, the bank gets a full share of blame.

The Parable of the Polish Plumber

In recent years unemployment rates in France and Germany have been over 10 percent. When unemployment is high, it highlights immigration issues. One of the reasons that French voters said *non* to the EU constitution was fear of migrants from the new EU countries in eastern Europe. When Polish plumbers arrived in France to find plenty of work, citizens were fearful that the 150 plumbers represented a new wave of immigrants. With Polish panache, the national tourist office placed ads in French newspapers, featuring a glamorous plumber, inviting the French to come to Poland! (Photo 6.11)

And the summer of 2005 rumbled from one bizarre episode to another. In early July, President Chirac of France declared that only Finland had worse food than Britain, adding you "cannot trust people [the British] with such bad cuisine" and looked thunderous when a reporter asked if he liked *rosbif?* The next day it was announced that London had narrowly beaten the long-standing favorite Paris to host the 2012 Olympic games. Chirac smiled bravely and the European leaders went off to Gleneagles in Scotland for the G8 meeting to discuss global warming, aid to Africa, and world agricultural trade. Agricultural trade led back to CAP and the way it closed European markets to third-world producers.

The summer of 2005, with the failure to approve a constitution, did not weaken the EU. All the free trade, central bank, and regulatory organizations remained in place. The summer did find some of the original members, including France, Germany, and The Netherlands, for differing reasons, expressing dissatisfaction with parts of EU operations.

The EU is absorbing the new eastern European members, which, with low labor costs, are attracting much capital and many new enterprises. How much further Europe can be widened is problematic. If the EU allows an application from Turkey, France says it will hold a referendum on the issue of Turkish membership. In the present mood,

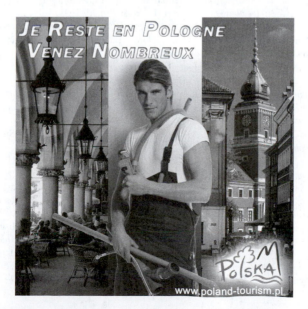

Photo 6.11 French newspaper ad inserted by the Polish Tourist Board inviting the French to visit Poland. The ad was a reaction to an overheated response to Polish plumbers arriving in France.

Source: LASKI DIFFUSION/GAMMA.

the French electorate would vote against the admission of Turkey, which would halt the application.

Bulgaria and Romania joined EU in 2007 but in terms of democracy and human rights they still have room for improvement. Events in Ukraine in 2004, where the old guard communists tried to steal an election and poison the opposition candidate (who eventually won the rerun election), demonstrate the fragility of democratic institutions. Ukranians probably want union with Europe. The ethnic Russians, in the east of the country, look to Russia. As the *New York Times* (November 24, 2004, p. A10) put it, "Europe's new dividing line seems to run somewhere east of the Dnieper River." North of the Ukraine the dividing line jags sharply to the west because the rulers of Belarus are authoritarian and still look to Russia.

▶ SUMMARY

With the 2004 and 2007 enlargements, the EU spatially contains most of the elements of Charlemagne's Holy Roman Empire (Le Goff, 2005). Switzerland, Catholic in the Middle Ages, has not joined but the Catholic west Slavs—the Poles, Czechs, Slovaks, and Slovenes—are in along with the Hungarians (but not the Croats, yet).

Orthodox east Slavs in Romania and Bulgaria joined in 2007. The EU is at the cultural divide between the Church of Rome, the Orthodox Church of Byzantium, and the secular state of Turkey, populated largely by Muslims. The rewards of crossing the cultural boundaries successfully may be large in global terms.

▶ FURTHER READING

Agnew, J. *Hegemony, The New Shape of Global Power*. Philadelphia: Temple University Press, 2005.

Agnew, J. "Bounding the European Project," *Geopolitics* 10(3) (2005): 575–581.

Agnew, J., and J. Nicholas Entrikin (eds.). *The Marshall Plan Today, Model and Metaphor*. New York: Routledge, 2004.

Bainbridge, T. *The Penguin Companion to European Union*. Third edition, New York: Penguin Books, 2002.

de Blij, *Wartime Encounter*. New York, NY: Hudson River Enterprises, 2006.

Blouet, B.W. *Geopolitics and Globalization in the Twentieth Century*. London: Reaktion Books, 2001.

Blouet, B.W. (ed.). *Global Geostrategy: Mackinder and the Defense of the West*. London: Frank Cass, 2005.

Blouet, B.W. *Halford Mackinder: A Biography*. College Station: Texas A&M University Press, 1987.

Bowman, I. *The New World, Problems in Political Geography*. Yonkers-on-Hudson, New York: World Book Company, 1921.

Cohen, S. B. *Geopolitics of the World System*. Lanham, MD: Rowan and Littlefiled, 2002.

Dinan, D. *Ever Closer Union*. Boulder, CO: Lynne Rienner, 2005.

Ekman, I. and T. Fuller. "Cheap EU liquor: Many Swedes drink to that." *International Herald Tribune,* October 5, 2005. (*www.iht.com*)

EFTA Bulletin: EFTA and EU Enlargement. Brussels: European Free Trade Association, September 2004.

Fischer, F. *Germany's Aims in the First World War*. New York: Norton, 1967.

Heffernan, M. *The Meaning of Europe: Geography and Geopolitics*. London: Arnold, 1998.

Hugill, P. *Global Communications since 1844: Geopolitics and Technology*. Baltimore: Johns Hopkins University Press, 1999.

Judt, T. *Postwar: A History of Europe since 1945*. New York: Penguin Press, 2005.

Kaplan, D.H., and J. Häkli (eds.). *Boundaries and Place: European Borderlands in Geographical Context*. Lanham, MD: Rowman and Littlefield, 2002.

Kissinger, H. *Diplomacy*. New York: Simon and Schuster, 1994.

Le Goff, J. *The Birth of Europe*. Oxford, UK: Blackwell, 2005.

Leonard, D. *The Economist Guide to the European Union*. 9th edition. London: Profile Books, 2005.

Mackinder, H.J. *Democratic Ideals and Reality*. London: Constable, 1919.

Mackinder, H.J. "Geographical Pivot of History," *Geographical Journal*, 23(4) (1904): 421–444. Reprinted in Special Issue: Halford Mackinder and the Geographical Pivot of History, *Geographical Journal*, 170(4) (2004): 298–321.

Martel, G. *A Companion to Europe 1900–1945*. Oxford, UK: Blackwell, 2005.

McEvedy, C. *The New Penguin Atlas of Europe*. New York: Penguin, 2002.

Minghi, J.V. "Common Cause for Borderland Minorities?: Shared Status among Italy's Ethnic Communities," in N. Kilot and D. Newman, Geopolitics at the End of the Twentieth Century, *The Changing World Political Map*. London: Frank Cass, 2000, p. 199–208.

Murphy, A.B. "The Changing Geography of Europeaness." *Geopolitics* 10(3) (2005): 586–591.

O'Loughlin, J. "Ordering the 'Crush Zone': Geopolitical Games in Post-Cold War Eastern Europe," in N. Kilot and D. Newman, *Geopolitics at the End of the Twentieth Century, The Changing World Political Map*. London: Frank Cass, 2000, p. 34–56.

Overy, R.J. *The Dictators: Hitler's Germany, Stalin's Russia*. New York: W. W. Norton, 2004.

Overy, R.J. *Russia's War*. New York: Penguin Putnam, 1998.

Phillips, J. *Macedonia: Warlords and Rebels in the Balkans*. New Haven, CT: Yale University Press, 2004.

Pittaway, M. *Eastern Europe 1939–2000*. London: Arnold, 2004.

Smith, N. *American Empire, Roosevelt's Geographer and the Prelude to Globalization*. Berkeley, University of California Press, 2003.

Tunander, O. "Swedish Geopolitics: From Rudolf Kjellén to a Swedish 'Dual State.'" *Geopolitics* 10(3) (2005): 546–566.

White, G.W. *Nation, State, and Territory: Origins, Evolutions, and Relationships*. Lanham, MD: Rowman and Littlefield, 2004.

Winks, R.W., and J. Neuberger. *Europe and the Making of Modernity 1815–1914*. New York: Oxford University Press, 2005.

Winks, R.W., and J.E. Talbot. *Europe: 1945 to the Present*. New York: Oxford University Press, 2005.

II

THE CORE OF THE EUROPEAN UNION

7

FRANCE AND THE BENELUX COUNTRIES

▶ INTRODUCTION

In 1945, the economy of Europe was worn out by war. Manufacturing plants were bombed or shelled, and surviving factories were starved of resources and workers. Housing stocks were depleted by aerial bombardment and street fighting. Agriculture had run down, overused railroads needed maintenance, and the few remaining private cars were off the roads for lack of fuel.

France tried to reassert itself as a great power, on the old model, wanting to detach the Rhineland and Ruhr from Germany and sending troops to Algeria and French Indo-China (Vietnam) to regain colonial control. Elsewhere in Europe there was a sense that reconciliation was needed; never again should the region be allowed to devastate itself. Before and during the war, there was contact between clerics in Europe. A leader of this movement, and a critic of allied obliteration bombing during the war, was Bishop Bell of Chichester. Speaking to Germans after the war he talked of the "spirit of Europe":

> We want to see Europe as Christendom. . . . No nation, no church, no individual is guiltless. Without repentance, and without forgiveness, there can be no regeneration.

The above passage is quoted by Professor Norman Davies (1996, p. 923), who comments that the moral dimensions of the postwar European movement were strong before it was "hijacked by economists." Of course, the politicians had to create the environment in which cooperation was possible.

On September 19, 1946, a few months after his Iron Curtain speech, Winston Churchill proposed a United States of Europe, the first step being an understanding between France and Germany. Britain and the Commonwealth would be a separate but cooperative entity linking Europe and the United States. In 1948, Churchill chaired a Congress of Europe at The Hague, in The Netherlands. Eight hundred experts and statesmen attended, including Konrad Adenauer. Adenauer became the first Chancellor of the Federal Republic of Germany from 1949 to 1963. Adenauer, leader of the Christian Democrats, was a Rhinelander and, like many from the region, had a strong sense of the need to cooperate internationally. The Chancellor took Germany into NATO and made the understandings with France that laid the foundations of the EU.

On January 1, 1948, Belgium, Luxembourg, and The Netherlands created the Benelux Customs Union, producing a model of how Europe might remove tariffs and other restrictions between trade partners. The Marshall Plan, providing U.S. aid for European Reconstruction, led to the establishment of the Organization of European Economic Co-operation in 1948, which evolved into OECD in 1960. A defense pact, the Treaty of Brussels, was signed by several western European countries in 1948. The pact was a forerunner of the North Atlantic Treaty Organization (NATO). Fear of the Soviet Union was a force pushing Europe toward integration.

The Frenchman Jean Monnet (1888–1979), an advocate of European economic cooperation since World War I, developed a plan to integrate the coal, iron, and steel industries of Europe under a supranational authority, with the aim of ending conflicts over access to iron and coal. Under the Treaty of Paris (1951), France, Germany, the Benelux countries, and Italy agreed to establish a European Coal and Steel Community (ECSC). The treaty entered into force in 1952. The same partners signed the Treaty of Rome (1957), creating the European Economic Community (EEC).

The economic growth of the 1950s and 1960s reduced strains between EEC countries. By the early 1950s French industry was growing rapidly, exceeding prewar production. The German "economic miracle" was well established as industry was rebuilt and urban areas reconstructed. In The Netherlands, central Rotterdam was a bomb site in 1945 and weeds were growing on the docks. With the reopening of Rhine river trade, Rotterdam reestablished itself as Europe's major port. Antwerp, in Belgium, resurged with the reopening of Europe to overseas trade. Luxembourg benefited from increased demand for iron and steel. Italy had a pool of cheap labor and many workers went to Germany where labor was in demand in an expanding economy. By the late 1950s, low labor costs were making Italian cars, household appliances, and office machinery competitive in export markets.

In the period of economic growth, which lasted from the late 1940s through the 1960s, the establishment of the EEC was relatively smooth with the institution creating new trade channels and attracting investment by multinational corporations into the community.

▶ FRANCE

Introduction

In the view of Jean Gottmann (1969), France constitutes the western hub of Europe. In many ways the creation of the EU centered around France as it rose from the disastrous decade of 1935 to 1945 and German occupation from 1940 to 1944. France is situated on the crossroads of western Europe, linking the Atlantic side of the region to the Rhine valley and central Europe (Figure 7.1). From a locational perspective, France lies at the core of the original six EEC countries, and it is difficult to conceive of a European community without France in a central role, even though French perspectives frequently differ from those of neighboring states.

From north to south France stretches from the English Channel to the Mediterranean Sea. From east to west the country runs from the Alps to the Atlantic. The distance from Calais on the Strait of Dover, to Marseille on the Mediterranean, is approximately

Figure 7.1 France and Benelux—Topography.

Photo 7.1   Mont Blanc Massif, in the French Alps.

Source: AgenceImages/AgeFotostock.

500 miles. The straight line distance from Mont Blanc (15,771 feet) in the French Alps to the Atlantic port of Brest is around 500 miles. France is a compact country, and within its boundaries—defined by the Alps, the Jura, the Rhine, the Ardennes, the Atlantic, the Pyrenees, and the Mediterranean—are many resources, environments, and ecological niches. (Photo 7.1)

Climate

The Alps and the Pyrenees contain high-altitude environments. The Hercynian uplands of Brittany, the Massif Central, the Ardennes, and Vosges are high enough to be exposed and bleak in winter. The Channel and North Sea coasts of France receive maritime polar air in the winter, and in summer, as land surfaces warm, a cool sea-to-shore breeze moderates temperatures. In winter, the southwest of France—Gascogne and Aquitaine—receives Atlantic depressions blowing in from the Bay of Biscay. In summer, as the Azores high-pressure cell strengthens over the eastern Atlantic, and the depressions take a northerly track, the southwest comes under the influence of the Iberian heat island and the warm temperatures of south-central France. Eastern and central France are far enough into the European landmass to have a continental climate with cold winters and hot summers. However, like all of western Europe, France can experience the passage of depressions coming from the Atlantic at any time of the year. In summers when the Azores high-pressure cell is weak, depressions bring cool and wet weather spells all across western Europe.

The south coast of France has a Mediterranean climate. Winters are mild and moist, summers generally warm and dry. But in a wet, western European summer, depressions track along the north face of the Pyrenees, through the Carcassonne gap into Provence and out into the Gulf of Lion, as they do frequently in winter months. In winter, a deep low-pressure system, in the Gulf of Lion, can establish the conditions for the cold *mistral* to blow down the Rhône valley, damaging crops and disrupting shipping in the Gulf.

Regions

France contains many environmental regions and cultural landscapes. French geographers developed the concept of *Pays*—literally the word means country—but in cultural geography the term refers to landscape and lifestyle regions. France, and other European countries, contain many small, distinctive landscapes within larger physiographic regions. The Paris Basin is a large physiographic region containing many *Pays*, including the Pay de Calais and the Champagne country. Fernand Braudel (1998, pp. 41–57) identified, within the Burgundy (Bourgogne) region, over twenty *Pays*, including Côte D'or, Beaujolais, Maconnais, and others associated with the production of distinctive wines.

Here France is not described at the Pays scale. We outline the features of the major physiographic regions as set out in Figure 7.1, describing the major river basins, uplands, and the coasts.

Physiographic Regions: River Basins

The great river basins of the Seine, the Loire, the Rhône, the Dordogne, and the Garonne draining to the Gironde, are all agriculturally rich.

The Seine and the Paris Basin The rocks of the Paris Basin have been warped into a saucer shape, and the rivers have cut through the surface exposing the underlying limestones, sandstones, and clays in escarpments and ridges. Where clay underlies the limestone near the surface, springs emerge as water percolating down through the limestone encounters impermeable clay. The Seine is the major river draining the basin, but many tributaries, including the Marne, converge on the main stream in the region of Paris. As a result, many valley-floor routeways come together close to Paris. Paris enjoys a central situation in the middle of France's most extensive area of agricultural land. The initial site of Paris was an island, Île-de-La-Cité, which became a bridging point, linking routeways across the river. (Photo 7.2)

Much of the Paris Basin lies north of the limit of viticulture in France, and the major crops are wheat, oats, sugar beet, and vegetables for sale in Paris markets. There are rich vineyards in the south of the basin, and to the east, in the Champagne area, the limestone soils produce remarkable white sparkling wines that are matured in caves cut into the exposed limestone escarpments.

Before entering the English Channel the Seine passes through Normandy, a region where Norsemen (Vikings) were established by 911 A.D. Normandy, which contains several distinct Pays, focuses on Rouen, an important medieval port and today a center of manufacturing activity. Downstream of Rouen is the deep-water port of Le Havre. The coasts of Normandy contain fashionable beach resorts such as Deauville and, of course, the World War II beaches attract many visitors. The Normandy coast also exposes the limestone of the Paris Basin in impressive white cliffs, including Pointe D'Hoc, which Rudders Raiders scaled to attack an artillery position covering a beach on which allied troops were landing. (Photo 7.3)

Photo 7.2   Paris, Île de la Cité, Notre Dame, aerial view.

Source: Robert Cameron/Getty Images.

The Loire   The river Loire flows northward out of the Massif Central before turning westward to the Atlantic. As the river approaches the coast, it crosses the older, harder rocks of the southern part of the Breton Massif and the landscape becomes rugged. The port of Nantes, which flourished with colonial West Indian trade, lies at the head of the Loire estuary. The Loire valley lies south of the Paris Basin and is open to maritime Atlantic influences. The climate is milder and well suited to vineyards, which are extensive in the valley. (Photo 7.4) In the seventeenth and eighteenth centuries wealthy families built chateaux with fine formal gardens, for the Pays de la Loire were close enough to Paris to allow return to the capital for social and political functions. Although the river Loire is of limited use for navigation, the valley is a routeway to the Atlantic coast and contains important towns including Orléans, Tours, Angers, and the port of Nantes. At the mouth of the Loire estuary is the large, modern, deep-water port of St. Nazaire. The port is a major industrial center and French naval base. During World War II, it was a German submarine base engaged in the battle of the Atlantic.

Dordogne, the Garonne, and the Gironde Estuary   The major rivers draining the basin of Aquitaine are the Dordogne and the Garonne, which flow to the Gironde estuary. Both rivers receive tributaries from the east flank of the Massif Central and, in addition,

Photo 7.3 Normandy is famous for its beaches. The chalk cliffs are also impressive and mirror the white cliffs of Dover on the other side of the English Channel.

Source: SUPERSTOCK.

the Garonne drains the north slope of the Pyrenees. The upper courses of the Dordogne and Garonne, and their tributaries, are in steep valleys where arable land is limited, and much ground is used for pasture. On the upper course of the Garonne, close to the Carcassonne gap, sits the town of Toulouse and, through the gap, the Midi canal links the basin of Aquitaine to the Mediterranean lands of southern France. The Carcassone gap is an ancient routeway leading from the Atlantic to the Mediterranean lands. Holding the gap is Carcassonne, with fortifications dating to the Roman era.

Photo 7.4 Chateau with vineyard in Loire Valley, France.

Source: Cephas Picture Library/Alamy Images.

The lower courses of the Garonne and the Dordogne are characterized by extensive vineyards, as are the banks of the Gironde estuary. Bordeaux is a major port and a leading center of the wine trade; home to numerous merchants, blenders, and *négocients* who trade the Bordeaux red (claret) and white wines around the world. (Photo 7.5)

Like many major European ports, Bordeaux is on an estuary, the Gironde, and well inland. With the increasing size of vessels, in particular tankers, outports down the estuary in deeper water have developed. Near the mouth of the Gironde estuary is the port of Le Verdon-sur-Mer, with oil terminals, container-handling facilities, and warehouses to store tropical produce. Bordeaux remains a viable port and attracts many cruise vessels that can tie up close to town, with an easy walk to the gracious parks and buildings of the city. The coast, running from the mouth of the Gironde south to the resort of Biarritz and then the Pyrenees, is marked by dunes behind which are lagoons (*Étangs*). The Bassin d' Arcachon is connected to the sea at Cap Ferret and the port of Arcachon displays many features of a fishing settlement transformed by modern trends (Garner, 2005).

The Rhône   The Rhône rises in Switzerland and flows in a generally westerly direction, through the Alps and the Jura to Lyon where it makes a confluence with the south-flowing Saône, draining the limestone Burgundy country to the north. The Rhône is France's largest river in terms of volume of water, but levels vary seasonally. Much work

Photo 7.5   Barge traffic can progress upstream from Bordeaux on the Garonne River. Cruise boats come up the Gironde estuary and tie up at quay sides below the bridge. Passengers walk from the quay into the old town.

Source: S001/GAMMA.

has been done to build dams to regulate flow and generate hydroelectricity, with locks allowing the passage of large barges upstream to the industrial city of Lyon. (Photo 7.6)

South of Lyon, the Rhône flows to the Mediterranean between the Alpine foreland to the east and the Massif Central to the west. Higher ground is frequently close to the river as at the town of Avignon. At Arles the river divides into delta distributaries, the Grand and the Petit Rhône. Navigation through the muddy delta is difficult. The Romans built a canal from Arles to the sea to overcome the problem, and in the modern era canals link the lower Rhône with the major ports of Sète to the west and Marseille to the east. Neither port is close to the Rhône delta to avoid the mud. The Camargue lies south of Arles between the Grand and Petit Rhône. The Camargue is a large wetland region of lakes, marshes, lagoons, and grazing land famed for feral horses and the breeding of bulls. In Provence and Languedoc, bull fighting is still an entertainment and the Roman coliseum at Nîmes is a bull ring. In the north, toward the head of the Rhône delta, the alluvium of the Camargue has been drained for farming, with rice and other crops being harvested. Today, much of the Carmargue is a national park in which the wild horses graze along with other livestock. (Photo 7.7)

To the east of the Rhône delta is the Crau, a boulder-strewn region created in post-glacial times when the river Durance flowed with great force directly to the sea. Now the Durance joins the Rhône near Avignon.

The Mediterranean coastal region of France is known as the Midi. West of the Rhône the coast, leading to the Pyrenees, is characterized by dunes and lagoons. Inland

Photo 7.6 Tributaries of the Rhône have been dammed to provide hydroelectric power and to control flooding on the main stream.

Source: Targa/SUPERSTOCK.

Photo 7.7 Traditionally, fighting bulls have been raised in the Camargue. The herder is riding a native Camargue pony.

Source: Age Fotostock/SUPERSTOCK.

the plain contains vineyards producing over half the wine made in France. East of the Rhône is a Riviera coast in which limestone hills meet the sea in headlands, bays, and inlets. Marseille, a Greek foundation, was created in a rocky inlet. Now modern vessels are accommodated behind breakwaters built out in deep water. Toulon is the base of the French Mediterranean fleet and still retains the fortified walls. Innumerable resorts crowd the coast to the Italian border, including St. Tropez, Cannes, Antibes, Nice, and Monte Carlo in the independent principality of Monaco. (Photo 7.8)

Inland from the coast and the resorts, the landscape is dominated by limestone outcrops that are habitat for low-growing scrubs, herbs, and bees. The vegetation is drought resistant for the limestone does not hold surface water and in the summer months there are long periods of hot, dry weather. The vegetation type is known as *Maquis* in France and *Macchia* in Italy. Agriculture is found in valley bottoms where water is available.

Physiographic Regions: The Uplands

Two major mountain-building periods are evident in the landscapes of France. The Alpine orogeny thrust up the Alps, the Jura, and the Pyrenees. The Hercynian (Allegheny)

Photo 7.8   The fortified coastal town of Antibes with the Maritime Alps in the background.

Source: AgenceImages/AgeFotostock.

orogeny produced the Vosges, the Ardennes, the Brittany Massif, and the Massif Central. The former high peaks of Brittany, the Ardennes, and the Massif Central long ago eroded, and the harder igneous and metamorphic rocks, characteristic of mountain cores, are exposed at the surface (Figure 7.1).

To suggest that the Alpine and Hercynian mountain-building orogenies were discreet events is misleading. The Alpine orogeny caused buckling, warping, faulting, and volcanic activity in areas north and south of the Alps. The eroded surfaces of the Massif Central, the Ardennes, and Brittany were uplifted, and the rejuvenated rivers cut down into the plateaux surfaces in deep, steep-sided valleys where erosional debris has become the parent material of soils suitable for agriculture in places. On the plateaux surfaces, soils are thin and the uplands are exposed and cold in winter. Much land is in pasture. The Vosges, with steep slopes facing the Rhine valley and gentle inclines to the west, was uplifted, faulted, and tilted in the Alpine mountain-building era.

We now describe the uplands of France in a counterclockwise sequence: the French Alps, Jura, Vosges, Ardennes, Brittany peninsula, Massif Central, and Pyrenees, but as we pass along the eastern and northeastern boundary of the country we will examine the gaps in the natural defenses.

French Alps   The French Alps, rising to 15,771 feet at Mt. Blanc, run from the Swiss border, near Lake Geneva in the north, to the Mediterranean Sea, west of Nice. The high, snow-covered mountains create a barrier between France and Italy through which run a number of passes, including the Mount Cenis Pass.

To visit part of the French Alps we could start in the Rhône valley, at the town of Valence, where the river Isère, flowing from the Alps, joins the Rhône. Traveling east, up the Isère to Grenoble, the valley narrows as it passes through the Alpine foreland. Agriculture changes from an emphasis on Mediterranean crops to temperate crops as altitude lowers temperatures and shortens growing seasons.

Photo 7.9 Grenoble is in the valley of the Isère river which has provided power for industries since Medieval times. An alpine peak is in the background.

Source: Hemis/AgeFotostock.

The city of Grenoble, surrounded by snow-covered peaks, is an ancient urban center incorporated into France in 1349 with the surrounding Dauphine region. The university dates from 1339 and the city contains many research institutions. Water-wheels, turned by the fast-flowing Isère, powered Grenoble's medieval industries. Today the river drives hydroelectric turbines, which fuel the diversified industries of the city. (Photo 7.9)

Grenoble is a tourist and winter sports center. Above Grenoble, in the Isère valley, agriculture gives way to pasture. Conifers grow on the steep valley sides, and above the tree line are alpine pastures, to which cattle are driven to graze in the summer months.

The Isère receives tributaries flowing from the high Alpine massifs to the east. If we follow the river Arc into the high country, we come to the Mount Cenis pass (6,831 feet) leading to Italy. The pass is surrounded by high, snow-covered peaks and is liable to be closed in winter. This routeway across the Alps from France to Italy has been made easier with the construction of rail and road tunnels.

The Jura From the city of Lyon the river Rhône can be followed eastward into the Jura, a region of high, folded limestone ridges and valleys. The Rhône rises above Lake Geneva and flows through the lake before cutting through the Jura ridges to meet the river Saône at Lyon. The Jura is predominantely used for raising livestock, and the dairy industry produces some fine cheeses. In winter, snowfall is heavy in the Jura and the region can be isolated for months. Historically, winter isolation helped encourage workshop industry, including clock making and gem cutting. Today the valleys and small towns of the region are home to numerous light industry plants.

The Vosges   The Vosges Mountains consist of old, harder, faulted blocks of rock, rising up to 4,672 feet. In the east the mountains fall steeply toward the Rhine and Strasbourg, a seat of the European Parliament. In the west the decline to the plain of Lorraine is gentler. The forested Vosges produce timber but livestock farming on the hills is marginal and tourism has become a major activity.

To the south of the Vosges is the strategic Belfort Gap, or Burgundy gate. The gap is some 15 miles wide and provides a passageway between the Rhine and Rhône routeways. Through the gap runs the Rhône–Rhine canal, linking navigation on the two river systems. The fortified French city of Belfort protects the gap. Today it is a modern industrial center.

The Ardennes   The Ardennes is a region similar in geological age and topographic character to Appalachia, with uplifted peneplained surfaces and steep wooded valleys cut by rejuvenated streams. The greatest part of the Ardennes lies in Belgium. The Ardennes was seen as a barrier to the invasion of northeastern France, but in June 1940 and the Battle of the Bulge in December 1944, German tank columns came through the narrow valleys, taking Allied armies by surprise.

Moving westward from the Ardennes we come out onto the plain of Picardy and near the coast, low-lying Flanders, notoriously muddy in wet weather. This region is drained by the river Somme, along which some of the worst battles of World War I were fought. A British army was consumed in the fighting and 600,000 Allied soldiers died, together with a similar number of Germans.

By now peace-loving readers will be asking, do we need a tour of the battlefields? The battlefields were the graveyards on which sacrifices were made that eventually allowed Europe to reconcile. There was hell before there was reconciliation and regeneration. Amazingly a rich literature came out of the trenches of World War I: *All Quiet on the Western Front,* the poetry of Rupert Brook and Wilfred Owen, and postwar memoirs, including Siegfried Sasson's *Memories of an Infantry Officer* (1928), Robert Graves's *Good-bye to All That* (1929), and Vera Brittain's *Testament of Youth* (1933). The literature entered national consciousness, particularly in Britain, and contributed to both appeasement before World War II and postwar reconciliation.

The gaps south and north of the Ardennes are drained by the Moselle and the Meuse. In order to defend the gaps against threats from the east, France built fortress towns: on the Moselle, Nancy and Metz; on the Meuse, Sedan and Verdun. All the fortress towns were engaged in the Franco-Prussian War, World War I, and World War II. At Verdun there were bitter battles in 1916 that consumed over a million French and German men. Much of Verdun was destroyed by artillery shells, the civilian population fled, but French forces held the town.

Brittany   The Breton Peninsula extends westward into the Atlantic Ocean between the English Channel and the Bay of Biscay. In a drowned estuary at the west end of the peninsula is Brest, the home port of the French Atlantic fleet.

The Brittany region consists of eroded older, harder rocks, mostly granites, creating a mature landscape of rounded hills. The peninsula is highest in the west in the mountains

D'Arrée. Coming east, around Rennes, the capital of the region in the valley of the river Vilaine, is an area of arable land. Most of Brittany is *bocage* country with small, hedged fields and patches of woodland. The region is noted for livestock and dairying, although along the milder south coast early vegetables are raised for sale in French and English markets. (Photo 7.10)

The coasts of Brittany are rugged but possess numerous well-sheltered estuaries, bays, inlets, and fishing ports. Long ago the boats from the north coast port of St. Malo were fishing the Grand Banks of Newfoundland and the seas around Iceland. Jacques Cartier (1491–1557) was born at St. Malo. Cartier sailed into the St. Lawrence River in 1534. Fishermen from St. Malo gave their name to the south Atlantic islands of Les Melounes, the modern Falklands (Malvinas). St. Malo was heavily damaged in World War II, but the attractive fortified town has been restored.

Massif Central The extent of the Massif Central dwarfs the area of the Breton Peninsula, the Vosges, and the Ardennes. The Massif rises up, in the center, to the *Plomb du Cantal* (6,076 feet) and the *Puy de Sancy* (6,185 feet). The *Puy* name indicates a volcanic cone, of which there are many in the Auvergne area of the Massif. The geological history of the Massif is complex. There are extensive outcrops of Paleozoic rocks, but the older rocks were uplifted, tilted, and faulted in the Alpine orogeny. Faulting allowed volcanic rocks to flow to the surface, creating basaltic outflows as well as the now extinct cones (*puys*), common in the Auvergne. In the last Ice Age the higher valleys of the Massif filled with glaciers that deepened and widened them into characteristic U shapes. The higher regions of the Massif Central display both volcanic and glacial landscapes. (Photo 7.11)

Photo 7.10 Brittany has a bocage landscape of small fields surrounded by hedges and trees.

Source: Robert Harding Picture Library/Alamy Images.

Photo 7.11   Puy de Dome, Auvergne in the Massif Central. Puy are extinct volcanic cones. The Auvergne is a region of livestock raising and low population densities.

Source: AgenceImages/AgeFotostock.

Distribution of Population, Urban Areas, and Economic Activity

Paris is primate. The city is many times larger than any other place in France and contains a large share of the economic activity of the country (Figure 7.2). Of the 60 million inhabitants of France, 10 million live in the Paris conurbation at the center of which is the historic city, containing national symbols and institutions such as the Arc de Triomphe, the Eiffel Tower, Notre Dame, and the Louvre.

Paris and its region dominates the political, economic, cultural, and social life of France. The city sits at the center of the Paris Basin, a rich agricultural region where routeways converge on the good bridging point across the Seine, with the Île de la Cité providing a defensive site (Figure 7.3). There are no important deposits of raw materials or coal close by, nor is the city a major port. Paris, however, has been the capital of France since 987 A.D. and over time has accumulated cultural institutions, historic buildings, and government bureaucracies, enhancing the power and prestige of the place. In the first half of the twentieth century the population of France grew slowly but Paris attracted many young adult migrants who formed families and contributed to population growth. Similarly, service industries, the engine of economic growth in recent decades, have located in the capital because Paris possesses huge purchasing power. Many of the traditional craft and fashion industries remain, supported by a concentration of luxury consumers. The historic city is surrounded by suburbs often characterized by high-rise apartment buildings. Even when France tried to slow population growth in the Paris

TABLE 7.1   France: Hierarchy of Cities

Conurbation	Population
Paris	9,350,000
Marseille-Aix-en-Provence	1,350,000
Lyons	1,340,000
Lille	1,100,000
Nice	890,000
Toulouse	760,000
Bordeaux	750,000
Nantes	550,000
Toulon	520,000
Douai-Lens	510,000
Strasbourg	430,000
Grenoble	420,000
Rouen	390,000

region by creating new towns elsewhere in the country, the capital city had strings of new towns built to the north and south.

In 1900, the population of France was approximately 39 million. In 1945, at the end of World War II, the figure was slightly less. The situation changed rapidly as many French men and women who had left the country in 1940 returned, prisoners of war were released, and people forced to work in Germany came home. There was a postwar baby boom.

Then, in the 1950s and 1960s, decolonization resulted in another inflow of people. The war in French Indo China (Vietnam) was lost in 1954. Tunisia and other African colonies became independent in the late 1950s and early 1960s, resulting in the return of colonial administrators, their families, military men, merchants, and professionals who had made careers overseas. The numbers were not large but, combined with the baby boom, produced population growth. By 1959, France had a population of 45.2 million.

Then came the Algerian exodus. France took over Algeria in 1830 and made the North African territory a settlement colony. Large numbers of *colons* moved to Algeria establishing businesses, farms, and vineyards. In the years between World War I and World War II, Algeria was France's major trading partner, exporting cereals, sheep, tobacco, wine, iron ore, nonferrous metals, and phosphates across the Mediterranean. France and Algeria shared the French franc.

After World War II, agitation for independence developed, leading to a bitter war between indigenous forces and the French military. Torture and atrocities were employed by both sides. De Gaulle came back to power in 1958 as the government in Paris was losing control and the French military increasingly threatened civilian authority. De Gaulle demanded a new constitution giving more power to the President. The constitutional proposals were accepted in a referendum. France got the Fifth Republic and a president with wide powers, as in the United States.

De Gaulle insisted that remaining French colonies vote *oui* or *non* on attachment to France. Those voting *non* were to become independent. Algeria and all the remaining

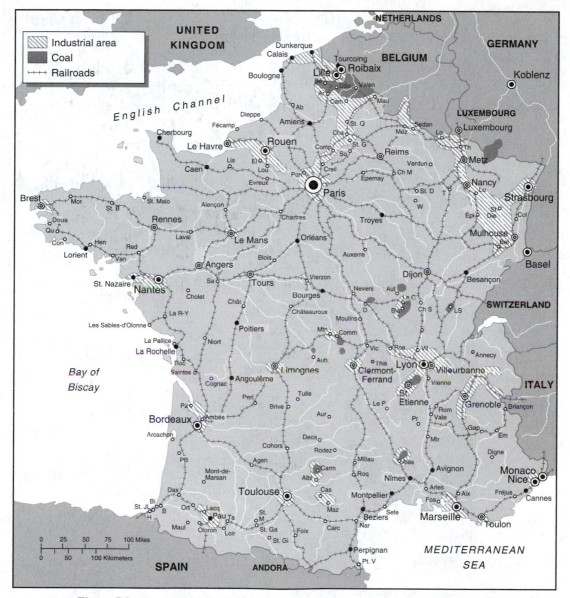

Figure 7.2 France: Distribution of Major Urban Places. Circles are proportional to city size.

African colonies voted *non*. Algeria moved toward independence in 1962, and the military tried to assassinate de Gaulle. (Read Frederick Forsythe, *The Day of the Jackal,* for a fictionalized account of a failed assassination attempt.)

As French rule ended in Algeria a million *colons* entered France. Many Algerians who had worked with the French administration also left. Those who stayed had a shortened life expectancy. Algerians coming to France did not have difficulty finding work, for the economy was expanding. Many jobs were low paying but there was also

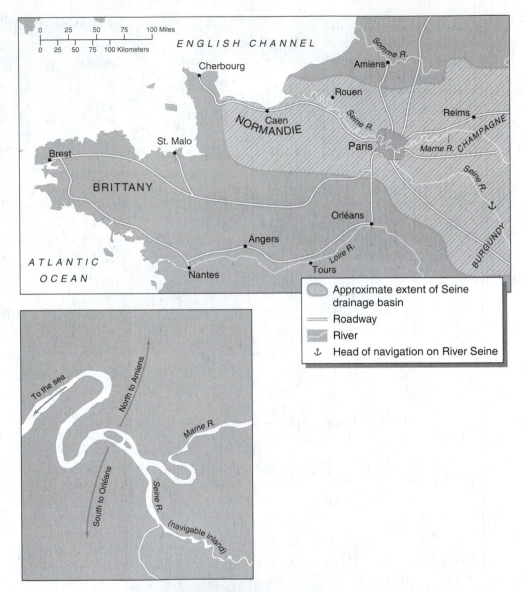

Figure 7.3 Paris: Site and Situation.

demand for workers in the better-paying car plants. Once Algerians found work in France, migration chains were set up as family and friends moved from North African towns and villages to immigrant communities in France. Immigrants were often accommodated in high-rise housing projects in dormitory suburbs around major cities.

Due to the birth rate and in-migration, the population of France increased by three million people between 1959 and 1964. And growth continued: 50 million by 1969, 54 million by 1980, and at a slower pace as birth rates dropped, to 61 million today.

Photo 7.12   Marseille, old fortified harbor. Modern commercial vessels use offshore quays.
Source: Hemis/AgeFotostock.

Much of the growth was promoted by in-migration from the Caribbean, West Africa, and North Africa. Today over a tenth of the population is foreign born or the offspring of immigrants.

Some cities, like Marseille, have absorbed the immigrant communities. (Photo 7.12) Ports always have immigrants, and people have the experience to respect the customs and cultures of newcomers. Many towns and cities were unable to absorb, let alone assimilate, people from foreign lands. In the fall of 2005 rioting broke out in the immigrant-filled suburbs and satellites of Paris. Disorder quickly spread to other towns and cities, although Marseille and *Le Cage,* where many immigrants lived, largely escaped. Unemployment in immigrant communities is often 25 percent. Grandparents and parents, who had made new lives in France, found their offspring unemployed and alienated. The generation born in France would not take the jobs their parents had, preferring to exist on France's generous unemployment allowances.

France is proud of its culture. It is not easy for most of the population to appreciate other cultures. President Chirac is rude about English cooking, seeing cross-channel cuisine as indicative of fundamental flaws in the character *Anglais.* American culture is both despised and seen as a threat with efforts to exclude films made in the United States and words with an American origin. Most French men and women think there are too many "foreigners" in the country. The attempt to prevent young Muslim women wearing traditional headwear to school is more likely to promote division than assimilation (Price, 2005, pp. 403–408).

Economic Activity

As in most of western Europe, the economy of France is dominated by service industries that generate nearly three quarters of the gross domestic product (GDP). Manufacturing industry earns a quarter of GDP and agriculture only about 3 percent, although agricultural goods account for nearly 15 percent of exports. France has the largest agricultural sector of any European country and is a leading wine producer. The wine industry has encountered competition in international markets from California, Chile, Argentina, South Africa, Australia, and New Zealand. In addition, per capita wine consumption in France has declined.

French farming has partially modernized over recent decades. From the early twentieth century until after World War II, France had approximately three million workers in the farming sector. Agriculture depended on a large labor input. There was little mechanization; traditional methods of ploughing with horses and sowing and harvesting by hand were widely used.

After World War II, with economic expansion, labor began to leave the land and today there are approximately three quarters of a million workers in farming. The whole process of modernization and labor reduction was given impetus by the CAP, which began to have an impact in the 1960s. Today, France receives 20 percent of all CAP payments and is highly resistant to any change, although CAP policies have largely achieved the original aims.

The structure of farming has altered in France. Half a century ago there were over two million farms in France. Today the figure is just over half a million. Land has not gone out of cultivation and farm size has increased, on average to one or two hundred acres. This is small by North American standards and the trend will continue toward fewer, larger farms. The criticism of CAP payments in France is that a large proportion go to big farms and a much smaller share to the many small farms.

France produces over four million cars a year, many being manufactured in the Paris region. At Toulouse, the European Airbus is assembled. In the north the older industrial region around Lille is less involved in coal mining and heavy industry but remains an important economic center, with high-speed rail connections to Brussels, Paris, and London.

France and the EU

France was the leading force in the creation of the Coal and Steel Community and then the Treaty of Rome in 1957. The form of the European Community reflected French organizational ideas—the original community was protectionist with a common external tariff, statist with a central authority at Brussels implementing policies, and bureaucratic.

No country has gained more politically, economically, and diplomatically from the creation of EU than France. After rapid defeat and occupation in World War II, France was able to rise in the postwar years to take the leading role in Europe. European policies favored France. The Coal and Steel Community integrated the western European iron and steel industry, removing the fear that German heavy industry would dominate again. The Common Agricultural Policy was highly favorable to France. As the largest producer of agricultural products, France receives the greatest part of CAP subsidies. Further, as France produced a food surplus, crops could be sold into the protected, high-priced

European market. France received both CAP subsidies and high prices for food exports. As a result, French farming has partially modernized, although some sectors, including wine production, require reorganization to address problems that have been apparent for decades (Crowley, 1993).

Although it is generally agreed that CAP payments and the EU budget need radical change, France resists revisions that would cut subsidies to French farmers, who are a vocal and disruptive force within the country when cuts are discussed. Presently there is dissatisfaction with Europe within France, as shown by the rejection, by French voters, of the European Constitution in 2005. France has an immigrant problem, having failed to absorb the second generation of North Africans. The widening of EU in 2004 produced the prospect of more migrants, even if they were from eastern Europe rather than North Africa. With unemployment exceeding 10 percent, more workers were unwelcome.

Summary

The French economy is sluggish, unemployment is high, particularly among young workers, and many immigrants feel alienated. Addressing the problems is difficult. France, like most western European countries, developed a social contract. Workers receive generous benefits and have security of employment. As a result, employers are reluctant to hire because shedding workers is difficult and costly.

To encourage employers to hire young workers, and reduce unemployment among young Muslims, the government of France proposed that when a business took on new workers under 26 years of age, the first two years would be probationary. Workers could be let go if company needs changed. The result was more riots across France in the spring of 2006. The proposed legislation was withdrawn.

▶ BELGIUM

Introduction

After the Napoleonic wars, the territory we now recognize as Belgium became a part of the Kingdom of The Netherlands. The arrangement did not last long, the Belgians revolting in 1830 and declaring independence. The neutrality of Belgium was guaranteed by the Treaty of London (1839), signed by Britain, France, Prussia, Russia, and Austria. Neutrality was respected in the Franco–Prussian War (1870), but in 1914 Germany dismissed the treaty as "a scrap of paper" and violated Belgian neutrality, bringing Britain into World War I. The restoration of Belgium became one of President Woodrow Wilson's war aims, specifically mentioned in the fourteen points speech.

Belgium resumed a neutral status after World War I but was attacked and occupied by Germany in World War II. Belgium joined NATO (1949) and became a founding member of the European Coal and Steel Community (1952) and the European Economic Community in 1957. The headquarters of NATO and the European Commission are in Brussels.

Although Belgium is at the center of the movement toward European integration, the country has problems with its own unity. Belgium is populated by Flemings and

Walloons. Flemish is a Germanic language; Walloon is akin to French. The Flemings, predominant in the north, are Protestant. The Walloons, found in the south, are Catholic. The regional tensions were reduced by constitutional reform (1993), which created a federal state.

The Environment

Geologically, topographically, and climatically Belgium can be divided into two: lowland Belgium and upland Belgium. The line of division is the Sambre-Meuse valley. Culturally, lowland Belgium and Flanders are populated by Flemings. Upland Belgium and the Ardennes is predominantly Walloon. Brussels is in lowland Belgium. In and around the capital are communities of Walloon and Flemish speakers. Brussels, however, is a Francophone city.

Physically, lowland Belgium, facing the North Sea, is low lying and composed of geologically recent sediments at the surface. Upland Belgium, to the south and east of the country, is composed of older, harder rocks folded in the Allegheny (Hercynian) mountain building era, then eroded to a plain that was uplifted in the Alpine orogeny and subsequently dissected by streams. Upland Belgium is cooler and wetter than lowland Belgium. The lowland region, with deposits of marine clays, loess, and sands, has fertile soils. The upland soils are thin, except in the valleys where streams have deposited alluvium (Figure 7.4).

A Transect across Belgium

We can cross Belgium by rail from the North Sea port of Oostende, to Brugge, (Photo 7.13) Gent, Brussels, Namur in the valley of Sambre-Meuse, over the upland Ardennes, and down to the town of Arlon in the Belgium province of Luxembourg.

Entering the port of Oostende we observe Belgium's extensive, sandy North Sea shore with excellent beaches backed by banks of sand dunes. (Photo 7.14) Inland of the dunes is Flanders, a low-lying region ten miles wide drained by human action to create polders (reclaimed land) used for grazing livestock and to grow horticultural products for urban markets. On the landward side of the polders is the port city of Brugge, a center of medieval trade and the wool industry, until the outlet to the North Sea sited up and fossilized the town, which today attracts tourists and pilgrims. Brugge was reconnected to the North Sea, at Zeebrugge, by a canal opened in 1907.

Inland from Brugge and the polder region, which is close to sea level, the plain of Flanders rises from 15 feet near the coast to 150 feet inland. The plain, predominately clay at the surface, is well cultivated, densely settled, and supports many towns, including the major medieval wool town of Gent (Photo 7.15) on the river Schelde. Gent contains medieval buildings, testifying to the prosperity of both the medieval guilds and merchants. Gent possesses modern industries and a canal link to the coast.

Moving inland from Gent to Brussels, we enter Belgium's dissected central low plateau. The low plateau region is again well cultivated, densely populated, and contains the important cities, of Brussels and Antwerp, linked by canal. Brussels displays site and situation features similar to Paris. The early settlement was an island in the river Zenne (a tributary of the Schelde) and Brussels occupied a situation of great centrality within

Figure 7.4 Belgium and Luxembourg.

the routeways of the Schelde basin, just as Paris does in the Seine river system. Brussels is the capital of Belgium, Brabant province, and in several ways is the capital of Europe. Many European organizations are headquartered in Brussels. (Photo 7.16) The city is a rich combination of medieval, renaissance, industrial, and global eras.

Traveling inland from Brussels, the central plateau increases in height to 600 feet before reaching the city of Namur where the Sambre and the Meuse make a confluence. In the valley of the Sambre-Meuse are coal seams of Pennsylvania age, and we enter a region transformed by nineteenth-century industrialization. The valley was one of the earliest industrial regions in mainland Europe with many manufacturing innovations being transferred from Britain, although Belgian inventors and entrepreneurs developed new manufacturing technologies. The valley contains the industrial towns of Liège, Namur, and Charleroi. As in all older industrial regions on coalfields, mines, blast furnaces, and steel works have closed as the industries have concentrated into fewer, larger, more efficient units.

The Sambre-Meuse is a physical transition zone between lowland and upland Belgium as well as a cultural divide. Northern Belgium is Flemish speaking, prosperous,

Photo 7.13 Brugge was an important center of the Medieval wool trade which is an attractive tourist center today.

Source: Masterfile Royalty-Free/Masterfile.

and well populated. In southern Belgium, the language is Walloon, population densities are lower, living standards poorer, and upland farming marginal.

From Namur the rail line takes us onto the Ardennes, a region of older, harder rocks rising up to 2,000 feet and displaying relatively flat upland erosion surfaces with thin soils. The region is cold and bleak in winter. The land is used for grazing and forestry. In favorable valleys, arable crops are produced but population densities are low compared with the rest of Belgium.

At the southern edge of the Ardennes, the land drops to the plain of Lorraine with its limestone scarps and deposits of sedimentary iron ore. The land is better suited to agriculture, and in favorable areas vineyards produce good wine.

Kempenland

The transect of Belgium running through Brugge, Gent, Brussels, Namur, to the Ardennes and Belgian Lorraine did not take in the Kempenland, in the northeast of the country. The region rises to just over 300 feet. At the surface the Kempenland consists of coarse sands and gravels, deposited by the Meuse at the end of the glacial era when the river spewed meltwater filled with erosional detritus. Much of the Kempenland is still in heath and little used for agriculture apart from livestock grazing. Below the more recent geological material is a concealed coalfield, and in the twentieth century colleries developed on the Kempenland together with industrial towns, but coal output has declined.

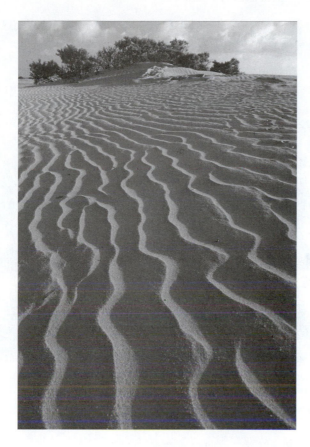

Photo 7.14 The North Sea coast of Belgium is protected by an almost continuous chain of sand dunes.

Source: Arco Images/AgeFotostock.

Urban and Economic Geography

Over 90 percent of Belgium's 10 million inhabitants live in urban areas. At the top of the urban hierarchy is Brussels, an international but Francophone city, surrounded by Flemish-speaking communities. Governments send three ambassadors to Brussels, one each for Belgium, NATO, and the EU. We might expect that Brussels would be a numerically primate city, but in the hierarchy of Belgian cities, Antwerp is approximately half the size of the capital.

Antwerp

The port of Antwerp is fifty-five miles inland on the tidal river Schelde at a site where the current swings against the right bank of the river, creating a deep channel.

The urban morphology of Antwerp displays the medieval and modern history of western European ports and commerce. The walled medieval city contained churches, markets, monasteries, and every trade of the time. Shops and workshops were up against churches, often in narrow lanes. In addition to the Saturday Groote market, held on the square surrounded by guild halls, there were specialized markets for commodities, including fish, cattle, horses, grain, oxen, milk, eggs, linen, gloves, and shoes. The meat

Photo 7.15 Leie canal and Graslei street, Gent, Belgium. Gent, like Brugge, was important in the Medieval wool trade.

Source: Age fotostock/SUPERSTOCK.

merchants were inside Butchers Hall and the cloth trading was conducted in Cloth Hall. In addition, street names designated specialized activities: Brewers' Street, Wool Street, Tanners Street, and Lombard Street, accommodating the bankers originally coming from northern Italy. The first stock exchange was founded at Antwerp in 1460.

In 1551, a start was made on building modern fortifications. (Photo 7.17) New streets and quays were constructed and canals were run into the heart of the city. The new walls did not save the city from being sacked twice in the late sixteenth century. Then, by the peace of Westphalia (1648), the Dutch restricted navigation on the Schelde to favor

TABLE 7.2 Population of Belgian Cities

Brussels	959,318
Antwerp	446,525
Gent	224,180
Charleroi	200,827
Liège	185,639
Brugge	116,240
Namur	105,419
Mons	90,935

Photo 7.16   There are European Parliament buildings in Brussels (above) and Strasbourg.

Source: Picture Finders/AgeFotostock.

Amsterdam at the expense of Antwerp. Navigation was reopened in the nineteenth century but the Dutch still levied tolls on the lower Schelde until 1863 when Belgium made a cash payment to eliminate the charges.

After the buyout, trade at Antwerp grew rapidly. The port was well linked, by rail and canal, to Belgian manufacturing and, being far inland, was closer to the Ruhr than Rotterdam. Ruhr products moved rapidly by rail for export through Antwerp, which acquired a reputation for efficient cargo handling that it retains to the present, as a result of sustained investment in modern facilities, in deeper waters downstream of the original port. The city handles cargo to and from northern France, Luxembourg, western Germany, and parts of The Netherlands.

Summary: Politics and Economics

After the Napoleonic wars, the Belgian provinces were attached to The Netherlands. Britain wanted France away from the lower courses of the Rhine and Schelde, with The Netherlands strong enough to act as a buffer between the French and German lands. The Netherlands was insensitive to the fact that the Belgian provinces were not Dutch. The Belgians broke away in 1830.

At first the Flemish/Walloon language divide was not a major issue. The ruling elite conducted the affairs of a unitary state in French and the universities taught in French.

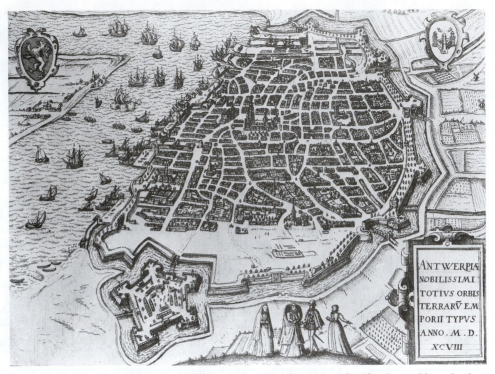

Photo 7.17 Engraving of Antwerp, 1598. Notice the 16th century fortifications with projecting bastions.

Source: The Bridgeman Art Library International.

The status of French was partially a reaction against Dutch rule. In the second half of the nineteenth century, the Flemish language gained in status and a Flemish-speaking elite emerged (Stephenson, 1972).

After World War II, at just the time Belgium was taking a leading role in Benelux and the founding of EU, the differing cultures in Belgium's two major regions became an issue. In the 1960s the country came close to falling to pieces with incessant student demonstrations. One student issue was the use of French in Belgian universities. A result of the protests was the creation of a university teaching in English! The regional issues were addressed in constitutional change. In 1993 Belgium evolved from a unitary to a federal state composed of three regions: the Flemish region, the Walloon region, and Brussels and surrounding territory. The country remains a constitutional monarchy.

Agriculture contributes less than 2 percent of GDP. There are still prosperous farming regions, but the majority of Belgians make a living in the service sector. Manufacturing, traditional and modern, is important. In the old Sambre-Meuse industrial region, cities like Liège, Namur, and Charleroi remain involved in heavy industry. Belgium produces over 10 million tons of steel a year. The old industries have downsized but not disappeared.

Brugge and Gent were prosperous medieval wool cities, which are well preserved and attract tourists. Both cities have modern manufacturing, as do Brussels and Antwerp.

A number of major transnational corporations are headquartered in Belgium, including Interbrew, which now controls many brands of European beer.

Per capita incomes in Belgium are slightly above the average for western Europe. Population numbers at 10.5 million are stable, with the total fertility rate standing at 1.6. Belgium used to be seen as the "cockpit of Europe," there was so much fighting over it and the surrounding territory. Today, Brussels serves as the administrative capital of the European Union, but tensions between Flemings and Walloons continue.

▶ THE NETHERLANDS
Introduction

The Dutch initiated many of the innovations that helped create the modern world. By the Treaty of Westphalia (1648) The Netherlands became independent, creating the model for the modern, sovereign nation-state. The Dutch East India Company and the Dutch West India Company created overseas colonies and trading stations and, unfashionably in the seventeenth century, pioneered free trade. It was the Dutch jurist Grotius who advanced the idea of freedom of the seas.

The Dutch had a large influence on English farming and finance. Capitalist Anglo-Dutch families were part of the Caribbean sugar revolution and held interests in Virginia tobacco plantations. There were Dutchmen and Dutch influence at Jamestown from 1607, the first sustained English settlement in the Americas. Many of the early Pilgrims at the Massachusetts Bay colonies came via Leiden in The Netherlands, and New Amsterdam was the forerunner of New York.

The Dutch, in the seventeenth century, made increasingly intense use of land expensively reclaimed from the sea, estuaries, and marshes. Dutchmen developed crop rotation, allowing land to be cultivated on an annual basis without fallowing. Crop rotation was made possible by growing nitrogen-fixing fodder crops like clover. The fodder allowed more cattle to be kept year round in a system of mixed farming. The larger livestock herds enhanced the production of butter, cheese, hides, and leather goods. Animals, fed in barns in the winter months, dropped manure that in spring was spread on fields to improve fertility.

The Netherlands is the most densely populated country in Europe with the exception of mini states like Monaco and Malta. High densities in the Dutch lands were achieved long before the rest of Europe. High rural densities were made possible by intensive farming and a system of canals to get produce to market, with urban "night soil" (human excrement) being a return cargo to fields close to waterways. The Netherlands had a dense network of towns early and the towns were markets for rural produce and distribution centers for imported Caribbean sugar and spices from the Dutch East Indies. Atop the urban hierarchy sat Amsterdam, a center of finance, trade, diamond cutting, fishing, and confectionary trades based on imported sugar. The city housed innumerable craft industry workshops and markets selling cheese, bulbs, leather, and every variety of agricultural product.

The Rhine runs north from Switzerland, through France and Germany, before turning west to pass through The Netherlands to join the North Sea and the Atlantic. The southern Netherlands lies on the distributaries of the Rhine. The Dutch developed ports, transshipment areas, and trading stations to link the continental commerce of the Rhine with the maritime trade of the Atlantic world (Figure 7.5).

No country was better equipped to take a lead in the European Union and on January 1, 1948, along with Belgium and Luxembourg, The Netherlands launched Benelux, creating the model for European integration. Once Germany and France decided on a European Economic Community, it was logical that the Benelux countries would join. From another perspective they had little choice, for economies were already linked to those of the bigger countries, and ports like Rotterdam and Antwerp served the trade arteries of the Rhine and northern France.

Topography

The greatest part of The Netherlands is close to sea level, and some of the country is below sea level. The surface of the land is composed of sands, esturine muds, marine clays, peat, and deposits of glacial detritus laid down at the end of the Ice Age.

As the Ice Age ended, the postglacial river Meuse (Maas) carried a huge volume of water to the North Sea, the additional flow allowing the transport of sands, gravels, and rocks. Much of the coarser material was deposited in what today is the Belgium Kempenland and the Southern Heathlands of The Netherlands. To the east, meltwater streams deposited sands and gravels, creating the Eastern Heathlands of The Netherlands. After the glacial retreat, surface drainage did not organize quickly, and many lakes, marshes, and bogs developed. As the postglacial sea level rose, low-lying coastal areas came under seawater.

Until around 1200 A.D. there was a continuous line of sand dunes running from the mouth of the Rhine along the east and north coasts of The Netherlands. Then, in the thirteenth century, the sea breached the dunes in the north, forming the Frisian Islands, and penetrated inland creating the Zuider Zee (the South Sea) in the heart of The Netherlands. Now, as a result of draining, the Zuider Zee is the freshwater IJsselmeer.

An early settlement strategy on low-lying land was to build mounds that would stand above floodwater in wet and stormy weather. These *terps,* as they are called in Dutch, are still a feature in the landscape today, often topped by a church and farmhouses. To control flooding and allow reclamation of land, embankments were built to keep out the sea, and on the landward side polders (reclaimed land) became agricultural assets. In the fifteenth century windmills were used to pump water into drainage channels. In the seventeenth century extensive areas were drained by a combination of embankments to keep out the sea, drainage canals to convey water to the coast, and windmills to pump water into canals. Dutch engineers became so expert at constructing drainage works that they were employed to reclaim fen land in eastern England.

By the mid-nineteenth century, steam pumps lifted water to the drainage canals. In the early twentieth century even more efficient diesel engines helped with land reclamation. There were, however, two major gaps in the seaward defenses. The delta lands were open to storms, and the northern end of the Zuider Zee connected directly with the North Sea. Devastating storms found the gaps.

Climate

The Netherlands is a low-lying country with a long coastline and no topographic barriers to air mass movement. Maritime influence on climate is marked. Along the coast, even

Figure 7.5 The Netherlands.

in fine summer weather, a cool breeze blows off the North Sea to the land. For much of the year The Netherlands experiences depressions, coming off the North Atlantic, bringing mild, moist, cloudy weather. In winter, continental polar air does, in cold years, extend across the country, bringing low temperatures and freezing inland water surfaces.

A major danger in winter is a deep Atlantic depression, accompanied by high winds and storm surges in the North Sea. Storm surges, combined with high tides, can cause flooding inland, as happened in 1916 and 1953. After the 1916 flood the government funded a huge project to close off the Zuider Zee tidal estuary by the construction of the Afsluitdijk, completed in 1932, running from Den Oever in the west to Zurich at the east end of the barrier dam (Figure 7.5). (Photo 7.18) Behind the dam a freshwater lake was created, fed by the river IJssel. Large areas of reclaimed land (polders) were established in the former south and east of the Zuider Zee. The land was used for agriculture, and new towns were created to provide services to farming communities.

In late January 1953, a deep depression accompanied by gale-force winds tracked toward Norway off the northeast coast of Scotland. The storm then changed course, entered the North Sea, and headed for the Jutland peninsula and the North German Plain. Strengthening winds pushed water southward in the North Sea, raising sea levels along the coasts of England, The Netherlands, Germany, and Denmark. On February 1, as the storm moved east, gale-force southwesterly winds hit the coast of Holland and blew into the Rhine delta. River water could not escape to the sea. Assaulted by the storm surge, high winds, and the weight of impounded water, the restraining dykes failed.

Photo 7.18 The Afsluitdijk in north Holland keeps the North Sea out of the IJsselmeer. A modern highway links communities formerly separated by a large body of water.

Source: Jon Arnold Images/Alamy Images.

Nearly two thousand people died; tens of thousands were homeless. Livestock losses were devastating.

The 1953 flood led to the 1958 Delta Act and the *Deltawerken,* a scheme of dams, sluices, and locks to prevent the sea from surging into the Rhine, the Maas, and the Schelde while allowing the passage of ships (Figure 7.5). Again land was reclaimed and the building of new roads and bridges helped make the delta region more productive.

The danger remains that heavy rain over the Rhine, the Maas, and their tributaries might result in the flooding of the delta, particularly if water came downstream at a time when the sluices were shut against a high sea. Building higher levees along the river banks is not the policy, for the higher the levees, the worse the flood if the levees fail. So far the strategy is to divert river floodwater into low-lying "bathtubs" to limit damage. Farmers in the "bathtubs" do not like the policy.

In the last two decades, policies regarding land reclamation and drainage have modified. As some farmland goes out of cultivation, and EU agricultural surpluses build up, the pressure to reclaim land has lessened. In 1991, the plan to reclaim Markermeer as a major polder in the Zuider Zee was abandoned. With the development of a system of interlinked nature reserves, lakes, and wetlands, open fresh water is now seen as an asset to be conserved rather than pumped out (Koomen et al., 2005).

Urban Geography

Because of overseas trade and the early commercialization of agriculture, The Netherlands was the first modern European country to urbanize. In the seventeenth century, Paris and London were larger than Amsterdam, but neither France nor England had such a well-developed urban hierarchy. In the seventeenth century, Paris and London were primate cities, many times larger than any other place in their kingdoms. In the Dutch case, in the seventeenth century, we can see a near rank–size relationship (Table 7.3). Under the rank–size rule, the largest town is twice as big as the second largest place, three times the size of the third largest town, and four times the size of the urban place that is fourth on the list by population.

Amsterdam in 1622 was more than twice the size of Leiden and, in the seventeenth century, Amsterdam grew faster than the average and moved toward primacy. However, other cities on the list expanded too, and in the modern maritime world, the port of Rotterdam, which can berth large vessels, grew to serve The Netherlands and much of western and central Europe. Today, Amsterdam has just over a million inhabitants, and Rotterdam a little less than a million. The Hague, housing the national parliament, the royal residence, and the International Court of Justice, has over half a million inhabitants. Utrecht (366,186), Eindhoven (302,274), and Leiden (250,302) are the other cities with over 250,000 inhabitants. Ninety percent of the population lives in urban places.

A note on the 1622 list of towns: The list records the towns that had a vote in the States of Holland, the ruling body. Each place on the list had one vote. Small places were equal to big places, helping to diffuse political power and prevent it from being concentrated in one place, as happened in France. Not all towns are on the list. The Hague did not have a vote, and Utrecht was outside the States of Holland. Other modern towns were either not in existence or outside the territory of Holland in 1622.

TABLE 7.3 Dutch Towns: 1622 and 2000

Towns	1622 (rank order)	2000
Amsterdam	104,932	1,002,868 (c)
Leiden	44,748	250,302 (c)
Haarlem	39,455	191,079 (c)
Delft	22,769	96,095
Enkhuizen	20,967	16,000
Rotterdam	19,532	989,956 (c)
Dordrecht	18,270	241,218 (c)
Hague	16,000 (estimate)	610,245 (c)
Gouda	14,627	71,918
Hoorn	14,139	64,604
Alkmaar	12,417	92,836
Schiedam	5,997	75,589
Gorinchem	5,913	34,623 (estimate)
Edam	5,647	46,951 (estimate)
Monnikendam	3,990	10,089 (estimate)
Medemblik	3,983	8,037 (estimate)
Brielle	3,632	16,000
Schoonhoven	2,891	12,303 (estimate)
Purmerend	2,556	70,284
Utrecht		366,186(c)
Eindhoven		302,274(c)

(c), population in conurbation. Source of 1622 data, Price, 1994, p. 16.

No country in western Europe has planned the use of land as carefully as The Netherlands. It would have been impossible to embank and drain land without an overall plan. The layout of Amsterdam indicates how new blocks of housing could only be created after additional canals were put in place. The Dutch planning tradition was well developed in the twentieth century. (Photo 7.19)

Looking at the map of towns, we can identify a cluster of Dutch cities lying close together (Figure 7.5). Starting at Amsterdam, and moving in a clockwise direction, the cities are Utrecht, Rotterdam, Delft, The Hague, Leiden, Haarlem, and back to Amsterdam. The city cluster is referred to collectively as the Randstad, literally rim city. To prevent the cities coalescing and forming one giant conurbation, urban growth was planned to expand away from the central part of the Randstad. As in several western European countries, more growth has been permitted in recent decades but the landscapes of The Netherlands retain a well-planned appearance.

The Economy

Since the seventeenth century, The Netherlands has linked the maritime and continental worlds, looking into Europe along the Rhine and outward to the West Indies, the East Indies, Africa, and America. In addition to trading globally, the Dutch made intensive use of the land at home. These characteristics persist into the present. The Netherlands is the headquarters of major multinational corporations, including Royal Dutch Shell,

Photo 7.19 Traditionally windmills were used to pump water from the land into drainage canals.

Source: AgeFotostock/SUPERSTOCK.

Photo 7.20 At the end of World War II, Rotterdam was a bombed city with weeds growing on the quays. Today it is a major world port and a symbol of Europe's commercial resurgence.

Source: AgeFotostock/SUPERSTOCK.

Photo 7.21 Heated greenhouses are extensively used in The Netherlands to raise vegetables and flowers which are exported all across Europe and to North America.

Source: Peter Horree/Alamy Images.

Philips electronics, and Unilever. Rotterdam is the home of Europort, and Amsterdam remains a major commercial and financial center. (Photo 7.20) Dutch agriculture is the most efficient in Europe, and it responds to market forces. In 1950, there were over 400,000 farms in The Netherlands; today there are less than 100,000. Poorer land has gone out of cultivation, farm size has increased, the agricultural labor force has decreased, and with more inputs, output has grown. (Photo 7.21) The Netherlands is a large exporter of agricultural produce, including cut flowers, bulbs, vegetables, butter, cheese (including Gouda and Edam), ham, poultry, and even potatoes. Agricultural products make up a fifth of exports, but agriculture contributes less than 3 percent of GDP. Like other modern economies, the Dutch gain nearly three quarters of domestic product from services, but manufacturing still contributes around a quarter. The Netherlands has one of the most global economies in Europe, and the country is a positive force in the effort to move the EU to greater efficiency.

▶ LUXEMBOURG

Introduction

Luxembourg is a small state with less than half a million inhabitants. It was occupied by France during the Napoleonic wars and after the war became part of The Netherlands. The Dutch king became Duke of Luxembourg, but when, in 1890, a woman ascended the Dutch throne she could not, by gender, become Duke of Luxembourg. Luxembourg became independent. In 1914, German forces occupied Luxembourg and had World War I ended in a German victory, or a negotiated peace, Luxembourg would have become a province of Germany. President Woodrow Wilson in his fourteen points speech insisted that Belgium be restored to independence by Germany, but did not mention Luxembourg.

At the end of World War I, Luxembourg left the German customs system and entered into a customs union with Belgium. In World War II, Luxembourg, and its steel works, was annexed by Germany, being part of the Reich from 1942 to 1945, in spite of a general strike against Germany in June 1942.

Luxembourg is bordered by France, Germany, and Belgium. Central position and the fact that many inhabitants speak French and German, the official languages, in addition to *Lëtzeburgesch* (*Luxembourgeoise*), give the country an importance far beyond its size. Luxembourg houses the headquarters of a number of European institutions, international organizations, and multinational corporations. The European institutions include the Court of Justice of the European Communities, the European Investment Bank, the European Investment Fund, and the Statistical Office of the European Communities (Eurostat).

Luxembourg has made itself attractive by avoiding the overregulation apparent in Germany and France. In the 1920s, as radio established itself, European countries set up state broadcasting monopolies. Luxembourg exploited this by allowing the transmission of radio programs, in several languages, into surrounding European countries. Radio Luxembourg provided relief from the content of the formal national stations. In financial services the laws are constructed to allow legitimate "off-shore" banking activities. More than 200 banking corporations have offices in Luxembourg, which has become a center of international finance.

Physical Environment

Luxembourg consists of two topographic regions. The north of the country is part of the Ardennes, known as *Oesling*. The southern part of the country is the *Gutland* or *Le Bon Pays* (Figure 7.4). The good land is a continuation of Lorraine with its escarpments, limestones, sandstones, and clays. The region is carefully cultivated and has excellent vineyards. The higher, cooler, and wetter Ardennes in Luxembourg, although better cultivated than the Belgium Ardennes, produces timber, livestock, and some arable crops. (Photo 7.22) The *Gutland* is lower, warmer, and drier than the Ardennes. The Ardennes upland provides some protection against winter incursions of maritime polar air but, being inland, Luxembourg gets elements of a continental winter, particularly when continental polar air predominates over central Europe. Average temperatures in January, at Luxembourg, the capital city, are around freezing.

Economy

In the south of Luxembourg are deposits of sedimentary iron ore similar to those found in Lorraine. Starting in the 1870s, the ore became the basis of a large iron and steel industry, although coal had to be imported. In the 1960s, Luxembourg produced over

Photo 7.22 The Ardennes region is characterized by steep sided, well-wooded valleys

Source: Masterfile Royalty-Free/Masterfile.

4 million tons of steel a year and the industry was the largest employer in the country. Of necessity, the industry was integrated into the European Coal and Steel Community. Coking coal came from Germany, additional iron ore from France, and the output of iron and steel was sold into Germany, Belgium, The Netherlands, and France for further processing. Luxembourg still produces approximately 2 million tons of steel annually but the low-grade sedimentary ore is no longer used, being replaced by higher iron content, imported igneous ores.

Today over 80 percent of the workforce, including many workers from other parts of Europe, are in the service sector. Major employers are banking, investment houses, European organizations, the tourist industry, and corporate administration. Ninety percent of the population lives in urban areas. The picturesque, fortified capital city of Luxembourg stands above the entrenched river Alzette. The city contains historic buildings and is an attractive tourist destination. The total population of the predominantly Roman Catholic country is 441,300, and just over 80,000 people live in the capital, which remains a pleasant pedestrian city. Per capita incomes are the highest in Europe, infant mortality is below four per thousand live births, the total fertility rate is 1.7, but the population grows slowly due to in-migration.

▶ FURTHER READING

Berensten, W.H. *Contemporary Europe: A Geographic Analysis*. Seventh edition. New York: John Wiley, 1997.

Davies, N. *A History of Europe*. Oxford, UK: Oxford University Press, 1996.

Judt, T. *Postwar: A History of Europe since 1945*. New York: Penguin Press, 2005.

FRANCE

Ardagh, J, and C. Jones. *Cultural Atlas of France*. New York: Facts on File, 1991. Contains good regional portraits.

Braudel, F. *The Identity of France: Vol. 1. History and Environment*. New York: Harper and Row, 1988.

Crowley, W.K. "Changes in the French Winescape." *Geographical Review* 83(3) (1993): 252–268.

Dreyfus, F.G., J. Morizet, and M. Peyrard (eds.). *France and the EC Membership Evaluated*. New York: St. Martin's Press, 1993.

"Europe's Farm Follies." *The Economist*. December 10, 2005, pp. 25–27.

Garner, A. *A Shifting Shore: Locals, Outsiders, and the Transformation of a French Fishing Town, 1823–2000*. Ithaca, NY: Cornell University Press, 2005. Study of Arcachon on the Bay of Biscay.

Gottmann, J. *A Geography of Europe*. Fourth edition. New York: Holt, Rhinehart, and Winston, 1969. See Chapter 8, "France."

Halbert, L. "The Decentralization of Intrametropolitan Business Services in the Paris Region: Patterns, Interpretation, Consequences." *Economic Geography* 80(4) (2004): 391–404.

Harvey, D. *Paris, Capital of Modernity*. New York, Routledge, 2003.

Jones, C. *Paris: A Biography of a City*. New York: Viking, 2004.

Mayle, P. *A Year in Provence*. New York: Vintage Books, 1989.

Price, P. *A Concise History of France*. 2nd edition. New York: Cambridge University Press, 2005.

"Survey of France." *The Economist*. October 28, 2006, pp. 3–16.

BENELUX

de Blij, H. *Wartime Encounter*. New York, NY: Hudson River Enterprises, 2006.

De Vries, J., and A.D. Van Der Woude. *The First Modern Economy: Success, Failure, and Perseverance of the Dutch Economy, 1500–1815*. New York: Cambridge University Press, 1997.

Kooman, E. et al. "Simulating the Future of Agricultural Land Use in the Netherlands." *Tijdschrift voor Economische en Sociale Geografie* 96(2) (2005): 218–224.

Lambert, A. *Making of the Dutch Landscape*. New York: Seminar Press, 1971.

Price, J.L. *Holland and the Dutch Republic in the Seventeenth Century*. Oxford, UK: Clarendon Press, 1994.

Stephenson, G.V. "Cultural Regionalism and the Unitary State Idea in Belgium." *Geographical Review* 62(4) (1972): 501–523.

8

GERMANY AND ITALY

▶ INTRODUCTION

Germany and Italy were late unifiers. The German Empire was created after the Franco-Prussian War, in 1871. Italian unification was achieved in the 1860s. Two new, major powers altered the balance of European international relations. Germany, in seeking alliances, did not calm the fears of neighboring states. Italian unification did not signal the end of territorial ambition. Italy wanted the South Tyrol, Trentino, and Istria, objectives achieved at the end of World War I. Broader, Mediterranean ambitions could only come at the expense of France and Britain. German and Italian unification helped create the conflicts that paved the path to European Union.

▶ GERMANY

Introduction

Modern Germany has its origins in the expansion of Prussia in the eighteenth century. Early in the century Sweden ruled part of Pomerania around the port of Stettin (now Szczecin, Poland). The territory was taken by Prussia in 1720. Frederick the Great (1740–86) took Silesia from Austria in 1740 and acquired part of Poland in 1772. In the Napoleonic wars, Prussia lost lands, for a time, but at the Congress of Vienna (1814–15) had territory restored and gained control of part of the Rhineland.

In the 1820s, Prussia began to create a customs union (*Zollverein*) among the numerous German states. Bismarck came to power in 1862, declaring that German unification would be achieved with "iron and blood." Wars were provoked against Denmark to take Schleswig-Holstein (1864), acquiring the strategic territory that eventually allowed Germany to link her Baltic and North Sea coasts via the Kiel canal (Figure 8.1). To establish Prussian leadership in German unification, Austria, German in speech and culture, was defeated in 1866. Around the same time Frankfurt-am-Main was occupied militarily and forced to become a part of Prussia. The Franco-Prussian war, 1870–71, resulted in the defeat of France, the transfer of Alsace-Lorraine to Germany, and the creation of the German Empire. King William of Prussia became Kaiser Wilhelm I of the German Empire, on January 18, 1871, in the Hall of Mirrors at Versailles, on French soil, in a ceremony attended by European royalty and diplomats.

Having created the German Empire, under Prussian leadership, Chancellor Bismarck worked to produce a stable Europe by isolating France and developing a system of alliances and understandings with Austria, Russia, and Italy. Bismarck was forced to

Figure 8.1   German Bundes Länder (Federal States).

resign in 1890 by Kaiser Wilhelm II. Germany then adopted an aggressive foreign policy and built a high seas fleet that could be used to challenge the Royal Navy. Britain was pushed toward France (*Entente Cordiale,* 1904), disturbing the balance that Bismarck had created. By 1914 Germany was at war with France, Belgium, Britain, and the Russian Empire. Germany's war aims were extensive:

- Annex Luxembourg and much of Belgium and take territory from France, including the Belfort gap at the south end of the Vosges.

- Control the Belgium channel ports, together with Calais and Boulogne in France, on the south shore of the Strait of Dover.

- Create a continuous central African empire running from the Atlantic to the Indian Ocean.
- Establish a European Economic Association, which would enlarge German economic space and create a Europe surrounded by protective tariffs.

Britain was to be specifically excluded from the EEA. In eastern Europe, the Russian hold on non-Russian Slav peoples was to be broken. With the exception of eastern Europe, where Germany did partially succeed in creating new states out of the Russian Empire (Poland, the Baltics, and briefly, the Ukraine and Belarus), few war aims were achieved. The harsh armistice of November 1918 and the punitive Treaty of Versailles, in 1919, stripped Germany of armaments and gave territory to France, Belgium, Denmark, and Poland. The Saar with coalfield went, for a time, to France (Figure 8.1). The Rhineland was demilitarized and allied garrisons placed at the crossing points on the river. Poland got the coalfield and industrial assets of Upper Silesia. Germany paid reparations.

By diplomacy (the Treaty of Locarno, 1925), and subterfuge with the Soviet Union, Germany worked to get allied troops out of the Rhineland and rebuild the armed forces. When Hitler came to power, Germany remilitarized the Rhineland (1936), absorbed Austria (1938), dismembered Czechoslovakia (1938–39), defeated Poland (1939), and overran western Europe in 1940. Yugoslavia and Greece were occupied April 1941, and in June, German forces attacked the Soviet Union.

At the end of World War II, Germany was divided into French, British, U.S., and Soviet occupation zones. The Soviet zone became the German Democratic Republic (GDR) in 1949. The French, British, and U.S. zones became West Germany, the Federal German Republic in 1949, with the capital at Bonn on the Rhine. West Germany joined NATO in 1955.

Berlin was divided into the occupation zones of East and West Berlin. West Berlin was enclosed by territory controlled by the Soviet Union and East Germany. The capital of east Prussia, Königsburg, and the surrounding territory became Kaliningrad, part of the Russian Soviet Socialist Republic. Poland got much of East Prussia, Danzig, and German territory up to the Oder Neisse line. Germans in Poland, the Czech lands, and Hungary were forced back into Germany, increasing the problems of a country where every city center was rubble. East Germany lost industrial assets to the Soviet Union.

In West Germany economic growth was rapid. In the east, which had less industry prewar, recovery was slow. With a Moscow-oriented communist regime introducing central planning and state control of the economy, the East could not match the economic performance of the West. In 1953 there were uprisings in East Berlin and as labor became scarce in the West German "economic miracle" (*Wirtshaftswunder*), young East German adults and skilled workers moved from the GDR to the FDR to find work and a higher living standard. The Berlin Wall went up in 1961 followed by a high barbed-wire fence with guard posts, machine guns, and dogs sealing East Germany off from the western part of the country.

The Berlin Wall came down in 1989. In May 1990, the GDR ceased to exist and was absorbed into the FDR. The cost was huge for Germany and the European Community, which had not been consulted. The Deutschemark replaced the East German currency, efforts were made to improve living standards, industrial efficiency, and reduce environmental pollution. East Germans became part of the former West German state pension scheme, immediately overstretching the system. To avoid inflation, the Bundesbank pushed up interest rates, which increased rates throughout Europe and North America. In the global recession of 1990–91, German policies made things worse economically. East Germany merged with the European Community without application. Quickly the region became eligible for EU grants and subsidies. Living standards in eastern Germany remain significantly lower than in the former West Germany. Industry remains less efficient and unemployment higher. Many people recall a more secure life under communism.

Physical Environment

The distance from Germany's northern border, on the Jutland Peninsula, to the south of the country in the Bavarian Alps is over five hundred miles. By air the journey would take us from the border with Denmark, across the Kiel Canal to Hamburg, south over Hannover, Nuremberg (Nürnberg), and Munich (München), to the Austrian border (Figure 8.2). On the route we would pass over, from north to south, four major topographical regions (Figure 8.3).

1. The North German Plain
2. The Central Uplands and Swabian Jura
3. The Danube valley
4. The Bavarian foreland and the Bavarian Alps

The North German Plain

The plain is part of the extensive European lowland, which runs from the north flank of the Pyrenees into Russia. The North German Plain is a product, at the surface, of glacial activity. Much of the plain is a rolling landscape of moraines, marshes, and lakes, the moraines and outwash materials being deposited in the last glaciation. In the south, the plain is rich in loess, fine-grained silt picked up and deposited by the wind in the glacial era.

In the west, the North German Plain is open to maritime influences. It has a climate similar to The Netherlands. As we go eastward, continental influence upon climate is apparent. At Cologne, on the Rhine, the mean monthly temperature for January is 36°F; at Berlin, further east, the figure is 30°F.

The glacial clays and the loess areas of the North German Plain have been turned into agricultural land producing good harvests of cereals and root crops. In the north, there are many lakes and marshes providing wetland environments. Glacial outwash materials, ranging from sands and gravels to boulder fields, remain in heathland providing rough grazing.

The major rivers draining the North German Plain are, from west to east, the Ems joining the North Sea near Emden, the Weser flowing through Bremen to the North Sea, and the Elbe, which enters the North Sea below Hamburg.

The Central Uplands

The Central Uplands consist of faulted blocks of Hercynian age, which have been weathered and dissected. Numerous, well-developed valleys penetrate the uplands. In the north, including the Harz Mountains, the valleys drain to the Elbe and the Weser. (Photo 8.1)

Figure 8.2 Germany. Rivers and Cities.

In the south and west the valleys are tributary to the Rhine, which cuts through the Central Uplands from south to north before trending westward to The Netherlands and the North Sea.

At the south of the Central Uplands, the limestone folds of the Jura Mountains, extending from France through Switzerland, appear in Germany as the Swabian Jura or Schwäbische Alb.

Figure 8.3 German Topographical Regions.

The Danube Valley

South of the Swabian Jura is the valley of the Danube, trending generally east and southeast toward the Black Sea. The Danube is navigable well into Germany with large barges reaching Regensburg and smaller boats navigating upstream to Ulm. A canal from Bamberg on the Main, a navigable tributary of the Rhine, links the Rhine–Danube navigation systems. Inland cruise companies have itineraries that start in The Netherlands

Photo 8.1 Harz Natural Reserve in the central uplands of Germany.

Source: Age fotostock/SUPERSTOCK.

and provide transport through Germany on the Rhine to the Danube and eventually the Black Sea.

The Bavarian Foreland and the Bavarian Alps

South of the valley of the Danube is the Bavarian Foreland, an easterly extension of the Swiss plateau, which rises up to 3,000 feet in the foothills of the Bavarian Alps. The foreland is a broad plain on which the largest city is Munich. Many soils in the foreland are composed of deposits from formerly extensive valley glaciers that protruded from the Bavarian Alps. Much of the foreland is in pasture but arable crops are grown around Munich, including the hops, barley, and wheat that helped make Bavarian brewing famous.

The Bavarian Alps rise up to 8,904 feet in the Berchtesgaden. The mountains are full of glacial features including U-shaped valleys and arêtes. The winter games of Hitler's 1936 Olympics were held in the region.

The Elbe and the Rhine

The north to south simple cross-section of topography presented is more complex in reality, as a journey along any of the major German rivers flowing to the North Sea illustrates (see Figure 8.2).

The Elbe

The Elbe rises in the Czech Republic (barge traffic can reach Prague), cuts through the Erzgebirge Mountains to Dresden, and then flows onto the North German Plain in a northwesterly direction, receiving many tributaries. (Photo 8.2) The northwesterly course of the river reflects the position of the Scandinavian ice cap in retreat. The glacial deposits to the east and west of the Elbe differ in character. To the east, the North German Plain is largely composed of moraines deposited by glaciers, containing clay and many boulders of igneous and metamorphic rock plucked from Scandinavian surfaces by the ice. Out of the glacier front ran streams carrying silts, sands, and gravels. The gravel deposits make poor soils. The clays of the moraines, if drained and fertilized, make good deep soils suited to cereals and root crops like potatoes and sugar beets. Less well-drained areas are livestock pasture.

To the west of the Elbe, as it enters its lower course, is Lüneburger Heath, a type of landscape encountered in the Kempenland of Belgium and the Eastern Heathlands of The Netherlands. As the ice melted, huge quantities of water were released into rivers, reestablishing lines of drainage and frequently altering course. The increased flow allowed large rocks, in addition to sands and gravels, to be moved downstream and deposited toward the west. The result was extensive heathlands, including Lüneburger, with thin rocky soil and extensive sands and gravels. Farming is only possible in limited areas, and much land is planted in coniferous forest.

Like many European ports, Hamburg, on the Elbe, is well inland. The city, and surrounding communities, with nearly two million people, is a major manufacturing

Photo 8.2 Baroque Dresden on the river Elbe. The city has been restored after World War II bombing.

Source: Age fotostock/SUPERSTOCK.

center and by far Germany's biggest port. In the medieval era, Hamburg helped found the merchant group called the Hanseatic League, allying with Lübeck in 1241. In the modern era, Hamburg connected Germany with the maritime world and many in the city were reluctant to join the German empire in 1871, seeing it as a tool of continental Prussia. Hamburg did break briefly with Germany after World War I, becoming a short-lived socialist republic. The city has burned twice, in 1842 and 1943, the latter a result of Royal Air Force raids, leading to the deaths of tens of thousands of inhabitants. The city has rebuilt in a modern, stylish manner.

The Rhine

The Rhine rises in Switzerland and enters Germany downstream of the Swiss river port of Basle. The Rhine links three major physiographic regions of central Europe: the Alps, the Central Uplands, and the Northern Plain, and provides a routeway through them (Figure 8.3). Between France and Germany, the river flows in the rift valley between the fault blocks of the Vosges and the Black Forest. At Mannheim, the scenic Neckar flowing through Stuttgart and Heidelberg joins the Rhine. (Photo 8.3) Then the Main from Frankfurt enters before the Rhine cuts through the Hunsrück upland to historic Koblenz (Photo 1.9) where the Mosel, from eastern France, makes a confluence with the main stream. Downstream of Koblenz, the Rhine cuts a gorge between the Eifel and Westerwald mountains, before entering the Northern Plain at Bonn and flowing to Cologne. The river passes Düsseldorf and Duisburg (where the river Ruhr enters) before swinging west to enter The Netherlands at Arnhem. The Rhine receives many tributaries and enjoys an even flow throughout the year. It is the most important river trade route in western and central Europe (Elkins, 1972, p. 40).

Location of Cities and Economic Activity

At German unification in 1871, the country included many important cities that had been state or regional capitals performing administrative and commercial functions. Berlin dominated the hierarchy, but Hamburg was by far the largest port. In the Rhineland, Cologne, Frankfurt am Main, and Düsseldorf were prestigious cities with a Rhinelander view of the world, which was different from the predominant Prussian continental perspective in Berlin. The hierarchy of German cities in 1880 is shown in Table 8.1. Berlin, which had a population of 900,000 in 1871, has grown rapidly to over a million people, benefiting from its role as the administrative center of the enlarged state.

After unification German industrialization accelerated. With industrialization came rural-to-urban migration. By 1900, Germany possessed 22 cities with over 150,000 inhabitants. Hamburg grew rapidly as a shipbuilding center and a port exporting German products and importing raw materials and tropical produce.

In 1900, none of the industrial cities of the Ruhr region was in the list of top twenty towns, which is dominated by old regional centers (Table 8.1). Düsseldorf is on the list on the basis of a long-standing importance as a Rhine port and commercial center. Notice the number of cities in the list that are no longer part of Germany, including Strasbourg (in France), Stettin (now Polish Szczecin), Breslau (now Wrocław in Poland), and Königsberg, today an exclave of Russia named Kaliningrad.

By 1935, (Table 8.2) several of the major Ruhr industrial centers had more than 300,000 inhabitants, including Essen, Dortmund, and Duisburg, which links navigation

Photo 8.3 Heidelberg on the river Neckar. The Neckar joins the Rhine at Mannheim.

Source: Age fotostock/SUPERSTOCK.

on the Rhine to the river Ruhr. In 1955 (Table 8.2), Germany consisted of West Germany and East Germany. Notice that of the top listed cities, apart from Berlin, few are in the east, and at Leipzig and Dresden population numbers are below prewar levels. In West Germany, by the mid-1950s, in spite of blitzing in World War II and deaths as a result of street fighting, urban population numbers exceeded the 1935 figures. Cologne is an exception. West German economic and urban recovery was rapid. Not so in the east. In 1935, Leipzig was the fifth-ranked city in Germany by population. Dresden was seventh. In 2000, Leipzig was thirteenth and Dresden fifteenth, a commentary on the decline of East German cities in the decades after World War II (Table 8.3). When Berlin was put back together again in 1989/90, it was easily the largest city in the new Germany, but from the 1950s to the late 1980s the total population of the city was in slow decline.

Berlin

Berlin reflects the rise of Prussia, the German Empire, the emergence of the Weimar Republic and Nazi Germany. Today the city mirrors modern Germany, the country with the largest economy in the European Union. Berlin is not on a major river or a great

TABLE 8.1 German Cities 1880 and 1900

City	1880	1900
Berlin	1,120,000	1,890,000
Hamburg	410,000	710,000
Breslau (Today Wrocław in Poland)	272,000	422,000
Munich	230,000	500,000
Dresden	222,000	400,000
Cologne	145,000	373,000
Frankfurt am Main	137,000	290,000
Magdeburg	137,000	230,000
Hannover	123,000	240,000
Stuttgart	117,000	180,000
Leipzig	115,000	460,000
Danzig (Today Gdańsk in Poland)	109,000	141,000
Nürnberg	99,000	261,000
Düsseldorf	95,000	220,000
Stettin (Today Szczecin in Poland)	92,000	211,000
Chemnitz		207,000
Königsberg (Today Kaliningrad in Russia)	141,000	190,000
Bremen	112,000	164,000
Halle	71,000	157,000
Strasbourg	105,000	151,000 (Returned to France after WWI)

Source: Statesman's Yearbook

natural routeway, nor does it possess a raw material base. The surrounding agricultural region is not particularly fertile, the better drained areas being suited to growing rye and potatoes. Why is it the capital?

The original settlements were beside the river Spree. On an island in the river was the fishing village of Kölln. On the north bank was the small settlement of Berlin. The settlements received town charters in the thirteenth century and merged in 1307 to form the *Altstadt* of Berlin. Association with the Hanseatic League improved trade in the fifteenth century, and the town became the seat of the electors of Brandenburg in 1486. In the seventeenth century Berlin was still a small provincial town but fortifications were built and in the following century, with the rise of Prussia, the city grew as a military and administrative center as the Prussian kings established a prestigious court, erected stylish buildings, and laid out Potsdam. By 1780, the city contained 100,000 people.

During the Napoleonic wars, Prussia was defeated, for a time, and Berlin was occupied by French troops from 1806 to 1808. However, Prussia ended the war on the winning side and gained additional territory at the Congress of Vienna. Between 1780 and 1820 the population of Berlin doubled, and by 1860 the city contained half a million people. The wars of the 1860s and the creation of the German Empire were good for Berlin, which attained a population of over one million in 1880 and two million at the beginning of the twentieth century.

Much of the urban population growth, in the second half of the nineteenth century, derived from Berlin's role as the administrative and military core of Prussia and then the

TABLE 8.2   Leading German Cities, 1935 and 1955

City	1935	1955
Berlin	4,200,000	3,300,000 (divided into occupied zones)
Hamburg	1,100,000	1,800,000
Cologne	750,000	700,000
Munich	730,000	950,000
Leipzig	710,000	620,000 (East Germany)
Essen	650,000	680,000
Dresden	640,000	500,000 (East Germany)
Breslau (Today Wrocław)	620,000	370,000 (Wrocław in Poland)
Frankfurt-am-Main	550,000	570,000
Dortmund	510,000	610,000
Düsseldorf	490,000	630,000
Hannover	440,000	520,000
Duisburg	440,000	470,000
Stuttgart	420,000	590,000
Nürnberg	410,000	410,000
Wuppertal	410,000	400,000
Chemnitz (Karl-Marx-Stadt, 1953–1990)	350,000	290,000 (East Germany)
Gelsenkirchen	330,000	360,000
Bremen	320,000	490,000
Königsberg (Today Kaliningrad)	310,000	202,000 (Kaliningrad in the Russian S.S.R.)
Bochum	310,000	340,000
Magdeburg	300,000	270,000 (East Germany)
Halle	209,000	290,000 (East Germany)
Mannheim	275,000	285,000
Stettin (Today Szczecin in Poland)	272,000	230,000 (Szczecin in Poland)

Source: Statesman's Yearbook

German Empire. There is another element in Berlin's growth. Throughout the nineteenth century, Berlin became increasingly important as a center of education in the humanities, sciences, and technical training. Berlin University was founded in 1809. When the second industrial revolution (the drive to maturity) arrived in the second half of the nineteenth century, Berlin was well placed to take the lead in technological industries such as electrical engineering. The way the elements of government, the military, and technology came together is illustrated by the career of Werner Siemens (1816–1892). A son of a farmer, Siemens could not afford an education at a university or school of technology. He joined the army. The training in artillery and engineering provided a good scientific education and Siemens became familiar with telegraph systems in the Berlin artillery workshops. In 1847 Siemens, with a partner, started a factory to make telegraphic equipment and built the first underground telegraphic line in 1849. For obvious reasons the German military was interested in communications, and orders for telegraphic cable and equipment were acquired. Siemens invented a dynamo, built an electric locomotive, and used electricity to power industrial equipment. Other major producers of electrical equipment, including

TABLE 8.3   Top Twenty German Cities by Population, 2000

Berlin	3,400,000
Hamburg	1,700,000
Munich	1,200,000
Cologne	900,000
Frankfurt am Main	650,000
Essen	600,000
Dortmund	590,000
Stuttgart	580,000
Düsseldorf	570,000
Bremen	540,000
Hannover	520,000
Duisberg	510,000
Leipzig	490,000
Nürnberg	480,000
Dresden	470,000
Bochum	390,000
Wuppertal	370,000
Bielefeld	320,000
Mannheim	310,000
Bonn	300,000

Source: Statesman's Yearbook.

AEG (*Allgemeine Elektrizitäts Gesellschaft*), which specialized in electric lighting and electricity-generating plants, also developed in Berlin. The city became a leading center of technological development through to World War II. New forms of manufacturing developed to supply the large and growing consumer markets in Berlin and other centers of population to which the city was well linked by railroads, canals, and river transport. Although Berlin grew rapidly in population numbers, it remained compact, with much of the population living in tenements with two-room family apartments.

Berlin continued to grow in population, manufacturing capacity, and administrative functions early in World War I. As the war progressed, birth rates dropped and death rates rose. Social unrest developed as defeat approached. The Kaiser went to The Netherlands for political asylum and abdicated. Workers councils attempted to take power after the armistice of November 1918. In early January 1919, the Spartacist Uprising resulted in street fighting, which troops put down.

The early years of the 1920s in much of Germany were marked by unemployment, food and fuel shortages, culminating in the great inflation and economic disruption of 1923–24, when the currency became nearly worthless. By 1925, greater Berlin contained four million people and the city underwent an amazing cultural renaissance in theater, literature, music, painting, and cinema. More than London or Paris, Berlin was a magnet for artistic talent, and Berliners supported the theaters, orchestras, and operas. Something of the atmosphere of the city in the late 1920s and early 1930s is portrayed in the novels of Christopher Isherwood, including *Goodbye to Berlin* (1939).

By the end of the 1920s, Berlin was in political turmoil as the communists and the fascists battled each other and the police. The Nazi Party gained supporters, not just among disaffected working people, who were out of work and impoverished again in the Depression, but amongst students and professors in Berlin's universities. Hitler came to power in 1933.

Under the Prussian rulers and the Kaisers, Berlin had developed elements of the absolutist city (Chapter 4) with monumental buildings and neoclassical triumphant structures like the Brandenburg Gate (1789–93). Now Hitler, commanding a team of architects led by Albert Speer, wanted Berlin to be the world capital, *Germania*. There was to be an extensive new north–south thoroughfare, deliberately grander than the Champs Élysées, with a triumphal archway, fifty times larger than the Arc de Triomphe. New buildings were to house the ministries of state, and a huge, domed Hall of the People, would dwarf all historic buildings in the city. Few of the plans were implemented and the mayor of Berlin was against many of them because they ignored the scale and character of the existing city. Speer and his team did erect some buildings including a new Reich Chancellery (1938), built in nine months, for Hitler to move into. Of the Nazi buildings, the most prominent survivor is the stadium built for the 1936 Olympics. The final of the 2006 World Cup was played there. (Photo 8.4)

At the beginning of World War II, Berlin was little inconvenienced by hostilities. The early German campaigns were *Blitzkrieg*, lightning wars after which the economy reverted to normal production. German living standards remained at a good level until 1944, and the population did not suffer the hunger and hardship of occupied Europe until Germany itself was occupied and defeated.

At first, although bombed, Berlin was protected from sustained attack by distance. Berlin was far to the east, and when allied bombing increased in 1942 the major targets were in the west, a relatively short distance from bases in eastern England. Rhine

Photo 8.4 The Olympic stadium at Berlin was one of a few fascist buildings to survive the Nazi era.

Source: Marko/AgeFotostock.

cities like Cologne were targeted as were Ruhr industrial centers, including Essen and Dortmund.

In 1944, Berlin became a major target but the greater distances from Britain allowed raids to be plotted and projected, defenses to be organized, and fighter squadrons deployed. The longer distances meant heavy fuel loads, more air time for malfunctions to develop, and fewer escorting fighters. In spite of heavy bomber losses, production in Berlin was damaged and companies like Siemens were forced to decentralize production to smaller, safer towns and cities. It was not until late in 1944 that armaments production declined as a result of Allied action.

Bombing did not destroy the fabric of Berlin. The Red Army artillery barrage of 40,000 shells and the house-to-house, street-to-street fighting reduced Berlin to rubble piles in the spring of 1945. In 1939, Berlin had a population of 4.3 million people. When a census was taken in 1946, 3.2 million inhabitants were recorded in the city.

Postwar the city was divided into East and West Berlin. In 1948–49 the Soviets blockaded West Berlin and supplies were flown in by allied planes. The erection of the Berlin Wall, starting in 1961, was a continuing cause of tension until it fell in 1989.

In the postwar era, East Berlin was part of the Soviet bloc and reconstruction was guided by socialist planning principles. On the Unter den Linden, running west from the River Spree to the Brandenburg Gate, buildings were restored, demolished, renamed, or reinterpreted to reflect a different ideology, as happened in "hundreds of cities across central and eastern Europe during the Cold War" (Stangl, 2006, p. 371).

The uncertainty surrounding Berlin hampered investment. Companies like Siemens moved manufacturing to other cities. In East Berlin, factories were often deprived of equipment to supply the Soviet Union with war reparations. BMW had a factory at Spandau, a suburb of Berlin. Postwar car production was concentrated in West Germany.

When the Wall came down, Berlin entered a new phase of urban development. The capital moved from Bonn on the Rhine to Berlin. New ministries and residential areas were rapidly built. (Photo 8.5) Territory that had formed the strategic zone between the two Berlins became available for construction, often obliterating the sites of tragic events where East Berliners had died trying to reach the West. (Photo 8.6)

Much building and rebuilding has taken place since unification but the visitor is powerfully reminded of Berlin's past in museums, walking tours, and the memory district, which holds the Holocaust memorial, the Topography of Terror space including a small segment of the Berlin Wall, Checkpoint Charlie, and the Jewish Museum (Till, 2005).

Once again, Berlin is flourishing as a political and cultural center with museums, art galleries, operas, orchestras, theater, and literature. Berlin has reclaimed its place as a major European city. (Table 8.4)

Hamburg

Hamburg, Germany's second largest city, has a different history and culture from Berlin. Hamburg was an important medieval port, a member of the Hanseatic League (Chapter 4) along with Bremen, Cologne, Lübeck, and Danzig. In the nineteenth century, Hamburg became a focus of overseas trade, a shipbuilding center, a processor of imported raw

Photo 8.5 A view of the Potsdamer Platz featuring many new buildings. The former line of the Berlin Wall is top center.

Source: Daniel Karmann/dpa/Landov LLC.

TABLE 8.4 Berlin Population Numbers

1800	172,000
1820	200,000
1850	419,000
1860	548,000
1870	826,000
1880	1,122,000
1890	1,579,000
1900	1,889,000
1910	2,071,000
1920	3,801,000
1930	4,243,000
1939	4,332,000
1946	3,200,000
1950	3,337,000
1960	3,261,000
1970	3,208,000
1980	3,057,000
1990	3,438,000
2000	3,382,000

Photo 8.6 Commemorative plaques at a remnant of the Berlin Wall. Many East Berliners died attempting to cross the wall to get to the West.

Source: Bernd Settnik/dpa/Landov LLC.

materials, and an entrepôt importing goods that were sold in European markets. Hamburg employed free trade, did not join the *Zollverein,* and, although part of the German Empire in 1871, insisted on staying outside the Prussian-dominated tariff system. Hamburg became more committed to the German Empire when Germany acquired overseas colonies and launched a navy. Hamburg shipyards built many of the cruisers and dreadnoughts. World War I stopped the port's overseas trade for several years.

In World War II, trade was again disrupted, and in 1943 the central city was destroyed in a huge Royal Air Force bombing raid. Few buildings survived apart from historic St. Nicholas Church. The city was rebuilt in a modern style, reclaiming its role as Germany's major port and commercial center. (Photo 8.7) With Berlin isolated in East Germany, many major corporations located headquarters in Hamburg. In recent years, the rivalry between Berlin and Hamburg has revived as they try to poach corporations from each other.

Summary

West Germany had a rapid recovery after World War II. The term "economic miracle" (*wirtschaftswunder*) is overused, for there was nothing miraculous about German reconstruction. Germany's cities and industries had been bombed and shelled but German expertise in science, technology, engineering, and construction survived. After the war, there was a pressing need to rebuild. Britain and the United States, in their occupation zones, rapidly restored economic activity. As Europe recovered, there was sustained

Photo 8.7 The port of Hamburg. The central city is all modern buildings with the exception of the churches, the only structures to survive the 1943 bombing.

Source: NewsCom.

demand for German goods. With the Cold War, and the perceived Soviet threat, NATO needed a strong West Germany with the capability to withstand attack from the Soviet bloc. West Germany was rearmed and brought into NATO in 1955.

In the postwar decades, West Germany outdeveloped stagnating East Germany. Poor economic performance in the East was partially a result of a cumbersome central economic planning system but the western part of Germany with the Rhine and the Ruhr was the more productive part of the country before the war, with the exception of Berlin. Postwar Berlin was isolated, blockaded, and divided.

Germany has been more successful than Britain and the U.S. in retaining a home-based manufacturing sector, although the strength of the sector has been threatened in the last decade as companies have moved operations to cheap labor regions. Workers in the manufacturing industry in Germany have high productivity and good working conditions. The work week is short and the number of vacation days large, compared with U.S. standards. The state provides an excellent social welfare safety net. Inevitably with high living standards, good fringe benefits, and a generous system of state support, manufacturing costs in Germany now appear high against those in eastern Europe and China.

Much German investment in manufacturing went into eastern Europe before the eastward EU expansion of 2004, and the trend continues. German unions have advanced the case against the transfer of manufacturing to China, or other parts of Asia, pointing out that labor is only one of the costs involved. The administrative costs of entering the Chinese economy are high, as a large state bureaucracy issues numerous permits without which it is impossible to operate. Acquiring the permits is costly and takes more than one year. Then a plant has to be built and personnel trained. As the unions point out, when you move you leave behind the accumulated knowledge and skills that have been responsible for the success of the manufactured product in the first place. Unions have commissioned studies portraying the full costs of transfer to China, and corporations have studied the results.

In Germany, life expectancy is high and infant mortality low but the total fertility rate does not reach replacement level. On the natural increase account the population of the country will shrink. Admitting immigrants is not popular because unemployment exceeds 10 percent and, in the past, it has been difficult to integrate newcomers, particularly workers from Turkey. Germany will attract migrants from eastern Europe and inhabitants of the new EU members will eventually have right of entry under the community's free movement of labor provisions.

Germany and EU

After World War II, the European movement, backed by leaders like Chancellor Adenhauer, allowed Germany to regain a place in Europe when the country might have been shunned after the Nazi collapse.

Economically, Germany, with large corporations and combines, is well suited to operate in the enlarged European market. German-manufactured products have done well in European markets and with a larger "home" market have a better base to compete in world markets.

EU has allowed Germany to exercise peacefully the economic and political influence it always wanted in *Mitteleuropa* (Middle Europe). German corporations have been leading investors in eastern Europe, since the collapse of communism and the integration of the region into EU.

▶ ITALY

Introduction

Italy, like Germany, was late to unify, doing so in the decade between 1860 and 1870. The pieces that came together to create the country had differing characteristics. Much of the North had been under Austrian rule. The popes had ruled central Italy and did not relinquish Rome to become the capital of Italy until 1870. The South, consisting of Naples, Sicily, and the lands between, was a Bourbon Kingdom, the area where Garibaldi launched the insurrection that linked with Piedmont and Lombardy, led by Cavour, to create the Kingdom of Italy in 1861. Veneto (1866) and Rome (1870) were later additions (Figure 8.4).

The new country was linguistically diverse, with each region having a well-developed dialect. In Sicilian *bello* became *beddu* and regions had so many dialect words that there were, for example, dictionaries of Milanese, Genoese, and Neopolitan. Florentine became the received pronunciation and national broadcasting systems have reduced the power of regional tongues.

Figure 8.4 Regions of Italy.

Italian unification created an extensive state, but the country did not possess a good mineral resource base. There was some iron ore, mercury, bauxite, and a few other minerals, but coal, the energy source of the nineteenth century, had to be imported. Political unification did not bring economic integration. Transport costs were high as a result of distance and rugged terrain in peninsular Italy. In the South, poverty was widespread and there was little demand for goods manufactured in northern cities, like Turin and Milan.

In spite of a poor resource base, limited infrastructure, and an incomplete manufacturing sector, Italy soon aspired to play as a power in international affairs. In 1882,

Italy entered into the Triple Alliance with Germany and Austria. Spending on armaments increased and the country developed imperial ambitions, establishing itself on the Red Sea in what became Italian Eritrea and attempting, disastrously, the conquest of Abyssinia (Ethiopia) in 1895. Early in the twentieth century, Italy was more successful in North Africa, getting Turkey to recognize Italian control over Tripolitania and Cyrenaica (modern Libya) but the local Senussi did not accept Italian rule.

In World War I, Italy left the Triple Alliance, joined the western allies and under the secret Treaty of London, in May 1915, was promised Austrian territory by Britain, France, and Tsarist Russia. President Woodrow Wilson was not bound by the secret Treaty of London and wanted Italian boundaries settled on ethnogeographic principles. Under the St. Germain peace treaty with Austria (September 1919), Italy did receive German-speaking areas in the Tyrol and Alto Adige, up to the Brenner Pass, but was disappointed on the Adriatic front, although the port of Trieste, with a substantial Italian-speaking population, did become part of Italy at the end of World War I.

Mussolini and Fascism

When Mussolini became Prime Minister in 1922 and established a full fascist regime (1928–43), he advanced foreign policy aims that could only come at the expense of Italy's World War I allies, France and Britain. Mussolini wanted to turn the Mediterranean into an Italian sphere, *Mare Nostrum*. Italians were told that Nice, Corsica, Tunis, and Malta belonged to them. Further Italy should have a role in running the Suez Canal, owned by a French company in which the British government was a large shareholder. Italy attacked Abyssinia (Ethiopia) late in 1935.

Mussolini realized that his territorial aims could be met if Hitler broke the power of France and threatened Britain. He declared the Rome–Berlin Axis in 1936. The Pact of Steel was signed with Germany in 1939, and Italy seized Albania, on the other side of the Adriatic, at Easter the same year. When Germany attacked Poland on September 1, Mussolini declared Italy a nonbelligerent but, as France fell in June 1940, "Hitler's jackal" entered the war. Italy did not have the raw material base and the industrial output to sustain a long war.

Italian industrialization began late, and it was not until the period 1890–1910 and the generation of hydroelectricity in the alpine valleys that modern units of production appeared in textiles, steel making, fertilizers, electrical and railroad equipment. Modern, industrial plants were concentrated in the Genoa, Turin, Milan triangle.

When Mussolini came to power, he believed in corporatism, with the state guiding overall economic policies and having a capital stake in industries. In the running of state corporations employees would have an input. The policy of state involvement was accelerated by the establishment of IRI, *Instituto per la Ricostruzione Industriale,* which held a capital stake in many corporations and encouraged the expansion of heavy industry. Mussolini aimed to make Italy as self-sufficient as possible. Land, including the malarial Pontine Marshes in the Tyrrhenian coastlands near Rome, were reclaimed and agricultural output increased. But self-sufficiency (autarky) was impossible for Italy because of the deficient resource base. By September 1943, Italy had surrendered, although German forces remained in control of north and central Italy.

Postwar Italy

At the end of the war, Italy lost territory in the northeast to Yugoslavia. The monarchy, retained under Fascism, was replaced. The electorate voted for a republic. Italian politics remained unstable after World War II, with one coalition after another attempting to govern the country for short periods of time. The political scene has rarely been stable through to the present. There may be some truth in Mrs. Thatcher's unkind view that Italians were keen on European institutions because they could not govern themselves, for the electoral system of proportional representation resulted in large numbers of parties and compromise coalitions that frequently split. Modification of proportional representation has brought more stability but has not eliminated political scandals.

Politics were volatile in the postwar years but economic recovery was rapid. By 1952, industrial production was 40 percent above the figure for 1938, and through the 1950s industrial output rose 8 to 9 percent per annum. A transformation was beginning. In 1950, over 40 percent of the workforce was in agriculture. In regions like the Abruzzi and Basilicata, approximately three quarters of the working population was making a poor living in inefficient, unprofitable agriculture that produced few cash crops (Figure 8.4). Efforts were made at land reform with some success with the establishment (1951) of the *Cassa per il Mezzogiorno* to provide funding for development in the South. The term *Mezzogiorno* literally translates as mid-day referring to the intensity of the sun at noon. The best individual cure for southern rural poverty was to leave the land, and move north to find work in Turin, Milan, Genoa, or Venice. Italy had an asset, a large pool of relatively cheap labor.

After World War II, the South (*Mezzogiorno*) still had third-world characteristics. Agriculture was small scale and unmechanized, incomes were below poverty levels, and infant mortality was high. People died from malaria. Roads in southern Italy and Sicily were poor, winding along the coast and dropping down steep valley sides to small seaside settlements. The rail system linked the region to central and northern Italy but south of Naples the roads did not provide the infrastructure for economic development.

In the North, industrial growth was rapid in the 1950s and 1960s. With labor costs low, Italian scooters, cars, agricultural equipment, typewriters, chemicals, and consumer goods were competitive. Entrepreneurs from Britain found that household goods, like washing machines, could be made in Italy and marketed cheaply but profitably in the United Kingdom.

Italy joined the Coal and Steel Community (1952) and the EEC (1957). The free movement of labor encouraged Italians to work in the expanding German economy, and the EEC provided a larger market for competitive Italian products.

The Coal and Steel Community created a fair market, within EEC, for iron and steel products. In the nineteenth century, iron and steel plants were located on an orefield or a coalfield. In the second half of the twentieth century, plants were built in coastal locations where raw materials could be cheaply assembled. Large integrated steel plants were built at Genoa, Naples, and Taranto. Today, Italy produces over twenty million tons of steel annually. Within EU, Italy is the second largest producer after Germany.

As prosperity grew, the South began to experience economic growth. Labor was cheap and some companies, including Alfa Romeo, set up plants in southern cities like Naples. The Autostrada system was completed down both sides of peninsular Italy and a Messina Strait bridge, to link Sicily to the mainland, is talked about. All over Italy, including the South, vegetables, flowers, fruit, and grapes are grown and sold into European markets. In this activity the *Mezzogiorno* has advantages, for the further south an operation is, the longer the growing season, although transportation costs more.

Regions known previously only for country wines, including Sicily and the Abruzzi, have become producers of low-priced wine that can be sold into world markets. Italy is one of the world's largest producers and exporters of wine. The forces of globalization have opened distant markets. Many relatively small Italian corporations have been quick to produce fashionable consumer products, including clothing, shoes, ceramics, jewelry, and furnishings that fill market niches.

Rapid economic growth brought inflation. To keep the price of Italian exports competitive, the Lira was frequently devalued. Inflation was to be contained by adoption of the Euro. The Euro slowed Italian inflation but production costs still rose, and with the increasing value of the Euro, Italian goods became expensive in world markets. Historically, when prices rose, Italy devalued. Now Italy cannot devalue the Euro nor can it make money "cheap" by lowering the interest rates set by the European Central Bank. Today, many in Italy, including cabinet members, question Italy's adoption of the Euro.

Tourism

Since the eighteenth century, Italy has been a tourist destination. Italy was on the grand tour by which the wealthy from northern Europe, particularly the British, visited the landscapes of Rome and purchased Italian art (Black, 2003). The large number of Italian paintings in British art galleries and country houses is a product of this era. By the second half of the nineteenth century, prosperous families were going for extended stays in the Italian lakes, the Dolomites, Tuscany, and Florence. Resorts such as San Remo and Rapallo, along the coast from the French Riviera, became popular for winter stays. (Photo 8.8) The Ligurian Alps cut out the cold north air and the sun shines on the south-facing coast. The mean monthly temperature for January at San Remo, on the Italian Riviera, is 46°F, the same figure as Naples hundreds of miles further south.

Mass market tourism was established by 1960. As European prosperity grew and discretionary income increased, growing numbers of working people had the means to buy "package" holidays. Tour companies block-booked rooms in purpose-built hotels and chartered aircraft to fly visitors to seaside locations. Bulk buying of rooms and airline seats, together with high occupancy rates, brought the costs of vacations down. Brits, Scandinavians, and Germans bought holidays in the sun. All of Mediterranean Europe benefitted economically from this north-to-south tourism, but Italy was the first to achieve substantial foreign currency earnings from the trend. Of course, many who did not want a mass market package holiday still went to Florence, Rome, Ravenna, Naples with Vesuvius, and Taormina (Sicily) with Etna. For the affluent, a vacation in a Tuscan villa is the desired summer experience.

The Climates of Italy

In the north, the Italian Alps rise to over 15,000 feet. Latitudinally, the country extends north to south over 10 degrees of latitude. Climates vary from alpine to subtropical, and in the southern summer, with the *Scirocco* blowing up from North Africa, it can be extremely hot.

Photo 8.8 Camogli, Italian Riviera town in Liguria. On many Mediterranean coasts, high density development overcrowds the landscape.

Source: Travelshots/AgeFotostock.

Typical Mediterranean climates with winter rainfall and summer drought are predominant in peninsular Italy, although inland, on higher ground, winters can be cold. In northern Italy there is rainfall in all months and in the Po Valley, which has enough landmass to create continental conditions, winters are cold and damp. Frosts are frequent.

Southern Italy and Sicily have a Mediterranean climate with mild winters and hot summers, but inland and upland, winter can be severe, as in the Abruzzi Mountains lying between Rome and the Adriatic Sea. In Sicily, Mount Etna, rising to 10,902 feet, is snow covered at the peak throughout the year.

In alpine environments in northern Italy, the sequence from valley floor to summit is cultivated lands, coniferous woodland, summer pasture, scree, bare rock, and snow and ice at the higher levels. A similar sequence is found in the French, Swiss, and Austrian Alps.

The Alps are an important factor in the climates of Italy. The mountain range is generally high enough to keep out much of the cold continental, central European air in winter. Around Italy's northern lakes, in sheltered south-facing valleys where the water bodies moderate temperatures, vineyards are widespread, and in favored places fruit-bearing citrus trees are grown. In the east, the Alps are not continuous enough to keep out all cold winter air. If there is a low-pressure system over the Adriatic Sea and a high-pressure pool of cold air over central Europe, a north wind, the *Bora,* the Italian equivalent of the French *Mistral,* blows north to south down the Adriatic.

Regions of Italy

Italy can be divided into five major regions.

1. The Alps
2. The Northern Plain largely drained by the river Po
3. Peninsula Italy with the Apennines and usually narrow coastal plains
4. Sicily
5. Sardinia

The Alps

The Alps form an arc running from Slovenia (formerly part of Yugoslavia) westward along the boundaries with Austria and Switzerland before turning south to form the Maritime Alps and the border with France. The mountains are highest in the west, and broadest in the north central sector, where extensive, south-facing, glaciated valleys are filled with finger lakes. The Italian lake district includes lakes Maggiore, Lugano, Como, and Garda. (Figure 8.5)

Geologically, the mountains contain more crystalline and metamorphic rocks in the west. Sedimentary rocks are best developed in the east, as in the limestone Dolomites where the rock has weathered into spires and, underground, dissolved to form caverns and caves.

The Alps surrounding northern Italy appear to be a barrier but there are numerous valleys leading up to passes that cross the mountain range. The valley of the Dora Riparia, a tributary of the river Po, cuts into the mountains west of Turin and leads to the Mt. Cenis Pass and France, with road and rail tunnels now speeding travel. The Dora Baltea, which flows southward from the mountains to join the Po, is a routeway to the St. Bernard Pass and Switzerland. Traveling north from Milan through the Italian lake district, we enter Switzerland and then cross the St. Gotthard Pass. Historically this has been an important routeway linking Italy to Switzerland and the Rhine, as has the Simplon Pass route into Switzerland, reached by traveling northwest from Milan.

The best-known pass into Austria is the Brenner. Traveling north from Verona, up the valley of the Adige, we reach Trento and then Bolzano (Bosen to the German-speaking inhabitants) in the Dolomites, before climbing over the Brenner into Austria and reaching Innsbruck. A 40-mile-long railroad tunnel is being dug under the Brenner from Innsbruck in Austria to Fortezza in Italy in an effort to reduce polluting road traffic, which fouls the alpine air. Historically, the alpine passes allowed merchants to trade the products of northern and central Europe for Mediterranean goods, including spices, coming to ports like Genoa and Venice from Asia.

The Northern Plain

Running nearly 300 miles from the Alps in the West to the Adriatic Sea in the east is the plain of Lombardy, drained largely by the river Po. Tributaries of the Po enter the plain from the Alps, to the north and west, and from the Apennines lying on the southern flank of the lowland (Figure 8.5).

The river Po fluctuates in level, being full of water in the spring and early summer, as snow melts in the mountains, but lower in late summer and fall. The river is used to

Figure 8.5 Italy.

generate hydroelectric power but is only navigable by smaller craft. Close to the river the plain is low lying and flood prone. The Po carries a sediment load, which is deposited in a delta extending out into the Adriatic Sea. Venice lies north of the Po delta on and around a group of islands and lagoons. South of the Po delta, the coast has extended into the Adriatic as a result of deposition. Ravenna, a Roman and Byzantine port, is now miles from the sea.

In the east of the plain, settlement is often linear along roads or embankments on low-lying land deposited by the river. Because of the flood risk there are only a few towns on the banks of the river Po. Settlements tend to be back from the water on low river terraces. Piacenza is an exception.

Cities on the plain form two lines. In the north, Turin, Milan, Bergamo, Brescia, Verona, and Vicenza have situations on the north–south routeways leading to alpine passes. In the south, on the line of a Roman road running inland from Ravenna, the towns include Bologna, Modena, Reggio nell'Emilia, Parma, and Piacenza at a crossing point on the Po. All the towns on the old Roman road are close to valleys running into the Apennines that provide routeways over the upland to Tuscany, Florence, and Rome.

The northern plain is the most extensive area of agricultural land in Italy. Close to the Po, on wetter ground, pasture is plentiful. Higher on the river terraces, wheat and corn are common. Parts of the flood plain support rice paddy fields. Tree crops include mulberry to provide fodder for silk worms and fruit for humans. Vineyards are important on higher ground, and in the west there are wine-producing areas including Alba, Asti, and Barolo, famous for full-bodied red wines. In the east, around Venice, is a large wine-producing region.

Peninsular Italy: The Apennines, The Coasts, and The Coastal Plains

The Apennines run from the border with France, along the shore of the Ligurian Sea, and then trend southeastward down the Italian peninsula through Tuscany, Umbria, the Abruzzi, and Calabria to Sicily. The Apennines are highest and widest in the region that lies between Rome and the Adriatic Sea. The high peaks include Monte Amaro (9,170 feet), Monte Greco (7,490 feet), and snow-capped Monte Corno (9,560 feet) in the Abruzzi region.

South of the Abruzzi, the Apennine summits are lower and the mountains break into separate masses, as on the Gargano peninsula, projecting into the Adriatic Sea. In Calabria, the toe of Italy, the Apennines trend southward before turning west to form a range along the north coast of Sicily. The Apennines consist predominately of sedimentary rocks of Triassic, Jurassic, and Cretaceous age pushed up in the alpine orogeny. Limestones are commonly exposed in characteristic Mediterranean landscapes. In the toe of Italy, older granite rocks are exposed at the surface.

As we come south down peninsular Italy, we enter the zone where the European and African plates come into contact. Plate movement and tectonic activity have caused changes in sea level. Raised shorelines are evident. Earthquakes and volcanic activity are present. In Umbria, there are a number of *calderas,* lakes formed in the craters of volcanoes. The best known example is Lago di Bolseno. Vesuvius, near Naples, is famous for overwhelming Pompeii and Herculaneum in 79 A.D. and has rumbled and erupted in the last half century. In eastern Sicily, Mt. Etna (10,902 feet) is an active volcano, although at times it is possible to climb the cone and look into the crater. The last major eruption, with lava flows threatening towns and villages on the flanks of the volcano, was in 1981.

North of Sicily in the Tyrrhenian Sea are the Eolie islands, including Isola Stromboli, Vulcano, and Lipari itself. Stromboli is in an almost continuous rumble, puffing out gases and debris several times a day. (Photo 8.9) Lipari was a major source of the obsidian used to make cutting tools in the Mediterranean Neolithic. Obsidian fractures with sharp edges. The rock was widely traded and, as the Lipari obsidian is chemically distinctive, trade routes can be mapped, pointing to wide Neolithic trade networks.

Earthquakes are common in the same region as the volcanoes. The Messina earthquake of 1908 destroyed the town. The quake produced a tsunami, the effect of which was made worse by the narrow straits of Messina that separate Sicily from Calabria in peninsular Italy. Further south in Sicily, an earthquake in 1693 destroyed the towns of Noto and Catania and did widespread damage in surrounding areas. An earthquake in 1980 near Naples killed over 3,000 people and left many homeless.

There are few extensive coastal plains in peninsular Italy. Frequently, the hills come to the coast, creating rugged shorelines.

The West Coast The coast running from the French border around the Gulf of Genoa to beyond La Spezia is a Riviera coast with the Maritime Alps and the Apennines meeting the sea in cliffs. There are few major inlets and most settlements are built in coves under

Photo 8.9 An aerial view of the Stromboli Island volcano during eruptive activity. The volcano erupts frequently, but people live permanently on the island. Notice the habitations bottom right.

Source: Protezione Civile Itlaia/©AP/Wide World Photos.

the cliffs, as at Portofino. From west to east along the coasts, the major places are San Remo, Savona, Genoa, Rapallo, and La Spezia. Genoa is a major European port and manufacturing center. Genoa does have a natural harbor but huge offshore breakwaters and quays have been built to protect vessels calling at the port. Behind Genoa there is a relatively low pass over the Ligurian Mountains to the plain of Lombardy, Milan, and the Alpine routeways into central Europe.

La Spezia, in a bay, serves as a port, naval base, and shipbuilding center. Southeast of La Spezia is the Carrara marble country, the material Michaelangelo used for his carving of *David* in Florence. The rock is still exploited and exported as a high-quality building material

The Arno Valley In narrow peninsular Italy there are few navigable rivers. An exception is the Arno, which flows from the Apennines through Tuscany to the Tyrrhenian Sea. In medieval times the Arno was navigable to Pisa and Florence, close to the head of navigation on the river. From Florence, routeways pass over the Apennines to Bologna and the northern plain. (Photo 8.10)

Downstream of Florence, the Arno valley broadens into a fertile lowland. Upstream, in the Apennine hill country, livestock raising is predominant. Down the valley, there is

Photo 8.10 Leaning Tower of Pisa with the Duomo (cathedral) behind.

Source: Age fotostock/SUPERSTOCK.

still plenty of pastureland but the main crops are grapes for wine making and olives for pressing, Lucca being a major center of the olive oil trade. Olives are grown toward the coast to avoid the frosts common on higher ground inland. Vines are grown widely in the Arno valley and Tuscany to produce well-known wines, including Chianti. The vine is frost tolerant and can be grown inland on hillsides.

South of Florence, on the routeway to Rome, is the Tuscan hill town of Siena, an early example of medieval town planning. In the thirteenth century the city imposed a building line to bring some regularity to the position of houses around the piazza and to stop encroachment on the large, central open space. The result was the *Piazza del Campo* where the *Corsa del Paglio* is run, with entrants riding horses representing the neighborhoods of the city.

Coming west down the Arno Valley from Florence, we come to Pisa. There is still some traffic on the river, mostly recreational, but silting has reduced navigation on the Arno. At the coast, a few miles below Pisa we turn south to reach the industrial center of Livorno (Leghorn), now a huge container port. Lower on the Tuscan coast road is Piombino and, offshore, the island of Elba where Napoleon was exiled from May 1814 to February 1815, with a pleasant house in the capital of Portoferraio and a nice summer villa. He grew restless, returned to France, raised an army, lost the Battle of Waterloo, and was dispatched to cold, bleak, isolated St. Helena in the south Atlantic.

Campagna di Roma   The coastal plain south of Livorno remains narrow until, south of Civitavecchia, it broadens into the *Campagna di Roma*. Civitavecchia is an ancient town that was heavily refortified in the sixteenth century. It still serves as a port for the Rome region and the inland industrial area around Terni. The Campagna is an extensive coastal plain that extends into the Tiber valley and south along the coast to Gaeta and the river Liri.

Although the legend of Romulus building Rome makes an attractive story, we should not forget that Romulus was good at locational analysis. Like London and Paris, Rome is not on the coast but upstream, on the Tiber, at a bridging point. Further a number of valleys and routeways come together upstream of Rome. The seven hills provided a defensive site, and downstream of Rome, the coastal plain broadens into a large area of flat land, with agricultural potential. If we were choosing a location for the capital of the Roman Empire, we might not favor the situation of Rome but, in terms of centrality within the Italian peninsula and routeways across the peninsula to the Adriatic Sea, the city has advantages when compared with other capital locations for Italy.

Rome became the capital of an empire and the center of the Christian church. Even with the decline of empire and the splitting of the Church into the Church at Rome and the Eastern Orthodox Church at Constantinople, the city retained powers of attraction. The present-day city has many consumer goods industries to supply the urban population. Vatican City is a sovereign state to which ambassadors are appointed from around the world. Rome is the legislative, executive, and bureaucratic center of the Italian state. The urban fabric is full of Roman buildings and great examples of Renaissance and Baroque architecture, including St. Peters and St. Peters Square in Vatican City. Rome, with the Vatican, is a center of pilgrimage, tourism, and modern, stylish living. (Photo 8.11) (Photo 8.12)

Photo 8.11   St. Peter's Basilica. Vatican City, Rome, Italy.

Source: Age fotostock/SUPERSTOCK.

Photo 8.12   St. Peter's Square and Via della Conciliazione from the Dome of the Basilica.

Source: Age fotostock/SUPERSTOCK.

Naples and the Campania   Naples is a Greek foundation (600 B.C.) in an extensive area of good agricultural land, the *Campania,* running from near Gaeta in the north to the Sorrento peninsula on the south side of the Bay of Naples. The advantages of the Naples site were obvious to colonists coming from the sea. The Gulf of Naples is protected by the Pozzuoli Peninsula in the north and Sorrento Peninsula to the south, together with the offshore islands of Ischia and Capri. Naples is sited on the north shore of the gulf, with its back protected from the north wind, the *Tramontana.* Vesuvius (4,190 feet) has provided rich volcanic soils, and to the north of Naples the valley of the Volturno gives access to the Apennines. Naples was taken over by the Romans, who built Pompeii and Herculaneum too close to Vesuvius. The eruption of A.D. 79. buried both places. (Photo 8.13)

Today, Naples is the capital of the *Mezzogiorno* (the South) and a major port and industrial center. South of Naples, Calabria is a region with eroded hillsides and agriculture only in favored places. At the toe of Italy is the city of Reggio Calabria and the Strait of Messina, a narrow, dangerous channel between the mainland and Sicily.

Southern and Eastern Coasts. The southern and eastern coasts of Peninsular Italy front onto the Ionian Sea and the Adriatic Sea. Within unified Italy, some of the poorest regions were in this zone, including Basilicata, Calabria, and, fronting onto the Adriatic, Apulia, Molise, the Abruzzi, and Marche (Figure 8.4). Frequently on the Adriatic coast, the Apennines come close to the coast and the regions are composed of hills and mountains with limited areas of low-lying agricultural land. The traditional olives and fruit trees are grown on hillsides, and in recent years, vineyards have become more extensive but

Photo 8.13   Temple of Jupiter, ruins of the old Roman city of Pompeii. Excavation has removed the volcanic debris which covered the city.

Source: Age fotostock/SUPERSTOCK.

historically much of the zone was pasture and poor-yielding arable land. The percentage of the workforce employed in agriculture is still relatively high, in Molise 15 percent, and unemployment is above the national average. Life has improved for many. Fifty years ago over half the workforce was in farming. Before that people emigrated to escape poverty. In the years before World War I, when emigration from Italy to the United States, Argentina, and Brazil was high, a quarter of the population of Basilicata left for the Americas. After World War II, people from all across the South and Sicily moved north to Turin and Milan or over the Alps to work in Germany. The Basilicata region has many hilltop villages, which have lost population. During the Fascist era, those suspected of disloyalty were exiled to the impoverished South. One of the exiles was a doctor, Carlo Levi, sent to Aliano in Basilicata, who wrote a book on his experiences, *Christ Stopped at Eboli* (1947), which describes, among other aspects of life, the incidence of malaria.

Taranto Parts of the east and south coast prospered after Italian unification. Taranto, Bari, and Brindisi got rail links and developed as ports, naval bases, and points where Italian influence could be projected across the Adriatic to Albania, Greece, and the Middle East. The city and port of Taranto, in Apulia, is a Greek foundation sited at the head of the Gulf of Taranto, with higher ground inland to protect against northerly and northeasterly winds in winter months. Like the coastal towns of Sicily, Taranto was occupied by most of the powers that controlled the central Mediterranean up to Italian unification (1860–70).

When Italy unified, Taranto become an Italian naval base with the capacity to build warships. An Italian navy was fine as long as Italy was allied with the major Mediterranean sea powers, France and Britain, as she was in World War I. As France fell, in June 1940, Mussolini joined the war hoping for territorial gains in North Africa, Corsica, Greece, and the British naval base on the island of Malta, south of Sicily. For fear of air attack the British fleet withdrew from Malta but air units remained, including photo reconnaissance planes, which identified the Italian battle fleet at Taranto, in November 1940. An aircraft carrier launched biplane torpedo bombers against the well-defended harbor, incapacitating battleships and heavy cruisers. The Japanese naval attaché to Italy sent a team to Taranto to examine the work of Britain's fleet air arm, which had solved the problem of torpedoes hitting harbor bottoms. The Japanese too solved the torpedo problem and a year later attacked Pearl Harbor, bringing the United States into the war. By 1943, Taranto was in Allied hands and what was left of the Italian fleet was tied up at Malta to avoid German takeover.

Tourism has had an impact in the Abruzzi, which contains the high Monte Corno (9,560 feet) in the Gran Sasso d'Italia. The *Parco Nazionale d'Abruzzi* was created in 1922–23. The region is high enough for winter snow, and winter sports attract visitors. In summer, hikers and fishers come to the higher, cooler region to enjoy the wild beauty of a formerly remote area. Along the Adriatic coast many visitors arrive for summer seaside vacations, but the regions fronting onto the Adriatic and the Ionian Sea have living standards below national averages.

Sicily

The Apennines trend west along the north of Sicily, leaving a narrow coastal plain except in the well-irrigated *Conca d'Oro,* around Palermo. (Photo 8.14) Much of interior Sicily, which rises up to several thousand feet, is composed of easily erodable sediments that do not produce good agricultural lands. In addition, the southern half of Sicily is climatically arid and semi-arid. Rich farming exists where irrigation water is available, but the region is not well supplied with surface streams. Rivers are dry for much of the year and many valleys are choked with erosional debris.

The major settlements of Sicily are coastal (Figure 8.5). Messina, on the narrow straits that separate the toe of Italy from Sicily and join the Tyrrhenian and Ionian Sea, is the entry point to the island from the mainland. The strait is rightly feared for strong currents and dangerous swirls of water. The ancients had the straits of Messina inhabited by *Charybdis* and *Scylla,* great dangers to mariners, including Odysseus. The 1908 earthquake and tsunami destroyed Messina.

Messina was a Greek foundation but along the north coast, Palermo, in the fertile *Conca d'Oro,* originated as a Phoenician settlement of the eighth-century B.C., passing to Rome in 254 B.C. in the First Punic War. After the fall of Rome, Sicily became part of the Byzantine Empire from 535 A.D. to 831 A.D. when the Arabs took Palermo and conquered Sicily and southern Italy. Palermo became the capital of the Emirate of Sicily in 948 A.D. until 1072 A.D. when the Normans took Palermo and control of Sicily. Roger and his son Roger II built an empire across the central Mediterranean to North Africa. Palermo became the center of a remarkable multicultural community. Byzantine, Norman, Italian, and Arab cultural traits flourished. The Arab geographer Idrisi (1099–1180) was one of many scholars working at the Norman court. The Arabs contributed architectural ideas, as can be seen in the cathedral at Palermo, and their knowledge of water usage and irrigation techniques improved Sicilian agriculture. (Photo 8.15)

Photo 8.14 Cefalù on the north coast of Sicily. The cathedral is Norman. The settlement is built on a raised shore beneath an old cliff line.

Source: Digital Vision/SUPERSTOCK.

Photo 8.15   The Duomo (cathedral) in Palermo, Sicily, displays Islamic architectural influence.

Source: Age fotostock/SUPERSTOCK.

On the west coast of Sicily is the town of Marsala, which has a rich wine-producing area in its hinterland. Marsala produces a fortified wine of the same name that is shipped all over Europe and North America. The dry south coast of Sicily is not marked by major towns. Licata is a port for shipping Sicilian sulphur but most places are small fishing settlements.

On the east coast, the cities of Siracusa and Catania are Greek foundations later absorbed into the Roman Empire. Siracusa was founded on a small, defensible island (Ortygia) in the Bay of Siracusa. Today, the island is attached by bridge to the mainland and the onshore city is an industrial center.

Catania, on the north shore of the Gulf of Catania, has an extensive hinterland, including the plain of Catania and the Simeto valley. The port ships sulphur and is a center of manufacturing. North of Catania is the small port of Naxos, an early Greek foundation in Sicily. Inland is Taormina, a famed winter resort where many writers including D.H. Lawrence stayed. In and around the town are Greek, Carthaginian, Roman, Arab, and Norman structures along with spectacular views of Mt. Etna.

Inland Sicily is a land of poor villages, few towns, and deserted hilltop forts. From Roman times to the sixteenth century, Sicily was an exporter of grains. Then in the southern European famine of 1591, the harvest in Sicily failed. The country never recovered as a wheat exporter. The island descended into poverty. Only recently has Sicily begun to develop economically.

Sardinia

Sardinia is similar in size to Sicily but geologically and culturally it is distinct. Sardinia is composed of massifs of older, harder granites and schists. The island was involved in the Caledonian, Hercynian, and Alpine orogenies with ancient surfaces being uplifted in the Alpine mountain-building phase. The island rises in the central Gennargentu range to 6,037 feet at Punta la Marmora. There are deposits of iron, lead, copper, zinc, and coal in the sedimentary rocks in the southwest of the island. Starting in the Fasicst era, some streams were harnessed to provide hydroelectric power and irrigation water.

In general, industries in Sardinia are basic, including metal processing, oil refining, woodworking, and foodstuffs. The island is not rich in agricultural resources, with igneous and metamorphic rocks providing thin soils and steep slopes. Rural resources have been depleted by deforestation and erosion as soils were washed downstream as trees were cleared. Much of the upland is covered by *macchia* vegetation, providing grazing for sheep and goats. In lowland areas crops include vines, olives, corn, wheat, and tobacco. The output of olive oil and wine has increased from low levels as a result of EU subsidies, via the Common Agricultural Policy. Fishing is important in coastal communities with tuna (tunny) and lobster being major catches. In general, living standards are low and unemployment high.

Tourism has grown to be a source of employment. The regional capital and port, Cagliari, holds approximately one-tenth of Sardinia's 1.65 million inhabitants. The city is a Phoenician foundation, taken by Rome in 238 B.C. Today, Cagliari is a modern port with manufacturing and processing industries. In the north Sassari, with around 120,000 inhabitants, is an agricultural center handling wine, olive oil, cheese, and fruits.

Urban Geography

Although Italy possesses Rome, a major city in world affairs, city-size distribution tends to follow a rank–size relationship. Rome is not a primate city in terms of size within the Italian city system. Italy is extraordinarily rich in vibrant cities. Partially this is a result of history, for the country possesses Phoenician, Greek, Roman, medieval, and modern urban foundations. However, the vibrancy of life in regional cities is also a function of late unification. Naples, Palermo, Milan, Florence, Venice, and Bologna were major regional centers in the nineteenth century before Italy unified. Naples, capital of the Kingdom of the Two Sicilies, was the largest city in the newly unified Italian Kingdom when the first census was taken on December 31, 1871. In 1881, both Naples and Milan were larger than Rome (Table 8.5).

Rome

Rome, because of an imperial past and the Holy See, became the capital of the new Italy and began to grow rapidly. Naples and other regional cities lost higher administrative functions to Rome. Rome did not exert powerful political and economic control until the Fascist era when decision making, and the allocation of national resources, were concentrated at the capital. By the 1920s, the northern cities were powerful economic engines but could not compete with Rome where the politicians and the bureaucracy created jobs at a much faster pace than manufacturing in the north. The growth of Rome was self-feeding for the suppliers of consumer goods, and services created additional work. Rome is a good example of how capital status, and the concentration of political power, allows a city to grow at a faster pace than the national average. (Table 8.6)

TABLE 8.5 Cities of Italy, 1881, 1911, and 1936

	1881	1911	1936
Naples	463,172	678,031	865,913
Milan	295,543	599,200	1,115,848
Rome	273,268	542,123	1,155,722
Turin	230,183	427,106	629,115
Palermo	205,712	341,088	411,879
Genoa	138, 081	272,221	634,646
Florence	134,992	232,860	322,535
Venice	129,445	160,719	264,027
Bologna	103,998	172,628	269,687
Catania	96,017	210,703	244,972
Trieste	Austria. Annexed by Italy 1919	245,000	248,379

Source: Statesman's Yearbook

Washington, D.C., and surrounding suburbs have displayed this characteristic over the last half century.

Venice

In the fifth century A.D., as the Roman Empire collapsed, settlers took refuge on islands in a lagoon at the northwest end of the Adriatic Sea. By the tenth century the craftsmen and traders on the islands had organized the city of Venice and taken control of the Adriatic Sea, a trade artery leading to the Levant (eastern Mediterranean). Large fortunes

TABLE 8.6 Cities over 200,000 in population (Census 2000)

Rome	2,655,970
Milan	1,301,551
Naples	1,000,470
Turin	900,987
Palermo	679,290
Genoa	632,366
Bologna	379,964
Florence	374,501
Catania	336,222
Bari	332,143
Venice	275,368
Verona	257,477
Messina	257,302
Trieste	215,096
Padua	209,641
Taranto	207,199

Source: Statesman's Yearbook.

were made in the spice trades and wealth was translated into churches, palaces, public buildings, and works of art. Military decline started in the sixteenth century as the emerging Ottoman Empire took Venetian Cyprus (1571), then Crete (1669), and positions in the Peloponnisos in southern Greece (1715). Trade was not helped by the new sea routes opened up in the Columbian age. Economic decline reduces urban redevelopment and thus the medieval and early modern buildings of Venice, built along the hundreds of canals, are largely preserved.

The major threats to Venice are subsidence and sea-level rise. The Italian government is funding project Moses, a system of barriers raised by compressed air, at times of high water. A scheme to pump seawater into the subsoil beneath Venice to reverse subsidence remains unfunded. Ecologists criticize the projects, claiming the works will cause more damage than conservation.

▶ SUMMARY

Italian unification brought together northern Italy, peninsula Italy, and the islands including Sicily and Sardinia. Few would claim that the process of unification is complete and the 2001 constitutional reforms gave Italy's twenty regions more power over taxes, education, and environmental issues. The Northern League still pushes for a Northern Republic of Padania, separate from the poorer South, Sicily, and Sardinia.

▶ FURTHER READING

GERMANY

Annesley, C. *Post Industrial Germany*. Manchester, UK: Manchester University Press, 2004.

Elkins, T.H. *Germany*. Third edition, London: Chatto and Windus, 1972.

Elkins, T.H., with B. Hofmeister. *Berlin: The Spatial Structure of a Divided City*. New York: Methuen, 1988.

Ferguson, N. *Paper and Iron: Hamburg Business and German Politics in the Era of Inflation, 1897–1927*. New York: Cambridge University Press, 1995.

Fest, J. *Speer: The Final Verdict*. New York: Harvest Books, 2003.

Fischer, F. *Germany's Aims in the First World War*. New York: Norton, 1967.

Fulbrook, M. *A Concise History of Germany*. 2nd edition. New York: Cambridge University Press, 2004.

Jones, A. *The New Germany: A Human Geography*. New York: John Wiley and Sons, 1994.

Large, D.C. *Berlin*. New York: Basic Books, 2000.

Mellor, R. *The Two Germanies: A Modern Geography*. London: Harper and Row, 1978.

Pounds, N.J.G. *The Ruhr, A Study in Historical and Economic Geography*. London: Faber and Faber, 1952. The classic study of the development of the Ruhr industrial region.

Spiegel Special International Edition: The Germans. No. 4, 2005.

Stangl, P. "Restoring Berlin's Unter den Linden: Ideology, World View Place and Space." *Journal of Historical Geography* 32 (2006): 352–376.

Till, K.E. *The New Berlin: Memory, Politics, Place*. Minneapolis: University of Minnesota Press, 2005.

Weinberg, G.L. *A World at Arms, A Global History of World War II*. Cambridge, UK: Cambridge University Press, 1994.

ITALY

Black, J. *Italy and the Grand Tour*. New Haven, CT: Yale University Press, 2003.

Bosworth, R.J.B. *Mussolini's Italy: Life Under the Fascist Dictatorship, 1915–1945*. New York: Penguin Press, 2006. Describes poor living conditions in Italy before World War II.

Domenico, R. *The Regions of Italy: A Reference Guide to History and Culture*. Westport, CT: Greenwood Press, 2002.

Hibbert, C. *Rome: The Biography of a City*. New York: Penguin Books, 1985.

King, R. *Sicily*. Newton Abbot, UK: David and Charles, 1973.

Levi, C. *Christ Stopped at Eboli: The Story of a Year*. New York: Farrar, Straus, and Giroux, 1947. Still available after many printings.

Phelps, D. *A House in Sicily*. New York: Carroll and Graf, 1999. Interesting commentary on the Taormina region and the Mafia in the decades after World War II.

PART

III

ENLARGEMENT OF EU

9

DENMARK, IRELAND, AND THE UNITED KINGDOM

▶ INTRODUCTION

The only major European economy outside the EEC, at foundation, was Britain (Figure 9.1). If the community were to expand, Britain had to be a member. However, Britain did not accept many EEC economic concepts. Britain was traditionally a free trader, and the EEC had a protective tariff to disadvantage imports. In 1950, the UK's top trading partners in rank order were the United States, Canada, Australia, Germany, and New Zealand. Many trading partners, large and small, were Commonwealth countries with preferential access to British markets. If Britain joined EEC, Commonwealth members, like Canada, would pay EEC tariffs to enter the UK market. In the aftermath of World War II, Britain felt a debt to Commonwealth countries, which, by June 1940, were the UK's only allies against the Axis powers.

Britain disliked the idea of centralized, common policies being administered by Brussels and was suspicious of the statist, protectionist ideas that drove the formation of the EEC. At the important Messina meeting in 1954, the Six (Benelux, France, Italy, and Germany) agreed to go ahead with the Common Market (EEC). Britain declined to be involved.

Britain's major political parties, the Labour Party, in power from 1945 to 1951, and the Conservatives, in power 1951 to 1964, had difficulty seeing how Britain could join a Europe designed to serve France and Germany. The Labour Government "nationalized" the rail, iron, steel, and coal industries of Britain, meaning the state took over industries, paying compensation to shareholders. Politically, it was impossible to join the European Coal and Steel Community. The trade unions, major Labour supporters, would have objected, and the Conservative opposition in Parliament would have suggested that Britain's basic industries had been taken from rightful owners, to be controlled by continental powers. Winston Churchill, leader of the Conservative party in Parliament, set out his view on Europe at Zurich in 1946. A united Europe was to be encouraged, but British interests lay with the Empire/Commonwealth.

The Conservatives and Winston Churchill returned to power in 1951. Anthony Eden became the Prime Minister as Churchill retired from office in 1954. Eden favored

Figure 9.1 The British Isles.

Commonwealth development. Britain would fund the economic diversification of colonies, as they moved to independence. Commonwealth trade would expand. Eden's premiership ended in 1957, after the failed Suez expedition (1956). Pro-Europe Harold Macmillan became Prime Minister. Macmillan accelerated decolonization (his critics said rushed) and prepared an application for EEC membership, with support from British industries and financial institutions that feared being "outside Europe."

To be successful, Britain's 1963 EEC application had to be accepted by France, Germany, Italy, and the Benelux countries. This requirement meant that one country could block British entry, which General de Gaulle and France did. Macmillan had worked closely with de Gaulle in World War II but this was no advantage. The General found it easier to work with Germany, the country he had fought against in World War II, than the Anglo-Saxons, whom he thought had not treated him with the dignity he deserved as leader of the Free French forces in Britain.

De Gaulle had a vision of a European community of countries (*Europe des Patries*), with a sovereign France in the leadership role. De Gaulle understood that the free trade UK would try to alter the nature of EEC, reducing protective tariffs and diluting common policies administered from Brussels. De Gaulle saw the UK as a threat to France's EEC leadership. France wanted major power status and had nuclear weapons, with the air power to drop the bombs (the *force du frappe*). De Gaulle was about to push NATO headquarters from French territory and remove French forces from NATO command (1966). In retrospect, there was little chance that Britain would enter the EEC with de Gaulle in command. He used the French veto again, in 1967, to exclude Britain, when a Labour administration applied for community membership. The story from Macmillan's perspective is well told in a volume of his memoirs, *At the End of the Day 1961–1963,* which includes an anecdote concerning the French Minister of Agriculture describing the rejection of Britain in barnyard terms: *Mon Cher. C'est très simple. Maintenant, avec les six, il y a cinq poules* (chickens) *et un coq.* He went on to say that it would not be agreeable for France to admit Britain and have *deux coqs* (two cocks) in the barnyard (1973, p. 365).

De Gaulle pushed France forward in international affairs and, using the country's large gold reserves, pressured the dollar as the world's leading currency. Unfortunately for de Gaulle, spending on armaments and Gaulist national aggrandizement policies left education and social services underfunded. In May and June 1968, students demonstrated continuously in Paris, and the country entered a series of debilitating strikes. De Gaulle resigned in 1969. It is easy to see de Gaulle as obstructionist but his perspective suggests there are different views of a European community. Differing national interests have to be accommodated within European institutions.

With de Gaulle gone—he died in 1971—those with a broader vision of Europe advanced the case for enlargement and the entry of Britain. And if the UK joined, then adjoining countries that traded heavily with Britain would also enter. The countries included Ireland, Denmark, and Norway. Ireland had low per capita incomes, little manufacturing industry, and traded few products into international markets (except Guinness). Ireland was independent but closely tied to British markets for agricultural exports, and Irish citizens could enter Britain to work and settle, if they wished.

For Norway and Denmark, the EEC decision was difficult. Norway was a maritime country with a small population, a massive merchant fleet, and worldwide connections. Denmark, with a Parliamentary tradition stretching back a thousand years, was wary of great powers controlling Europe. Germany had taken territory (Schleswig–Holstein) from Denmark in 1863–64 and occupied the country from 1940 to 1945. However, the UK was a major export market for meat, poultry, and dairy products. Denmark also sold these products into Germany, and CAP policies had become a hindrance to trade. Norway had markets in Britain for timber products and fish, besides extensive maritime interests. Sweden, a major trade partner of Britain, Denmark, Norway, and Germany, did not apply to EEC in 1972 because of long-standing neutrality. If a country joined the EEC, it transferred some sovereign powers to a central authority, which compromised neutrality. Finland was strictly neutral, to appease the Soviet Union. It was not practical for the Finns to apply.

In 1972, the EEC accepted the applications of Denmark, Ireland, Norway, and the UK. Entry was set for January 1, 1973. Norway had a referendum on the issue. By a narrow majority the electorate rejected EEC membership. The government resigned. Norway did not enter the Community. The electorate in Denmark voted for EEC membership. Had there been a referendum in Britain, it is unlikely the country would have joined, the electorate fearing loss of independence in a Europe dominated by France and Germany.

▶ DENMARK

Looking at the map, Denmark appears to have two major components: the Jutland Peninsula and, to the east, a group of Baltic Sea islands including Fyn, Sjaelland, Amager, Møn, Langeland, and Lolland (Figure 9.2). When we examine the distribution of towns, villages, roads, and arable farmland, our eyes are drawn to Århus and Aalborg in eastern Jutland, to Odense on Fyn, and Copenhagen spreading across from the island of Sjaelland to Amager. In western Jutland, towns and villages are fewer, smaller, and further apart. The road network is less dense. Clearly agriculture in west Jutland supports fewer people than the densely populated, intensively cultivated lands of eastern Jutland and the Baltic islands.

To explain differences in agricultural intensity and settlement density we have to go back to the end of the Ice Age with glaciers melting and retreating. The western edge of the Scandinavian ice cap ran roughly north to south down the Jutland Peninsula. Out of the ice front flowed numerous meltwater streams with quickly changing courses, carrying silts, sands, and gravels that were deposited over western Jutland.

The ice retreat left a large end moraine that runs as a ridge down much of eastern Jutland. The ridge, composed mostly of clay, is rolling, hummocky country rising up to 568 feet at Yding Hill, the highest point in Denmark. East of the ridge, and in the Baltic islands, glaciers left ground moraine and end moraine materials, for glaciers readvanced in colder spells. The topography of the islands replicates the rolling landscape of eastern Jutland but not as high, with Fyn rising to a few hundred feet.

There is a little solid rock at the surface in Denmark, although underlying Cretaceous chalk is exposed as at Møns Klint; the white cliffs on Møn. The Danish island of Bornholm, lying off southeast Sweden, is composed of igneous and metamorphic rocks, but there is little solid geology in Denmark to provide building materials. The white crumbly Cretaceous chalk is too soft for construction, but is used to make cement.

Lacking outcrops of solid rock suitable for building, the inhabitants of rural areas used the materials available. Glacial moraines contain rocks brought by the ice from Norway and Sweden. The rocks are seen in foundations. Glacial silts and clays are used to make tiles and bricks. Woodland provided timber for framing. In rural Denmark, half-timbered houses incorporating brick in the walls are common. Building stone was also imported, as seen in Copenhagen and other cities. (Photo 9.1)

Figure 9.2 Denmark and the Regions of Denmark.

Land Use

Western Jutland—composed of sands, gravels, silts, and peat—supports marshes, heathland, grazing, and plantings of conifers on sandy soils. The clayey moraines of eastern
Jutland and the Baltic islands make good farmland, with deep ploughing and the rotation
of cereals and root crops. Root crops, particularly potatoes, grow deep in the soil and
break it into finer particles, making the ground easier to work and better for plants to
root in. Overall in Denmark, the major arable crops are barley, wheat, and potatoes.
Just over 10 percent of the land is in timber, primarily in western Jutland. The main
agricultural activity is feeding pigs, cattle, and poultry to produce pork, bacon, ham,
milk, cheese, butter, eggs, and beef. Agriculture now contributes less than 5 percent of

Photo 9.1 Traditional Danish farmhouse.

Source: Leslie Garland Picture Library/Alamy Images.

Denmark's gross domestic product but accounts for over 10 percent of exports. However, the processing of foodstuffs counts as manufacturing not farming.

Regions of Denmark

Western Jutland Dune Coast

On the west Jutland shore, the sea sorted and shifted glacial sands to produce a dune coast along which sand spits enclose shallow lagoons. Formerly, dunes migrated inland, driven by the westerly wind. Planting of conifers and deep-rooting grasses has anchored the sand. There are few natural harbors on the western shore, which fronts onto the North Sea and the Atlantic world. At Esbjerg a natural inlet was used in the late nineteenth century to create a port to facilitate the export of produce to industrial Britain. Today, Esbjerg, with a population of 73,046, handles exports, fishing fleets, and North Sea Oil exploration.

Western Jutland Sand Plains

Inland of the dune coast, soils are still predominantly sandy and gravels are common. Two hundred years ago western Jutland was a heath with heather and other low-growing shrubs tolerant of acidic soils being the predominant vegetation. The poet Blicher saw the heath as a wilderness that was "a repository of natural virtues" in the way John Muir saw wilderness in the United States as a place to recharge the foundations of life. Other forces were at work and reclamation started, given impetus by the loss of agricultural land in Schleswig–Holstein (to Germany) and the founding of The Heath Society in 1866. Hans Christian Anderson declared "Hurry, Come! In a few years the heath a grainfield will be." The Heath Society encouraged reclamation. Conifers were planted, soils improved, and farms laid out. Today, the heath and marsh have been reduced to patches and productive land is in timber, arable, and pasture. However, the population densities remain much lower than in the east, and rural settlements are less dense.

Northern Lowlands of Jutland

When the ice retreated from the northern end of the Jutland Peninsula, a weight was lifted from the land, which rose from the sea to become a low-lying fertile plain with fjords, including Limfjorden, penetrating deeply inland. The uplift more than compensated for the rise in sea level brought about by glacial melting. The sea bed deposits became the parent material of fertile soils.

Eastern Jutland

Eastern Jutland consists of moraines and outwash material. Soils are predominately clay, but there are deposits of sand, gravels, and, of course, boulders in the moraines. The hummocky landscape is more fertile than western Jutland, and the density of farms, villages, and small towns is high compared with the west. The east coast of Jutland is marked by fjords, which provide sheltered, navigable inlets. The towns and cities are mostly inland on the banks of fjords, including, from north to south, Aalborg (120,000) on Limfjorden, Hadsund on Mariager Fjord, Randers (56,000) on Randers Fjord, and Åarhus (228,547) on the sea at Åarhus Bugt. Fredericia is on the sound between Jutland and Fyn. All the towns have access to the Kattegat and the Baltic, reminding us that many of the small ports are, or were, homes to fishing fleets. Denmark still has a large fishing fleet and a considerable merchant marine.

The Islands

Denmark has a system of modern highways linking all regions of the country and connecting with Germany and Sweden. Ferries sail to smaller islands and ports in Britain, Norway, and Sweden.

A major west-to-east highway from the port of Esbjerg on the North Sea coast of Jutland crosses the peninsula, intersects the north–south Jutland highway near Kolding, and proceeds east to the islands of Fyn, Sjaelland, and Amager where a bridge tunnel crosses The Sound to Malmö in Sweden.

On the island of Fyn is Odense (145,000), founded as a Viking trading settlement, the third largest place in Denmark. Hans Christian Anderson grew up here before moving to Copenhagen. The land of Fyn is intensively cultivated, with grains, root crops, clover, and good grassland frequently ploughed up as a part of crop rotations. The countryside of Fyn is full of fields, flowering hedgerows, and gardens around cottages, farms, and old manor houses. In the south of the island, on the fjord of the same name, is Fåaborg, a town of half-timbered houses with tiled roofs. (Photo 9.2)

Fyn is not flat and a number of hills rise to around 400 feet above sea level. Close to the coast they look impressive and give fine views of countryside and smaller offshore islands. A modern roadbridge links Fyn and Sjaelland. Eastern Sjaelland with Copenhagen

Photo 9.2 Half-timbered, tile roofed house, Faåborg on the island of Fyn, Denmark.

Source: Peter Horree/Alamy Images.

is heavily urbanized but, in the rural west, intensive, productive Danish agriculture is displayed in a well-ordered landscape. Denmark possesses some of the most efficient European farming units supported by cooperatives.

Agricultural Development

The change in Danish agriculture from traditional patterns to modern systems of production was stimulated in the nineteenth century. As Britain, and later Germany, industrialized, major urban markets for butter, eggs, cheese, poultry, beef, bacon, pork, and ham grew. The loss of Holstein (1863) and Schleswig (1864) pointed to the need to make remaining land more productive. Then, as cheap grain from the prairies and plains of North America, the Pampas of South America, and the steppes of the Ukraine reduced prices for grains in Europe, a farming slump developed. The state made loans available to allow small farmers to buy land, not to grow unprofitable grains but to produce dairy, pig, and poultry products. Butter, ham, and cheese require equipment and, in the 1880s, farmers were encouraged to join cooperatives that spread capital costs between members and standardized quality, making products easier to sell.

In recent decades, larger farms and bigger processing plants have emerged. In 1900 there were 200,000 farms. By the end of the twentieth century there were a quarter of that number as economic forces resulted in labor loss, mechanization, and larger agricultural units.

Copenhagen

The eastern edge of Sjaelland, looking into The Sound separating Denmark from Sweden, has a number of towns and cities. From north to south the towns include Helsingør, and Copenhagen, which spreads onto the smaller island of Amager.

Within Denmark, Copenhagen does not enjoy a central location. Entering the country by sea at Esbjerg, or by land from Germany, we drive all across Denmark to reach the capital. Conversely, coming across The Sound from Sweden, Copenhagen is reached almost immediately. The location of Copenhagen is explained by the existence of the Baltic narrows where the sea routes of the North Sea, the English Channel, and the Norwegian fjords connect with the Baltic Sea. From a situational point of view, Copenhagen was on the Baltic Sea lanes and enjoyed a sheltered site between the islands of Sjaelland and Amager, which provided a natural harbor.

By the twelfth century, trading towns were growing or being founded along the Baltic sea routes. Copenhagen was often in conflict with the Hanseatic League but stayed independent and in control of The Sound. Control was enhanced under Margrete of Denmark, who married Haakon VI of Norway in 1363. On Haakon's death, Margrete became ruler of both Denmark and Norway and was elected to the throne of Sweden. The unification lasted until the sixteenth century and allowed the Scandinavians to control the Baltic passage. Within the unified kingdom, Copenhagen had a central position.

To the north of Copenhagen at Helsingør, where the strait between Denmark and Sweden is at its narrowest, a fortress was built to control the sea route and extract tolls. Helsingør is Shakespeare's Elsinore, but he no more visited the castle when writing Hamlet than he did Sommer's Island (Bermuda) when writing The Tempest.

Copenhagen (661,000) is a historic city with castles, palaces, impressive public buildings, parks, monuments, museums, and statues. The urban landscape declares capital city status. Copenhagen is an industrial port city. In the modern world the glacial clays

Photo 9.3   The bridge over The Sound connecting Copenhagen, Denmark and Malmö, Sweden.

Source: Age fotostock/SUPERSTOCK.

were easily excavated to allow the construction of quays and docks to take large vessels. Shipbuilding and marine engineering developed, along with consumer goods industries, including the large Carlsburg brewery that shipped into overseas markets long before international beer brands were commonplace.

Copenhagen is a modern, clean, efficient, environmentally aware city that houses the European Environment Agency. The large electricity-generating windmills announce a commitment to alternative energy sources, and Denmark produces approximately one-sixth of its electricity using the wind. On all the indicators of quality of life and living standards, the 5.5 million Danes score highly. The capital is a reflection of a modern society, conscious of national heritage. (Photo 9.3)

▶ IRELAND

Ireland was neutral in World War II and sold all surplus agricultural produce to Britain at good prices. However, agriculture was generally inefficient, and many farms were so small they could not take advantage of improved market opportunities. At the end of the war, Ireland still had large pockets of rural poverty, particularly in the west where small farms predominated (Figure 9.3).

Ireland's peripheral position on Europe's far Atlantic fringe became an asset. As international travel revived in the postwar era, the new trans-Atlantic airliners did not have the range to fly nonstop from the United States to European capitals. In the west of Ireland, Shannon airport was developed as a refueling stop, and around Shannon a free port attracted investment. Further, the Irish government encouraged foreign direct investment in the 1950s. Although the number of international corporations investing in Ireland, mostly in Dublin, was small at first, interest increased when Ireland became an EEC member in 1973. Now U.S., Canadian, and Japanese corporations could build a plant in Ireland, where labor was cheap, plentiful, and trainable, and sell products into the territory of all EEC members, without paying tariffs. In the 1970s with oil costly and in short supply, European economic growth was slow, but Irish economic expansion came in the 1980s and by the early 1990s Ireland was transformed. New factories and service centers created jobs, wages rose, and those with Irish roots found they could return to develop careers in a diversified economy. In the 1990s, Ireland had strong net in migration (Jones, 2003). Dublin experienced renewed rural-to-urban migration, suburbanization, increased commuting, and peripheral office development. House prices rose, the gracious but grimey city was cleaned up, and more tourists arrived to enjoy Dublin as a base for travel into rural Ireland and the wild west uplands of Donegal, Mayo, Connemara, and Macguillycuddy's Reeks in the southwest, near Dingle Bay. (Photo 9.4)

The growth of Ireland's economy continued into the twenty-first century. When the EU expanded into east Europe in 2004, most of the existing members placed restraints on the free movement of labor from the new members. Ireland, Sweden, and the UK did not. Ireland needed workers and now Dublin has vibrant communities of Poles and others migrating from the east. House prices have soared. Local authorities strain to provide services to migrant families.

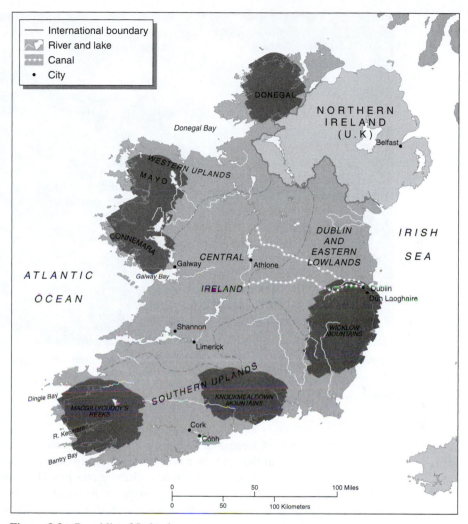

Figure 9.3 Republic of Ireland.

Physical Background

During the last glacial age, most of Ireland, except the southern rim, was covered by ice. When the ice melted, low-lying central Ireland was covered in glacial clays, outwash material, lakes, and marshes. Drainage was disorganized and in ill-drained areas extensive peat bogs developed.

The abundance of surface water created by the glacial retreat was re-inforced by climate. Ireland protrudes into the North Atlantic westerly wind system. The climate is mild and moist with rain on many days. From Dublin going to the west and the southwest, annual rainfall increases, the number of days with frost decreases, the mean monthly temperatures for January rise, wind speeds quicken, and the peat bogs multiply until the higher, rocky, wet west coast is reached.

Photo 9.4   The rugged west coast of Ireland supports tiny farms with vegetable plots, hayfields, enclosed pastures and rough grazing.

Source: SUPERSTOCK.

The mildest part of Ireland lies south of a line running from Dublin to the estuary of the river Shannon in the west. In much of this southern zone, mean monthly January temperatures are 41–42°F and frosts are few. Grass grows into December and rests until March. The region is well suited to the grazing of livestock, which only need to be provided with hay and fodder for a short period. Of all agricultural land in Ireland, only about 5 percent is arable and the rest is pasture, hay field, or rough grazing. With a long growing season and a high percentage of land in grass, Ireland is the greenest country in Europe, with greenness enhanced by verdant hedgerows containing a variety of trees, except on hard rock uplands.

We can speculate that the draining Dutch and reclaiming Danes would have made more intensive use of the damp, bogey postglacial environment, but in Celtic culture livestock raising was a central part of life and Ireland was well suited to grazing animals. Many traditional farms were more nature reserve than commercial operations, as described in the novel *By the Lake*:

The small fields around the house were enclosed with thick whitethorn hedges, with ash and rowan and green oak and sycamore, the fields overgrown with rushes. Then the screens of whitethorn suddenly gave way and they stood high over another lake. The wooded island where the herons bred was far out, and on the other shore the pale sedge and stunted birch trees of Gloria Bog ran towards the shrouded mountains. (McGahern, 2002, p. 23)

The lake supported breeding colonies of otters, herons, swans, wildfowl, pike, and bream. The farm had raised some cattle despite trouble with fences. When the farmer died the place was not sold to any neighboring farm but to a successful couple returning to rural roots in Ireland.

The Regions of Ireland

We can divide Ireland into four major regions (Figure 9.3):

1. Dublin and the Eastern Lowlands
2. Central Ireland
3. Western Uplands
4. Southern Uplands

Dublin and the Eastern Lowlands

Dublin does not occupy a central position in Ireland but it does face into the narrow Irish Sea that connects England, Scotland, Wales, and Ireland. The port and city stand on the banks of the river Liffey, with many bridges linking the north and south of the city. (Photo 9.5) South of the river is the Bank of Ireland, Trinity College (f. 1591 by Queen Elizabeth I), University College (1909), the National Museum, the National Gallery, the Irish Parliament, the castle, and cathedrals. North of the river is the Custom House, the Law Courts, O'Connell Street with the General Post Office so heavily involved in the Easter Uprising (1916), and beautiful, extensive Phoenix Park.

Today, the rapidly modernizing city has motorways radiating from it with a modern beltway to allow traffic to avoid the city center. Roads south lead to Dun Laoghaire, where the ferry leaves for Wales. The road north leads past the international airport, to the border with Northern Ireland, just fifty miles from Dublin.

Agriculture in the hinterland of Dublin has advantages. The eastern lowlands, being on the less wet side of Ireland, can grow grain crops, including barley for Dublin brewers, and fruit, vegetable, dairy, and poultry products can be sold into Dublin markets.

Central Ireland

Central Ireland is underlain by carboniferous (Pennsylvanian) limestone but the surface is composed of glacial deposits, lakes, marshes, and peat bogs. The central region is drained by the Shannon and smaller rivers with slight gradients. The French geographer Albert Demangeon described the drainage of the region:

> The centre of Ireland is occupied by a flat, slightly undulating plain in which the streams wander from lake to lake and from bog to bog so sluggishly that the eye cannot distinguish between running and stagnant water. (1958, p. 121)

Where the limestone is exposed in the west of Central Ireland, there is karst landscape with sink holes, underground caverns, and caves.

The market towns of the central regions are not large and include Athlone, Limerick, and Galway. The landscape is dotted with individual farmhouses for the *clackens* (clusters of houses) common in the seventeenth century have gone. Farm size in Ireland is small.

Photo 9.5 River Liffey and O'Connell Bridge, Dublin, Ireland.

Source: Age fotostock/SUPERSTOCK.

EU agricultural subsidies have brought more income to rural areas but the land market is inflexible. Families are reluctant to sell small farms, even when they no longer farm full time. Farm enlargement has been slow and costly.

The Shannon frequently widens into *callows,* marshy meadows that flood in winter. In summer, hay is made on the slightly higher meadows, and beside the river cattle graze on ground that rarely dries out. The Shannon *callows* are the largest area of annually flooded grassland in the British Isles. Past attempts at drainage were limited, and today, EU payments help preserve habitat for wild fowl.

Western Uplands

The uplands of Western Ireland, remnants of the former Caledonian Mountains, run from Donegal in the northwest to Mayo and Connemara in the west. The region ends at Galway Bay where the lower central region meets the sea. The uplands are not high; in Donegal, Mt. Errigal rises to 2,466 feet, Nephin in Mayo attains 2,644 feet, and the Twelve Pins 2,395 feet, but the wet, exposed, thin-soiled landscape supports little agriculture and appears bleak for much of the year. To quote Demangeon again (1958, p. 120):

Nothing wilder or more desolate can be imagined than the uplands of Connemara and Mayo. With their dark mantles of heather, their deserted valleys, their bogs, their innumerable lakes ... their bare slopes so polished by ice that they shine like glass when the sun strikes them.

There are few towns in the region, population density is low, and settlements are sparse. Much of the west of Ireland, in general, is classified by the EU as less favored for agriculture. The small farms receive some EU subsidies but are too small to benefit from improvement schemes. The western uplands still lose population, and as those who stay are older, the population is aging.

Southern Uplands

The Southern Uplands contain the following mountain areas from east to west: the Wicklow and Knockmealdown mountains, and several parallel folds of Old Red Sandstone in the southwest, running to the bays of Dingle, and Bantry. The Southern uplands are not so massive as those in the west, and in valleys there are farming communities. The major city is Cork with a University College. South of Cork is Cóbh (Queenstown) in a deep, well-protected bay at which trans-Atlantic liners used to pick up migrants to the United States. Here, in 1915, the survivors of the torpedoed *Lusitania* were brought ashore. The liner was British registered but many American passengers were on board when the vessel sailed from New York. The incident was a step toward bringing the U.S. into World War I.

Ireland and EU

Because Ireland was tightly linked economically to the UK and Irish citizens could enter Britain to find work, the country had little choice but to join the European Community with its neighbor in 1973. The 1970s were economically stagnant in Europe and North America. When economies prospered in the mid 1980s, Ireland attracted outside investment in manufacturing and information industries. The economy grew, in-migration developed, prosperity for urban areas followed. Ireland has the problems of rapid economic growth. The greater Dublin region now contains 40 percent of the population of Ireland and the city sprawls into surrounding rural areas.

In the 2004 EU expansion, Ireland, along with Sweden and the UK, were the only countries to immediately implement the free movement of labor provisions. The expanding Irish economy needed labor. Dublin is becoming more cosmopolitan with growing Polish, Lithuanian, and Latvian communities.

Summary

After World War I, Ireland left the UK, becoming independent in 1922, with the same dominion status as Canada, Australia, and New Zealand. Although many Irishman volunteered for service with the British forces, Ireland was neutral in World War II, not allowing Allied aircraft to operate in the Battle of the Atlantic against German submarines and unable to keep U boats out of coastal waters, west of Ireland. In 1949, the Republic of Ireland left the Commonwealth, the same year in which NATO membership was rejected. The six counties of Ulster, which opted to stay in the United Kingdom when Ireland became independent in 1922, remain a problem in relations between Dublin and the UK. However, a dispute born in the Catholic–Protestant wars of the seventeenth century has calmed as the Republic has become increasingly modern and prosperous as a result of EU membership. The population of the Republic of Ireland is 4.2 million, with a birth rate of 15, a death rate of 7 per thousand, total fertility at 1.9, and with in-migration grows at 1 percent a year.

▶ THE UNITED KINGDOM

Introduction

Britain has undergone transformations (Figure 9.4). After World War II, war damage was repaired, industries were nationalized, and a social safety network put in place, including a National Health Service. The swinging 60s, symbolized by the mini skirt and the mini car, saw living standards rise, new universities founded, and the greatest part of the decolonization process completed. In the 1970s, Britain entered Europe, suffered stagflation, and with trade unions demanding more in a time of inflation and stagnant growth, a long period of Labour Party rule came to an end in the 1979 winter of discontent. The 1980s were dominated by Prime Minister Margaret Thatcher and Thatcherism, with the selling of state assets including the telephone system, the railroads, the coal mines, and the national airline, to promote a competitive environment and an end to subsidized older industries. Thatcher was strong and certain in her political views. Inevitably she made mistakes, for example, cutting funding for higher education at just the time economic policies were creating an increased demand for an educated workforce. In foreign affairs Thatcher was a force, as the military *caudillos* who ruled Argentina discovered when they ordered the invasion of the Falkland Islands (1982) to distract attention from the political prisoners they had "disappeared." The short war, in the South Atlantic, ended in defeat for Argentina and the fall of the military junta. In Europe Thatcher insisted on more efficiency and was highly critical of the Brussels bureaucracy that lacked accountability.

Mrs. Thatcher was deposed from power by her own Conservative Party, which, after a few years, lost a general election to the Labour Party led by Tony Blair. The Labour Party no longer had the backing of a huge trade union movement. The old industries of shipbuilding, coal mining, and steel making had been privatized or had shrunk through lack of competitiveness. Mr. Blair introduced New Labour policies that took the middle political ground by preserving many of the middle-class gains achieved under Thatcherism.

Britain's first industrial revolution industries, including the coal mines, textile mills, blast furnaces, steel plants, and metal-working factories, are all much reduced in size, workforce, and importance. Even the products of the second industrial revolution—cars, motor bikes, and electrical goods—are now imported or are manufactured in Britain by foreign corporations. The greatest part of Britain's workforce is in the service sector, and financial services are particularly important in London, which, along with New York and Tokyo, is a global financial center.

Topography

In 1902, the Oxford geographer Halford Mackinder published *Britain and the British Seas*. Mackinder suggested that a line drawn from the river Tees in northeast England to the estuary of the river Exe, in the southwest, divided highland Britain from lowland Britain.

Figure 9.4   United Kingdom.

The lands north and west of the Tees–Exe line were older geologically, contained many of the coalfields, and agriculture was predominately hill farming and livestock rearing. The living Celtic cultures were in the highland region (Figure 9.5). (Photo 9.6)

Highland Britain is largely a land of stone-walled pastures, stone barns for livestock, and deciduous trees in sheltered places. Lowland England is a land of hawthorn

Figure 9.5 Highland and Lowland Britain.

Photo 9.6 Llangorse lake and the Brecon Beacons, Powys. In the Welsh uplands the valley floors support arable fields and improved pastures. As we go up the hillsides, pasture gives way to rough grazing with altitude.

Source: Graham Morley/Alamy Images.

hedges enclosing arable fields. The lowland towns prosper on service industries. The old coal-based industries of highland Britain have declined.

Like much of Mackinder's writing, *Britain and the British Seas* had a prophetic quality. A century after he wrote it, a study spoke of a northeast-to-southwest Humber–Severn line that divided Britain economically and demographically. Between 1991 and 2001 a million new jobs were created in the city of London. People moved to the southeast of England; Birmingham, Manchester, and Liverpool lost population. Everywhere south and east of the Humber–Severn line had become part of a greater London.

▶ REGIONS OF BRITAIN: SCOTLAND, WALES, NORTHERN IRELAND, AND ENGLAND

Scotland

Scotland, with a population of 5 million, can be divided into three major parts: the Highlands and Islands, the Central Valley where most Scots live, and the Southern Uplands, all of which display distinctive landscapes and ways of life (Figure 9.6).

The highlands and islands are largely composed of granitic rocks. (Photo 9.7) The uplands consist of igneous masses that were intruded into the base of the Caledonian mountains. The higher sedimentary layers of the mountains have long eroded away, leaving an igneous surface that has been cut into by glaciers and rivers. The weathering of granitic rocks is slow, soils are thin, slopes often steep and exposed. Agriculture in the highlands is found in favored valleys and on lower coastal lands. The traditional

Figure 9.6 Scotland.

crofting way of life, with clans of Gaelic-speaking families living in long houses heated with peat fires and by livestock living under the same roof, is largely gone, as are the plots growing oats and potatoes. The cottage crafts producing handwoven woolen textiles in distinctive patterns have mostly ceased, although the Hebridian island of Harris still produces the Harris tweed, with horsehair fortifying the woolen cloth.

The Highlands are an area of Europe with deserted rural places. The demand for holiday homes is not matching the outmigration. Much of the upland environment is used

Photo 9.7 Ben Nevis and Fort William from Loch Linnhe.

Source: David Robertson/Alamy Images.

for shooting, deer stalking, and in favorable areas, skiing. Dams and windmills produce electricity.

The west coast of highland Scotland has a remarkably mild climate due to the Gulf Stream and the westerly winds. Winters in the area are not as cold as in southeast England, around London, although midwinter days are noticeably shorter.

The major towns in the highland region are close to the coast and include Wick, Inverness, Aberdeen (population 212,000), and many smaller places, formerly important for fishing fleets. Aberdeen is an ancient town with a university founded in 1494. In recent decades it has prospered as a service center for North Sea oil.

The Central Lowlands of Scotland run from the Firth of Clyde and Glasgow in the west to the Firth of Forth and Edinburgh in the east. South of the Firth of Forth, East Lothian is a rich farming area, and north of the Forth, stretching to St. Andrews, is surprisingly rich farmland. St. Andrews, on the North Sea coast, is noted for golf courses and a university founded in 1412 A.D.

Beneath the central valley are the coalfields of Ayrshire, Lanark, and Fife. Although mining is much reduced, the proximity of the Lanarkshire coalfield to the Clyde was an important factor in developing early industry around Glasgow. Clydeside became a major producer of coal, iron, steel, textiles, and ships. The town of Paisley gave its name to a distinctive pattern used on cotton textiles. Famous shipyards lined Clydeside including John Brown's yard, which built the *Queen Mary,* the *Queen Elizabeth,* and the *QEII.* The *Queen Mary II* was built in a continental yard, a commentary on the decline of British shipbuilding.

Glasgow (population 580,000) is the center of Clydeside. As the traditional industries have declined, Glasgow has flourished. Glasgow was an industrial city and a design center. Local architects produced many fine Victorian and early twentieth-century buildings. The industrial grime has gone and Glasgow is now a tourist, shipping, cultural, financial, and high-tech center.

Edinburgh (population 450,000), capital of Scotland, and the seat of Parliament, sits on the south shore of the Firth of Forth, on and around the plugs of extinct volcanoes. The city is a mix of medieval and gracious eighteenth-century streets containing the castle, art galleries, universities, the Scottish National Library, and the National Archive. Like many smaller European capitals, Edinburgh's major sights can be reached on foot.

The Southern Uplands are less high and less exposed than the Highlands. Moorland and rough grazing dominate at higher levels, but in the valleys, particularly along the Tweed, farming is rich and productive. The region produces sheep, cattle, and dairy products. In valleys suitable for arable, grains and vegetables are grown. Along the coasts are fishing ports, but British inshore fishing fleets are much reduced as EU fisheries policy has opened up traditional fishing grounds to other fleets.

The relationship between Scotland and England is complex. The crowns of England and Scotland were united in 1603 when James VI of Scotland became James I of England. When the countries merged parliaments by the Act of Union (1707) they did so with commercial considerations in mind, to create a larger market and increase trade. Glasgow merchants, on the Clyde, got access to trade, on equal terms, with the North American colonies. Scottish financial institutions could operate in the English market. Scotland retained its legal system and education system along with the Presbyterian Church of Scotland.

There were armed uprisings in 1715 and 1745. During the Forty-Five, Bonnie Prince Charlie's army got to Derby in the English Midlands before running out of supplies. The uprisings were dynastic and nationalistic. One aim, in 1715 and 1745, was to reclaim the crowns of England and Scotland for the Stuarts who had lost the thrones when William and Mary took power in 1689. William and Mary were Protestants; the Stuarts were Catholics, receiving support from France on religious and strategic grounds. An independent Scotland, a friend of France to the north of England, could be a useful ally in conflicts between Britain and France.

The Scottish Parliament was reestablished in 1999, (Photo 9.8) creating constitutional complexities. Scots now elect members to sit in the Parliament at Edinburgh and, in Britain's national elections, elect members to represent Scottish constituencies in the Parliament at Westminster, where they often vote on purely English issues. For example, Scotland administers its own National Health Service. The Labour government, led by Tony Blair, wanted to reform the English National Health Service. The legislation only passed at Westminster with the help of Scottish members voting in favor.

Scotland elects many Labour members to the Parliament at Westminster. Much of Labour's electoral success has depended on the strong performance of the party in Scotland, and there are people in England who think that independence for Scotland would change the political balance. An English reaction to a separate Scottish parliament has been the emotive appearance of the flag of St. George, England's patron saint.

Photo 9.8   New Scottish Parliament buildings, Edinburgh, with Calton Hill in the background. The building costs escalated ten fold during construction.

Source: David Cairns/Alamy Images.

Whether Scotland becomes independent will be partly driven by a romantic recalling of a heroic "Braveheart" past. Not everyone in Scotland thinks independence a good thing, and the Scottish Nationalist Party, which favors independence, does not yet command a majority in the Scottish Parliament. Many Scots understand that an independent Scotland would be governed by a left-wing party, with higher taxes to support the ideology. On the other hand, an independent Scotland would become an EU member with a voice in Brussels and, as England and Scotland think alike on issues like fishing and agriculture, overall British influence at Brussels would strengthen.

Wales

Within the United Kingdom, Wales (2.9 million) has a special personality, linking the Celtic world to the modern era. In central and west Wales, a substantial part of the population speak Welsh as their first language. All of Wales is covered in Celtic place names: *afon,* river; *cae,* field; *ryn,* one of several words for hill; *glyn,* valley; *llyn,* lake; *môr,* sea; and *llan,* literally church, but usually signifying a settlement around a church.

The Welsh language is part of a living culture. Approximately 25 percent of elementary schools in Wales use Welsh as the main medium of instruction, the Welsh language schools being clustered in central and northwest Wales. Radio and TV programs are broadcast in Welsh, although, in the major cities of South Wales, English language channels are more popular.

Welsh literary culture is kept alive by bards (from the Welsh *bardd*) who recite poetry, retell legends, and sing. (Photo 9.9) Regional and national *Eisteddfods* (gatherings for the performance of music and literature) are events that maintain literary traditions traced to sixth-century Arthurian legends and the "Navigation of St. Brendon." St. Brendon (484–577 A.D.) visited many places around the Irish Sea and Celtic Brittany. Some argue that Brendan made voyages into the Atlantic.

The Romans occupied parts of Wales to exploit minerals, but the Roman presence in terms of road building and town founding was not as strong as in lowland Britain. When the Romans left, Wales was still divided into tribal territories, and there was little Anglo-Saxon settlement. Raids into England were common and to contain hostilities the Saxon King of Mercia, Offa, had a ditch and earthwork constructed from north to south along the Welsh borderland starting around 784 A.D. Offa's Dyke is still visible in the landscape today.

The Normans penetrated Wales in 1282 and established some castles, but Wales was not subdued. In 1277, Edward I campaigned in north Wales, establishing a string of fortified places including Conwy, Caernarfon, and Harlech. The medieval towns, with regular street patterns within well-fortified walls, are an attractive part of the North Wales landscape.

Edward I made his heir to the throne, Prince of Wales, a title that is still held by Charles, the present heir to the crown. Of course, Edward's gesture was intended to indicate that Wales was very much part of his kingdom but uprisings in Wales did not stop and powerful feudal lords were installed at Chester and Shrewsbury to keep peace

Photo 9.9   Bards at the National Eisteddfod Wales.

Source: The Photolibrary Wales/Alamy Images.

along the Welsh borderlands. Welsh union with England did not come until 1536 but Wales remained a culture separate from England.

The industrial revolution brought economic and cultural change. In South Wales was a large coalfield, with seams exposed in the sides of the valleys that ran to the sea at the Bristol Channel. In the nineteenth century, coal mines were opened in the valleys. South Wales coal was excellent for steel making and for firing steam-powered vessels. Coal was exported from South Wales ports like Cardiff, Swansea, and Barry to bunkers around the world to fuel steam vessels. Shipbuilding developed in the south coast ports and eventually huge iron and steel works were constructed at Port Talbot on the coast and at Ebbw Vale in one of the coal valleys.

Industrial development in the nineteenth century attracted capital and migrants. People from traditional Welsh-speaking communities of central Wales moved to the valleys of South Wales and the growing, coastal industrial centers of Cardiff and Swansea. Migrants arrived from England that, together with linkage to the commercial world, resulted in English becoming widely spoken in South Wales. Industrialization produced cultural changes, but the result was distinctly Welsh. All across Wales communities took to Methodism. Chapel and the associated choirs became a part of Welsh life. In South Wales not everyone spoke Welsh but people still sang in unison on any occasion. Every town in the south had a rugby club, supported as fervently on Saturday as Chapel on Sunday. The total population of Wales is less than three million people but the numerous rugby clubs in the south produced a Welsh national side that could defeat the teams of Scotland and England. In England the origins of rugby were in the fee-paying boarding schools. In Wales, rugby was a working-class game with players coming from docks, shipyards, mines, and steelworks. Today, much of the old heavy industry has closed. Wales has been successful in attracting new industries. Swansea and Cardiff are major shopping and service centers. The national rugby side is not the force it was.

Physical Environment

Wales consists of a central massif of the Cambrian mountains surrounded to the north, west, and south by coastal plains. In the east, on the landward side of Wales, the rivers Dee, Severn, Wye, and their tributaries drain the Welsh borderland (Figure 9.7).

The Cambrian mountains, running north–south, are the backbone of Wales. In the south, at Brecon Beacons, the land surface rises to 2,906 feet. In the north, Snowdon attains 3,560 feet, the highest point in Wales.

Climate

For much of the year Wales experiences southwesterly air streams coming from the Atlantic. In western coastal areas including the Isle of Anglesey, Cardigan Bay, and the Pembroke peninsula, the growing season exceeds nine months of the year and rainfall is plentiful. At Cardiff, close to the south coast of Wales, the mean monthly temperature for January is 40°F, in July the figure is 61°F. Annual rainfall averages 43 inches. With mild temperatures, long growing seasons, and frequent rainfall, grass grows for much of the year, providing grazing and meadows to make hay as winter fodder for livestock. On well-drained soils, arable crops yield well but the main emphasis of farming, in coastal areas and inland valleys, is livestock rearing and the production of dairy products. Traditionally, Wales exports cattle, sheep, milk, cheese, and butter to markets in England.

Figure 9.7 Wales.

In the mountainous interior of Wales, altitude and relief influence temperatures and rainfall. At 1,500 feet above sea level, average temperatures are five degrees lower than in the coastal areas and growing seasons are shorter, meaning that livestock have to be fed for a longer period of time, if kept through the winter. On the seaward slopes of mountains, where the westerlies are forced upward, rainfall totals are heavy. In northwest Wales, in the Snowdonia region, rainfall exceeds eighty inches a year, with maximum precipitation in the winter months.

Geology and Soils

Except on the Pembroke peninsula, in the southwest, Wales was glaciated in the last Ice Age. Glacial deposits of clays and outwash materials are widespread, as are U-shaped valleys, arêtes, and cirques (*cwm* in Welsh).

The Cambrian mountains are part of the ancient Caledonian mountain chain, which ran from Scandinavia, through Scotland to Wales. Northwest Wales is classic geological country with extensive exposures of lower Paleozoic rocks, which were first classified in the region. The pre-Cambrian and Cambrian rocks are named after the Roman term for Wales—Cambria. The Ordivician and Silurian rocks carry names from ancient Welsh tribes, the Ordovices and the Silures. The lower Paleozoic formations contain many igneous and metamorphic rocks, in addition to sedimentary deposits holding distinctive fossils, which allowed identification of geological periods.

In the Welsh upland, geology and climate combine to produce leached, acidic, podzol soils, the high rainfall in the mountains washing out many minerals. On well-drained hillsides and valleys, grassland thrives. Where drainage is poor a moorland of heather, bracken, mosses, and peat bogs is common.

Devolution

Devolution is the transfer of powers from central government to regional authorities. Both Wales and Scotland now administer their domestic affairs, with the Parliament at Westminster retaining control over defense and foreign affairs. In 1999, Welsh voters elected a sixty-seat National Assembly, which sits in Cardiff. In the same year Scotland elected 129 MPs to the Parliament at Edinburgh. The Scottish body has the power to impose taxes; the Welsh Assembly, by decision of the voters, does not. Wales shares the general revenue raised by the Treasury in London and allocates money to economic planning, the environment, health, urban problems, education, tourism, welfare, and rural development.

Several decades ago there was a Welsh nationalist movement with militant tendencies. Vacation homes in Wales, owned by English people, were burned or vandalized and visitors were made to feel unwelcome by a few. At the time, the number of Welsh speakers was declining and there was a sense that the culture was under threat. Official forms were available in Welsh but the Welsh Language Act, passed at Westminster in 1967, gave Welsh the same legal standing as English, reducing tensions. Devolution was offered to voters in Scotland and Wales in the 1970s and rejected. The result was reversed in 1997 but voter turnout was low and, in recent elections to the Scottish Parliament and the Welsh Assembly, turnout has been less than fifty percent. In the 2007 assembly elections 40% of the Welsh electorate voted. For the time being Welsh issues are not exciting Welsh voters. Like Scotland, Wales continues to send MPs to the Parliament at Westminster, but Welsh political culture is changing. Traditionally, South Wales was a Labour Party stronghold. Now candidates from other parties successfully contest seats.

Northern Ireland

In the seventeenth century, migrants from England and Scotland crossed the Atlantic to settle in North America. As part of the westerly movement there was settlement of English and Scots into Ireland. Settlement was marked in the northeast where many protestant Scots came to live and form the majority of the population. The remainder of Ireland was largely Roman Catholic. A long period of conflict started between the Irish and the English and Scottish settlers over land and religion.

When Ireland (Eire) became independent in 1922, the six northeastern counties, commonly referred to as Ulster, remained within the United Kingdom (Figure 9.8). Northern Ireland was created with a bicameral parliament, which met at Stormont near Belfast,

Figure 9.8 Northern Ireland.

and voters continued to elect members to the Parliament at Westminster. Stormont ran all aspects of life in the province except defense, foreign affairs, and taxation. Religiously, the province was deeply divided, with a Protestant majority and a Catholic minority. The Catholics had little say in the running of affairs and were discriminated against in the job market.

In the 1960s a civil rights campaign started that inflamed sectarian loyalties. At the time the constitution of Ireland stated that the north and the south should be unified under Dublin. Sinn Féin ("Ourselves Alone") was the political wing of the movement to unify. The IRA (Irish Republican Army) was the armed force. Both organizations had support in the Catholic populations of the north and the south. Protestants supported the Ulster Unionist Party and Protestant paramilitary groups appeared, including the Provisionals.

The Parliament at Stormont lost control. Troops were brought from the mainland, and in 1972 Northern Ireland was directly ruled from London. Violence was widespread, with the IRA and the Provisionals targeting each other. The violence spread to mainland Britain with the IRA planting bombs in shopping centers, office blocks, and transport facilities. The violence continued for thirty years, during which over 3,000 people were killed in Northern Ireland. Eventually, Britain, the Republic of Ireland, and Ulster politicians cooperated in a political process that has led to a rejection of paramilitary campaigns. Elections have been held. The majority Ulster Unionist Party and the minority Sinn Féin, have agreed to form a government.

Physical Environment

Northern Ireland is covered in many places by glacial deposits, but the older, harder rocks of the Caledonian mountains are exposed, as are lava flows associated with Tertiary volcanics. Having been eroded for long periods of geological time, the Caledonian mountains are no longer high but at the coast they often rise steeply, and impressively, from the sea. The granitic Mourne mountains, at the coast south of Belfast, rise ruggedly to 2,789 feet, the highest point in Northern Ireland.

North of Belfast the plateau-like Mountains of Antrim attain 1,807 feet and contain extensive outflows of basaltic lava. In the east, the Mountains of Antrim drain to the Irish Sea via the Glens, a series of deep, glaciated valleys that fall to the sea between cliffs rising sharply from the water.

To the west of Belfast is Lough (lake) Neagh. To reach the Lough from Belfast we travel inland along the valley of the river Lagan before it turns south, not far from the lake shore. The river Bann flows into Lough Neagh from the south and exits to the north, flowing via Coleraine to the Atlantic.

Much of the west of the province is drained by the river Foyle, which flows through Londonderry, Derry to the Catholic inhabitants, to enter Lough Foyle, an inlet of the Atlantic. Like a Scottish loch, a lough can be a lake or an arm of the sea.

Northern Ireland has a maritime climate. In Belfast the mean monthly temperature for January is 38°F. In July the mean is barely 60°F. Rainfall approaches forty inches a year. The climate is cool and damp. Prolonged frost is rare and, although winter snow showers occur, snow usually does not remain on the ground for long. On the mountains of Antrim and Mourne conditions are cooler, wetter, and more exposed.

As in the rest of Ireland grass grows well, with the growing season a little shorter in the north than the south. Much land is sown to grass and there are considerable areas of rough grazing. Livestock farming is the basis of rural life and the province produces pigs, sheep, cattle, and dairy products. In well-drained areas, grains and root crops are grown and Northern Ireland exports seed potatoes.

The Economy

Northern Ireland has few mineral resources other than rocks used in construction, including sandstone, granite, basalt, sands, and gravels. The province is linked to electricity systems in Scotland and Ireland. Natural gas comes by pipeline via Scotland.

A traditional textile industry produced linen. Flax was grown, harvested, and retted, a rotting process that separated the fibers, which were spun into yarn and woven into a fabric. Linen was produced in a cottage industry environment until the nineteenth century when the industry mechanized and moved into factories. Derry and Belfast were major centers of the linen industry, which has been much reduced as a result of international competition.

The other major nineteenth-century industry was shipbuilding, principally at Belfast. The *Titanic* and several other large passenger vessels were built at Belfast before World War I. (Photo 9.10) Shipbuilding declined in the Great Depression when world trade halved. It recovered in World War II when submarine warfare enhanced demand for new vessels. All shipbuilding in the economically developed world, with high labor costs, has suffered in competition with lower cost manufacturers.

Since the 1960s, efforts have been made to promote economic diversity in Northern Ireland. So keen was the UK government to attract new activities that it did not look closely at schemes to be subsidized. The outstanding example was the Delorean plan of the 1970s to build a futuristic sports car in Northern Ireland in a government-funded facility. By the time the scheme collapsed, the British government had supplied tens of millions of dollars. Only a few cars were produced, which are now collector's items.

Whatever the government did, it could not counteract the economic impact of terrorism, which disrupted daily life and deterred investors. Living standards were lower and unemployment high when compared with UK averages. Not until violence subsided, in recent years, has unemployment been reduced below 10 percent. The economy has ceased to depend upon security spending and new enterprises have been attracted. Much new investment has gone to Belfast (pop. 278,000) and nearby Lisburn (110,000). Derry (105,000), in a more isolated location and a much smaller market, is less attractive. Economic growth in Éire, and the increase of tourism there, has a positive impact on Northern Ireland now that violence is less likely to deter investment and visitors. Living standards remain below UK averages in Ulster.

Photo 9.10 The White Star liner, RMS Titanic, built at Harland and Wolff Shipyard, Belfast, struck an iceberg and sank on her maiden Atlantic voyage in 1912.

Source: Central Press/Hulton Archive/Getty Images.

England

The Uplands

Northern England is divided by the north-to-south Pennine upland, of Pennsylvanian age, composed largely of limestone but with significant outcrops of the sandstone known as millstone grit. (Figure 9.9) Igneous activity injected veins of minerals into the region but most of the economic deposits of lead and copper have been dug out. On the margins of the Pennines, to the west, the east, and the south, are coalfields.

Figure 9.9 England.

At the higher levels peaks and erosion surfaces of over 2,000 feet are common, with vegetation of grasses, sedges, bracken, heather, and many areas of ill-drained bog. At lower levels the valleys are largely in grassland for the raising of livestock. Cattle predominate in the valleys. Sheep live on the moorland higher up, until brought down for lambing in the spring.

On the North Sea side of the Pennines, northeastern England is underlain by the Northumberland–Durham coalfield. The region came to economic prominence shipping coal from Newcastle-upon-Tyne and Sunderland, on the river Wear, to London. Wagonways brought the coal from pits to staithes for loading onto ships. With the installation of iron rails, the wagonways became railways. Then an inventive mining engineer, George Stephenson, made a steam engine into a locomotive to pull coal wagons at a coal mine north of Newcastle. Quaker bankers wanted to connect the inaccessible south Durham coalfield to navigable water on the river Tees, and contracted Stephenson to build a railroad from Shildon on the coalfield to Stockton-on-Tees via Darlington. The Stockton and Darlington railway opened in 1825, the first public railroad, although several other mining engineers, in addition to George Stephenson, had built private railroads with locomotives to serve mines.

On the basis of coal the Northeast became a center of iron, steel, and shipbuilding. The landscape was covered with spoil heaps, coal mine winding gear, and lit by blazing blast furnaces. Huge workshops built locomotives and railroad equipment for Britain's railways and exported products worldwide, ensuring that Stephenson's 4 foot, $8\frac{1}{2}$-inch railroad gauge became the standard. On the tidal courses of the Tyne, Wear, and Tees the estuaries were crowded with shipyards.

Today, the mines, with their spoil heaps, have gone. Many shipyards have closed and steel-making capacity is reduced. Newcastle and other towns along the Tyne have turned to other things. Newcastle, reviving the well-designed early nineteenth-century streets, serves as an administrative, shopping, and cultural center with art galleries and vibrant clubs.

South of the River Tees is the Vale of York and Yorkshire. The fertile Vale is flanked in the west by the Yorkshire Dales, draining from the Pennines into the Yorkshire river Ouse. To the east of the Vale of York are the North York Moors and the Yorkshire Wolds. On the coast are the historic fishing towns of Whitby (Captain Cook's home port) and Scarborough. (Photo 9.11)

South and west of the ancient city of York, with British, Roman, Saxon, Norman, and medieval structures in the walled city, (Photo 9.12) are the great industrial centers associated with the woolen industry, which grew rapidly on the Yorkshire, Derbyshire, and Nottingham coalfield, in the nineteenth century, as steam engines were used to power mechanized factories. Leeds (716,000) is the commercial capital of the Yorkshire woolen region. Bradford (485,000), with surrounding towns, still produces woolen textiles, one of the few businesses of the first industrial revolution that are profitable in Britain.

In the postwar era, with demand for woolens high, mill owners wanted to run factories twenty-four hours a day but lacked the labor to do so. Migrants from Pakistan arrived to work the night shift. Soon Urdu was the mother tongue in central Bradford. Integration of Muslim Pakistanis was difficult and conflict grew between the multiculturalists, who wanted schools using Urdu for migrants, and those who argued that if migrants were to

Photo 9.11   A view of the upper harbor in Whitby, circa 1880 (with the Abbey on the hill). The small row boats fished close to the shore. The larger sailing vessels went out to the North Sea. Traditional inshore fishing has largely disappeared as modern efficient trawlers have depleted fish stocks.

Source: Frank Meadow Sutcliffe/Hulton Archive/Getty Images.

be upwardly mobile in Britain they had to learn English. In July 2005, militant Islamic suicide bombers blew themselves and other passengers up on London transport. Several of the bombers came from Pakistani communities in the Leeds area.

Southwest of Leeds, and still on the Yorkshire coalfield, is the metal working city of Sheffield (513,000). Cutlery and tools are made in the town, but not in the previous quantities. Major engineering companies operate in Sheffield but, like other industrial cities, Sheffield has become a regional shopping and cultural center, which houses large technology-oriented universities and teaching hospitals. The movie *The Full Monty,* set in Sheffield, is a commentary on the decline of steel.

The coalfield stretches into Derbyshire and Nottingham, but few coal mines now operate. The influence of coal continues, for it was a factor in the location of industries and the growth of towns in the nineteenth century. Derby, close to the limestone Peak District, at the south end of the Pennines remains a major railroad and engineering center. Derby was the original home of Rolls Royce. Nottingham (267,000), to the east, is a regional center with modern industries, large universities, and the castle overlooking the valley of the River Trent helping to attract tourists.

Photo 9.12 Medieval walls surround the city of York and York Minster.

Source: Bill Wymar/Alamy Images.

The Northwest and the English Midlands

In the north, on the west side of the Pennines, is the English Lake District, a region of natural beauty that inspired the Romantic poets, including Wordsworth, who lived at Dove cottage, at Grasmere. (Photo 9.13)

The core of the Lake District is a mass of older harder rocks rising up at Scafell Pike to 3,210 feet. The region was high enough to generate its own ice cap in the Ice Age. From the ice cap glaciers flowed down the valleys, widening them to U shapes and deepening valley floors. When the ice melted, water accumulated in the overdeepened valleys creating finger lakes, similar to those found in upper New York State. The Lake District is a National Park, still with much private ownership of land and towns serving as markets and tourist centers. Planning restrictions on development within the national park ensure that any new buildings incorporate traditional stone materials and display the architectural characteristics of the regional building style.

Much land is held by the National Trust, a private body that takes in trust historic houses and areas of natural beauty bequeathed by people wishing to preserve Britain's landscape heritage. In the Lake District, Beatrix Potter, author of the Peter Rabbit books, was a large benefactor. During the interwar years Potter, rich on Peter Rabbit royalties, bought farms in a period when hill farming was depressed. A typical hill farm had enclosed pasture lands running down to a lake and, on the upland moors and fells, the right to run livestock on open rough grazing. During the summer months hay is made on the permanent pasture and stored in barns to feed cattle through the winter months. The sheep are left out on the hillsides to look after themselves for much of the year but in the spring are brought to the enclosed fields to produce lambs, be sheared of wool, and

Photo 9.13   View of Grasmere. Wordsworth lived at Dove Cottage close to the lake.

Source: David Martin Hughes/Alamy Images.

dipped to kill ticks before being stamped with identifying dye and put out again onto open hillsides. This type of hill farming is common across highland Britain and in other European upland areas.

Beatrix Potter (1866–1943) acquired many hill farms, which passed to the National Trust on her death. The object was to preserve the working farms as part of a traditional way of life. But hill farming is marginal economically and even the Potter farms are facing amalgamation to create larger, more viable units.

South of the Lake District is Lancaster; the county town of Lancashire. The county is flanked by the Pennines in the east but the lower-lying west is underlain by a coal-field. In the eighteenth and nineteenth centuries the coalfield was exploited to transform Lancashire cotton manufacturing from a cottage industry to a factory-based production system in industrial towns like Preston, Bolton, Oldham, and Bury. Manchester (432,000) became the commercial and financial center of the cotton industry and exerted national political influence in the free trade debate. The Lancashire cotton industry imported raw cotton, much of it from the U.S. south, and exported cotton textiles. Workers needed cheap imported food, not bread made with protected, expensive, home-grown wheat. The city and the region, like other industrial areas in the nineteenth century, wanted free trade. If you wish to imagine a Victorian city in prosperity and pomp, go to Manchester to admire the nineteenth-century buildings erected in a neo-Gothic or classical style: the town hall, the library with theater, the Whitworth art gallery, the Victoria University of Manchester, the John Ryland's Library, the Free Trade Hall used for concerts by the Hallé Orchestra (f. 1852), and the commercial buildings including the railroad sta-tions, the station hotels, the banks, and the corporate headquarters. To advance the free trade views of the city, the *Manchester Guardian*, today the *Guardian*, newspaper was founded in 1821. The Lancashire cotton industry port was Liverpool (442,000), linked to Manchester by a railroad in 1830 and a canal opened in 1894.

Modern Manchester successfully hosts sporting events, including the Commonwealth Games, and major conventions. Canal street is full of clubs. There is a highly successful museum of the modern city, Urbis, and the renovated city enjoys a trendy image, with many young folk living close to the central city.

South of Lancashire is the county of Cheshire with the county town at Chester where the Roman walls and street plan are well preserved and Tudor townhouses project upper stories over the street. The English county towns were frequently bypassed by heavy industry and many still contain Tudor and medieval buildings within ancient walls. In northern England, York, Lancaster, and Chester are examples of well-preserved county towns. (Photo 9.14)

In lowland England, Norwich (Norfolk), Colchester (Essex), Canterbury (Kent), and Exeter (Devon) display the characteristics of the English county town with medieval fortifications, cathedral, castle, markets, craft streets, ancient narrow alleys, riverside quays, guild halls, county hall, and a range of building styles including medieval, Tudor, Georgian, Victorian gothic, and some modern.

South of Cheshire the lowland narrows between the Welsh uplands and the Pennines. Drainage is to the south via the river Severn. On the flanks of the Pennines is the Staffordshire coalfield, Stoke on Trent, and the Potteries industrial region. In the Potteries during the eighteenth century, industrialists, including Josiah Wedgewood (1730–1795), developed the mass production of pottery and china. The industry, now in competition

Photo 9.14   Chester has large numbers of Tudor style houses.

Source: Howard Barlow/Alamy Images.

with cheap labor products from Asia, is much reduced and the traditional pot-bellied firing kilns are obsolete but some of the long-standing manufacturers, including Wedgewood, still operate. (Photo 9.15)

South of Stoke we enter the English Midlands and manufacturing towns like Wolverhampton, Birmingham, and Coventry, lying on or close to the Birmingham coalfield. Metalworking in and around Birmingham was established in medieval times. With eighteenth-century industrialization, manufacturers adopted mass production techniques, supplying markets at home and abroad with everyday metal goods: pots, pans, and cheap tin trays. The city became a center of heavy engineering, producing the machinery needed to equip emerging industries. The industrialist Matthew Boulton (1728–1809), in partnership with James Watt, manufactured the Boulton–Watt steam engine in Birmingham. The steam engines were exported around the world along with other machinery from Birmingham.

Lowland England

Lowland England is a land of vales, ridges, escarpments, and rolling landscape. South of London, the Weald is framed by the chalk outcrops of the North Downs and the South Downs, the latter culminating in the White Cliffs of Dover. North of London, the

Photo 9.15 Two technicians applying relief molding to pots in a workshop at the Wedgewood pottery, Stoke-on-Trent, Staffordshire.

Source: Fox Photos/Getty Images/NewsCom.

Chilterns form part of a limestone ridge running from the south coast to East Anglia. Stretching from the Dorset coast, in the southwest, to Lincolnshire and east Yorkshire is the Jurassic ridge, rich in fossils on the Dorset coast and formerly exploited for iron ore further north.

Southeastern England

Lowland England, from the Humber to the Severn estuary, is dominated by London (Figure 9.9). Within the region all major towns are linked to London by rail and express bus services. London can be reached in two hours or less. People live in Oxford, Salisbury, Winchester, Ipswich, Canterbury, or the coasts of Kent and Sussex and commute to London. The highest densities of population are found in the southeast of England. London, after Paris, Moscow, and Istanbul, is the fourth largest city in Europe and one of the world's great financial and cultural centers. Institutions like the British Museum, the National Gallery, the Imperial War Museum, the British Library, and the Tate Gallery have major collections that attract visitors worldwide. International sporting events are held at Wembley, Wimbledon, Lords, and the Oval cricket grounds. Multinational corporations have head or regional offices in London, which contains the Parliament at Westminster, the ministries, Buckingham Palace, Westminster Abbey, and St. Paul's. All the major functions of commerce, finance, the church, and state are concentrated in London.

London has a capacity to constantly remake itself. After the Great Fire in 1666, which destroyed much of the urban area north of the Thames, together with the rats and fleas responsible for an outbreak of bubonic plague, the city was rebuilt, with Christopher Wren designing St. Paul's and some twenty other churches. In the nineteenth century, as London emerged as a global commercial and financial center, segments of the city were rebuilt with railroad stations, hotels, banks, office blocks, and of course, the Houses of Parliament in the neo-Gothic style, characteristic of Victorian architecture (Summerson, 1976). (Photo 9.16)

The bombing blitz of World War II, lasting from 1940 to 1945, demolished parts of the urban structure but did not destroy the functioning institutions of the city. After the war, bomb sites were redeveloped, often in a controversial way, as illustrated by the buildings erected around St. Paul's.

As London docks went out of use, a result of containerization and new patterns of trade and transport, the East End was redeveloped for housing and offices. The great tower at Canary Wharf symbolizes the transformation of London's docklands.

Around London are three international airports, Gatwick, Heathrow, and Stansted (near Cambridge). London has high-speed rail links, via the Channel Tunnel, with Brussels and Paris. In the era of globalization, London is home to innumerable corporations and businesses, owned by worldwide interests. In many ways, with a cosmopolitan population, London is no longer an English city but a world city situated in England.

After World War II town and country planning was introduced. London was given a green belt to contain urban sprawl. New or expanded towns were created as satellite centers beyond the green belt to house London commuters. For the country as a whole the use of all land was recorded and classified as residential, agricultural, industrial, and recreational. Changing the use of land from, for example, agricultural to residential requires a planning inquiry. On the one hand, the planning process has preserved much rural landscape in the United Kingdom and forced the reuse of formerly derelict land.

Photo 9.16   Big Ben and the Houses of Parliament are an outstanding example of Victorian Neo-gothic architecture.

Source: Jean-Pierre Lescourret/SUPERSTOCK.

On the negative side, the building of new homes has not met demand. House prices in Britain are high, particularly in the southeast.

The Eastern Lowlands of England

The eastern lowlands extend north from the Thames estuary, through Essex and East Anglia to the Fenlands draining to the Wash. Beyond the Wash is Lincolnshire running to the Humber estuary, north of which is the port of Kingston upon Hull and the Yorkshire Wolds.

Inland of the Wolds is the Vale of York, flanked in the west by the Yorkshire Dales, in highland Britain. Coming south of the Humber estuary, along the valley of the river Trent, we come to Nottingham sitting on the boundary between highland and lowland Britain. Further south, cities like Peterborough, Northampton, Bedford, and Cambridge are lowland places, with relatively short journey times to London.

Most of the Eastern lowland is intensively farmed. Around the Wash and the Fens, land has been drained to create rich farmland producing horticultural products and high-yield crops of grain. The Wash has historic connections with New England, many of the Puritan settlers coming from the original Boston on the river Witham and from the towns and villages of Norfolk and Suffolk.

The South Central Lowlands

West of London, up the valley of the river Thames, is an ancient routeway that runs between the Berkshire and Hampshire Downs into the drainage basin of the river Avon that flows to Bath and Bristol. The Thames and the Avon are linked by canal. South of this routeway, on Salisbury plain, is Stonehenge and the county town of Wiltshire, Salisbury, with a cathedral, a medieval street plan, and half-timbered houses.

The modern rail and road (M4) routes pass through Reading, steer north of the Berkshire Downs, then pass through the Cotswolds to Bath, Bristol, and cross the Severn estuary to South Wales. The region from west London to Bristol supports modern technological industries and major academic institutions. Along the south coast cities include Portsmouth, Southampton, and Bournemouth, all prosperous places with fast links to London.

The Southwest Peninsula

The peninsula consists of Cornwall, Devon, and parts of the counties of Dorset and Somerset. In Devon and Cornwall we are back in highland Britain with the great granitic uplands of Exmoor, Dartmoor, and Bodmin Moor. The moors rise via steep slopes to rugged tors (crags) on the skyline. Dartmoor, the setting for Conan Doyle's *The Hound of the Baskervilles,* houses a high-security prison dating to the Napoleonic wars and the War of 1812. The Dartmoor National Park is populated with sheep, ponies, and tourists in the summer. (Photo 9.17) The southwest peninsula stretches to the Lizard, Penzance, and Land's End, where ancient igneous and metamorphic rocks are exposed along the coast.

The south coast of the peninsula faces the English Channel, enjoying the mildest climate in Britain, with warm westerly winds prevalent for most of the year and the uplands serving as a buffer against maritime polar air coming from the north in winter. The north coast of the peninsula is open to the Atlantic and large breakers. In summer, there is surfing, but conditions are usually cool and can be dangerous.

Photo 9.17 Traditional Dartmoor ponies, Dartmoor Devon.

Source: Penny Tweedie/Alamy Images.

The major towns in the southwest peninsula are Exeter, the county town of Devon, Plymouth, the naval base (and home port of Francis Drake and Walter Raleigh), and Falmouth, which in the colonial era was a port from which mails and people left for the thirteen colonies. The major industry today is tourism, using the coasts, uplands, national parks, and historic towns to attract visitors. The mild climate results in many retirees settling in Devon and Cornwall.

► A UNITED KINGDOM?

There are regional tensions within the constituent parts of the United Kingdom that are not limited to the relationship between England, Scotland, Wales, and Northern Ireland. In Scotland, there is a historic cultural divide between what were the Gaelic-speaking highlands and islands, populated by the clans, and the lowlands running from Glasgow to Edinburgh, where English is spoken and even the distinct Glaswegian dialect is Anglo-Saxon at root. In Wales there is "Welsh Wales" where the first language is Welsh. In more prosperous North Wales and South Wales, English is widely used. Northern Ireland is bitterly divided into Protestant and Roman Catholic communities.

In 1855, Elizabeth Gaskell published *North and South,* a novel in which the differences between the two parts of England were explored. In 1855 the south of England was a region where wealth, social position, and political power were determined by land ownership. The landscape was controlled by families owning large estates whose income came from rents on agricultural land. The towns of the south, except London, a long-established commercial center, were service centers with agricultural markets, craft workshops, and artisans making consumer goods for the well to do.

On northern coalfields, new wealth was created at iron works, engineering factories, and textile mills. Northern towns were manufacturing centers, populated by growing numbers of low-paid factory workers living in overcrowded conditions lacking running water and sanitation. Inherently there was conflict between the industrial entrepreneurs accumulating wealth and workers paid subsistence wages.

The industrial centers of northern England, Scotland, Wales, and Northern Ireland were the growth engine of Britain in the nineteenth century. The power to govern, collect taxes, and spend revenue resided in London where Britain's national, imperial, and global agenda was determined in a political system controlled by the great land-owning families. Manchester, and the free traders, got protective tariffs removed on food imports in the 1840s, allowing imported cheap wheat from the Americas, but modern policies appeared slowly, particularly in education. As the drive to economic maturity arrived, Britain lacked technical colleges.

In the twentieth century, as industries consolidated into combines, head offices located in London, concentrating decision making in the capital. Banking crystallized into a few large nationwide organizations and banks, formerly serving Manchester, Birmingham, and Newcastle, became part of companies headquartered in London. When, after World War II, the Labour government nationalized coal mines, steel works, and rail networks, the headquarters of the National Coal Board, the British Steel Corporation, and British Rail were in London. A socialist government with more policy initiatives and regulations meant more government departments and an increasing number of bureaucrats who lived, worked, and consumed in and around the capital. London was not exceptional. The same process of the capital pulling people, power, and purchasing capacity was experienced at Paris, Berlin, Madrid, and most other capital cities. The booming suburbs around Washington, D.C., illustrate something of the process, although both the U.S. and Canada are fortunate that the major financial centers—New York and Toronto—are not capital cities.

For decades there has been a well-marked migration of people from the north to the southeast of England. When infrastructure investments were made the booming, growing south was first served. When motorways (equivalent to interstate highways) were built, the system concentrated on the southeast. Much of northern England and Scotland are still not linked by highways of interstate quality. The road traveler driving from Newcastle to Edinburgh along the east coast road (A1) is still likely to be slowed by agricultural equipment sharing the road.

After World War I, and the Great Depression, northern regions became increasingly peripheral. When the British car industry developed in the 1920s, it located in Birmingham, Oxford, and Dagenham (east of London) close to the more affluent consumer markets. The development of road haulage disadvantaged the northern regions (Crafts, 2005) as roads improved first in the south.

Southeastern England is now overcrowded and congested. Land available for home building is scarce. Groups like the Society for the Preservation of Rural England resist farmland becoming suburbs. Water supply is an increasing problem in dry summers and the main catchments are in highland Britain: Wales, Scotland, and the Pennines, regions that resist the construction of more reservoirs.

Northern regions draw some benefits from southern congestion. In a 2006 poll North Yorkshire was named the "Garden of England." Running from the Pennine upland to the North Sea shore, the region includes the Dales, the Vale of York, the North York Moors, and a Viking coast with scenic fishing ports including Whitby, Scarborough, and Robin Hoods Bay.

Denmark, the UK, and EU

Denmark and Britain did not sign the Treaty of Rome in 1957, preferring EFTA (1959), which simply made trade between members easier and did not involve merging sovereignty. Both Denmark and Britain were large importers of foodstuffs and agricultural inputs like feed grains for livestock. Joining the European Community, with its protectionism and high tariffs on agricultural imports, meant higher food prices.

However, Britain and Denmark had strong trade links with each other and with Germany. The Danes produced high-quality meat and dairy products and, as these goods crossed into Germany, Common Agricultural policies raised the price. British manufacturers found their goods disadvantaged on entering Europe, by the Common External Tariff.

After unsuccessful applications in the 1960s, Denmark and Britain entered the European Community in 1973. As later joiners, the new members paid a price, not getting a favorable budget deal and having to accept the newly created Common Fisheries Policy, which pooled fishing resources. Britain and Denmark had large fishing fleets.

Danes voted for EU entry in a 1972 referendum. There was no referendum in Britain, Parliament, controversially, making the entry decision without referring the issue to the electorate.

When it was time to ratify the Maastricht Treaty, which deepened EU and promoted the merging of sovereignty, the British Parliament again acted for the electorate. Denmark had a referendum and, by a narrow majority, rejected Maastricht. Denmark, like Britain, has a constitutional monarchy, and a parliamentary tradition stretching back a thousand years. Danes were fearful that Maastricht would allow the large countries to dominate the small countries of Europe. The Danish government negotiated opt-out clauses to the Maastricht Treaty and the electorate then voted for ratification in 1993. Denmark, like Britain, opted out of the monetary system and neither country has adopted the Euro.

In Britain a small *Better Off Out* campaign has appeared. In a letter to the right of center *Daily Telegraph,* dated July 20, 2006, five members of the House of Lords (the upper house in the British system) stated a case against the EU:

> In an age of emerging global markets, a protectionist EU surely does not offer the best future for Britain, or indeed for millions of others around the world who would . . . like to trade with us on an equal footing.
>
> Trade and co-operation with our European neighbors are . . . important, but as Norway and Switzerland ably demonstrate this can be easily achieved outside EU. Not only would this give us access to the booming economies of India and China, it would also free us from the growing erosion of parliamentary democracy and the many restrictions federal legislation places on British businesses.

The letter suggested, in an age of globalization, a regional organization like EU is outmoded. The opinion in the letter represents a minority view, but a significant proportion of the electorate in Britain and Denmark share the fear of the erosion of parliamentary democracy and the impact of EU legislation on economic activity.

▶ SUMMARY

Britain, Ireland, and Denmark are among the more dynamic economies in western Europe. In the United Kingdom, as a result of Thatcherite policies, old industries ran down as the financial and service sector grew. Unemployment is lower than in France and Germany. On the debit side manufacturing, which France and Germany have been more successful in preserving, is much less important than it was a few decades ago. Britain has encouraged investment from outside. France and Germany find ways to resist foreign capital taking over companies. Britain possesses an economy more open to the forces of globalization.

▶ FURTHER READING

DENMARK

Olwig, K. *Nature's Ideological Landscape*. London: Allen and Unwin, 1984.

IRELAND

Atlas of Ireland. Dublin: Royal Irish Academy, 1979.

Cabot, C. *Ireland*. London: HarperCollins, 1999.

Cahill, T. *How the Irish Saved Civilization*. New York: Doubleday, 1995.

Demangeon, A. *The British Isles*. London: Heinemann, 1958.

Hourihane, J. (ed.). *Ireland and the European Union: The First Thirty Years*. Dublin: Lilliput Press, 2004.

Johnson, J.H. *The Human Geography of Ireland*, Chichester, UK: John Wiley and Sons, 1994.

Jones, R.C. "Multinational Investment and Return Migration to Ireland in the 1990s—a county level analysis." *Irish Geographer*, 36(2) (2003): 153–169.

Maclaren, A. *Dublin*. New York: John Wiley and Sons, 1993.

McCarthy, M. *Ireland's Heritage: Critical Perspectives on Memory and Identity*, Alderhot, UK: Ashgate, 2005.

McGahern, J. *By the Lake*. New York: Alfred Knopf, 2002. A novel describing traditional rural life around Roscommon and Leitrim in the upper Shannon drainage system.

McManus, R. "Dublin's changing tourism geography." *Irish Geographer*, 34(2) (2001): 103–123.

Viney, M. *Ireland*. Washington, DC: Smithsonian Books, 2003.

UNITED KINGDOM

Ackroyd, P. *London: The Biography*. New York: Doubleday, 2001.

Addison, P., and H. Jones (eds.). *A Companion to Contemporary Britain 1939–2000*. Malden, MA: Blackwell, 2005

Aitchison, J., and H. Carter. *Language, Economy and Society: The Changing Fortunes of the Welsh Language in the Twentieth Century*. Cardiff, UK: University of Wales Press, 2000.

Annual Abstract of Statistics. New York: Palgrave Macmillan, 2005. Annual statistical source for UK.

Carter, H. *National Atlas of Wales*. Cardiff, UK: University of Wales Press, 1989.

Crafts, N. "Market Potential in British Regions, 1871–1931." *Regional Studies* 39(9) (2005): 1159–1166.

Daniels, P.W. "The Geography of Economic Change." in P. Addison and H. Jones (eds.), *A Companion to Contemporary Britain, 1939–2000*. Oxford UK: Blackwell Publishing, 2005.

Darby, H.C. (ed.). *A New Historical Geography of England*. Cambridge UK: Cambridge University Press, 1973.

Devine, T.M., C.H. Lee, and G.C. Peden. *The Transformation of Scotland: The Economy Since 1700*. Edinburgh, UK: Edinburgh, University Press, 2005.

Dorling, D. *Human Geography of the U.K*. Thousand Oaks, CA: Sage Publications, 2005.

Gilg, A. *Planning in Britain*. Thousand Oaks, CA: Sage Publications, 2005.

Gowland, D., and A. Turner. *Reluctant Europeans: Britain and European Integration 1945–1998*. London: Longman, 2000.

Lawton, R. and C.G. Pooley. *Britain: An Historical Geography*. London: Edward Arnold, 1992.

Mackinder, H. *Britain and the British Seas*. London: D. Appleton and Co., 1902.

Macmillan, H. *At the End of the Day 1961–1963*. London: Macmillan, 1973.

Manley, G. *Climate and the British Scene*. London: Collins, 1952.

O'Sullivan, Michael J. *Ireland and the Global Question*. Syracuse, NY: Syracuse University Press, 2006.

Pacione, M. *Glasgow: The Socio-Spatial Development of the City*. Chichester, UK: John Wiley and Sons, 1995.

Summerson, J. *The Architecture of Victorian London*. Charlottesville: University Press of Virginia, 1976.

Young, J.W. *Britain and European Unity, 1945–1999*. 2nd edition. New York: St. Martin's Press, 2000.

10

SOUTHERN EUROPE: GREECE, SPAIN, AND PORTUGAL

After World War I, southern Europe moved toward strong-man government and in some cases, fascism or communism. As discussed in Chapter 7, Mussolini embodied strong-man government with his march on Rome, leading to his appointment as Prime Minister (1922) and the establishment of a full fascist regime in the years 1928 to 1943. In Portugal the monarchy was forced out and after 1928 the country came under the rule of Salazar, who was determined to improve living standards in his "New State" (*Estado Novo*). Salazar was dictatorial and authoritarian but he was not military or totalitarian. Portugal did not employ the full terror apparatus to keep the population docile, but personal liberties were curtailed. Life was worse in Spain. A civil war between the fascists and the republicans lasted from 1936 until 1939. In the war battles were fierce, reprisals dreadful and, at Guernico, Europe witnessed the first terror bombing of a civilian population. At the end of the civil war, fascist General Franco was in charge, the monarchy suspended, republicans hunted down, political prisons filled, and armed guards on every street.

Greece came under General Metaxas in 1936. He restored the monarchy, adjourned parliament, and imposed a dictatorial rule that lasted until the German invasion in April 1941. The now defunct state of Yugoslavia, Mediterranean by possession of an Adriatic coastline, was under communist rule and Marshall Tito after World War II. Tito died in 1980. The country broke up in the 1990s.

Although under authoritarian rule the Mediterranean countries, with the exception of Italy, tried to stay out of World War II. Italy invaded Albania (1939) and then unsuccessfully attacked Greece. Hitler, coming to Mussolini's aid, invaded and occupied Yugoslavia and then Greece in April 1941. Hitler's efforts to get Spain into the war were unsuccessful. Franco understood that if he joined the war, Spain would be blockaded by the Royal Navy. If Germany wanted Spain as an ally, Hitler would have to commit supplies, raw materials, and armaments. Hitler could not provide the goods. German troops were denied permission to move through Spain to attack Gibraltar and close off the western Mediterranean to Britain. Portugal, traditionally a British ally, stayed out of the war.

Italy got rid of its fascist regime in 1943, surrendered to the Allies, declared war on Germany, and ended the war as part of the European movement. Neutral Spain and

Portugal ended the war with fascists in command. Until the fascists went in the 1970s the countries were not welcome in the European community. Greece suffered a post–World War II civil war, came under military rule in 1967, and did not get parliamentary democracy until the mid-1970s.

Lack of political stability was not the only disadvantage Greece, Spain, and Portugal carried. The countries had few industries, lacked diversified economies, and were unlikely to compete within the shared European economic space. Portugal's major exports were cork, wine, port, salted fish, and tins of sardines. Greece exported currants (dried grapes), olive oil, tobacco leaves, almonds, fruits, vegetables, and sheep skins. There was shipbuilding at Piraeus. The famed Greek shipping magnates operated internationally and invested profits overseas, the Greek currency, the dracma, not being an investment tool.

Spain did have coalfields, mineral resources, shipbuilding, and heavy industry. Industrial cities were Bilbao, Santander, Barcelona, and València. The main exports were citrus fruit, onions, and wine on the agricultural side. Iron ore, wolfram, and mercury were mineral exports. Greece, Portugal, and Spain, in the late 1950s, as the EEC was forming, had low living standards, high infant mortality, poor education, and lacked democratic political institutions.

The 1970s was the decade in which democracy returned. The Greek Colonels lost power in 1974. General Franco died in 1975. The first free elections in fifty years were held in Portugal in 1976. In the 1980s Greece (January 1981), Spain and Portugal (January 1986) joined the European community.

▶ GREECE

Greece is a young state in an ancient land. In the fifteenth and sixteenth centuries the lands that form the present Greek state came under Turkish rule as the Turks expanded into southeast Europe.

In the early nineteenth century there was increasing pressure for Greek independence, with Britain, France, and Russia pressuring Turkey into recognizing an independent Greek state. The Turks resisted, resulting in an allied naval force being deployed successfully at the Battle of Navarino (1827). By the Treaty of Adrianople (1829), Turkey recognized the autonomy of part of the Greek lands and Greece entered a phase of territorial expansion.

Territorial Growth of the Greek State

During the Napoleonic wars Britain had acquired the Ionian Islands (Figure 10.1) and ceded them to Greece in 1864. Britain was not supportive when Greece wanted to take Crete from Turkey in the late 1860s and the Greeks failed to take the island by force in 1897. As a result of the Balkan War of 1913, Greece acquired land in Macedonia and western Thrace, including the city of Salonica, now called Thessaloníki.

Eventually Greece joined the western allies in World War I against Turkey. Under the Treaty of Sèvres, negotiated in Paris after World War I, Greece was to be given the city of Smyrna (Izmir) and surrounding territory on the west coast of Anatolia, where there was a Greek population. The Turkish republican movement, under Atatürk, rejected the treaty signed by the Sultan and occupied Smyrna. Peace between Greece and Turkey came by the Treaty of Lausanne (1923). Greece gave up the claim to Smyrna (Izmir)

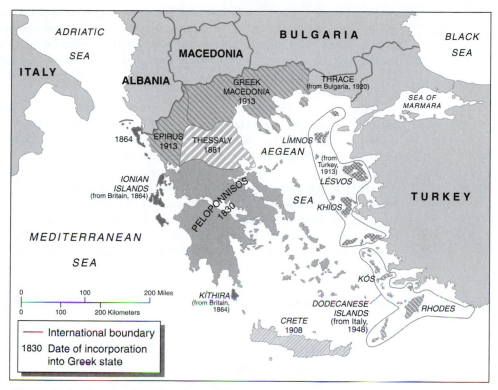

Figure 10.1 Growth of Modern State of Greece.

and relinquished part of Thrace. Populations were exchanged, partly under League of Nations supervision. In total over a million Greeks left Turkish territory. The number of ethnic Turks leaving Greece for Turkey was less but still involved hundreds of thousands of people. Bulgarians and Serbs were also forced out of Greece. Prior to World War I, Bulgaria had territory on the Mediterranean Sea, which was lost. As with many of the younger European countries, Greece was politically unstable during the 1920s and 1930s, swinging between a republic and a monarchy and coming under the rule of General Metaxas in 1936.

In October 1940 Italian troops, having seized Albania in the previous year, unsuccessfully attacked Greece. In 1941, prior to the attack on the Soviet Union, German divisions occupied Yugoslavia and then Greece. Before the German occupation was ended in 1944, Greece moved into a civil war between communists and monarchists. After the war the communists received Soviet bloc aid via Yugoslavia, prompting President Harry Truman (March 12, 1947) to issue the Truman Doctrine, which pledged support for "free peoples who are resisting attempted subjugation by armed minorities or by outside pressures." Greece received large U.S. financial and military aid. It helped when Marshall Tito of Yugoslavia, breaking with Moscow in 1948, ceased being a conduit for supplies to communist guerrillas. Greece joined NATO in 1951.

Political stability was slow in coming. The monarchy was restored after World War II but at the end of the civil war in 1949 political divisions were deep. In 1967 a military coup by the "Greek Colonels" led to the suppression of democracy, the exile of the king, censorship, imprisonment without trial, torture, and widespread abuses of human rights. The Colonels relinquished power in 1974 after failing to take over independent Cyprus. Cyprus had a substantial Turkish population, and the coup attempt led to a Turkish invasion and the division of the island into Greek and Turk sectors, separated by a UN Green line.

After the restoration of constitutional government in Greece, the Colonels were put on trial, the electorate voted for a republic, and parliament began to function. The country could now claim to be a democracy and it became possible to apply for EEC membership, although the economy was in no condition to compete within the European market. Greece was admitted to the European Community on January 1, 1981. The territorial disputes have not gone away. The Greeks and Turks still dispute small Aegean islands and Greece was resentful of Macedonian independence from Yugoslavia, thinking the name, at least, was Greek. It is difficult to reject the view that Greece came to EU membership as a result of European countries wanting to bring stability and European human rights standards to a country that had suffered so many internal and external disputes.

Regions and Environments of Greece

We can divide Greece into three major parts (Figures 10.2 and 10.3):

1. The Northern Landmass, extending down from mainland Europe
2. The Peloponnisos, almost but not wholly separated from the northern landmass by the Gulf of Corinth (Korinthiakós Kólpos)
3. The Greek islands, lying in the Ionian and Aegean seas

The Northern Landmass

Much of Greece is a land of mountains and uplands, which are young, steep-sided, and composed predominantly of limestone. The general trend of the mountains is north to south and the valleys lying in the folds of the mountains run north to south as well.

The major range is the Píndos mountains running into Greece from Albania and Macedonia. In the south, the mountains swing to the east along the Gulf of Corinth. The highest peak in the range is Mt. Smólikas, 8,650 feet, near the Albanian border, but peaks exceeding 8,000 feet are found along the range to the Gulf of Corinth. Much of the mountain environment is still forested. Rainfall can be heavy on the western Adriatic side of the mountains. The east is in a rain shadow.

Limestone is porous; water sinks in dissolving the rock, creating sink holes, caves, and systems of underground drainage. Occasional layers of clay impede the downward movement of water and result in lines of springs along the limestone–clay boundary, where villages are sited.

Frequently mountains and higher ground extend to the shore and coastal plains are not extensive. The plain of Thessalía is enclosed by mountains. In the plain are several agricultural centers including Lárisa, which is linked to the port of Vólos on the Aegean. The major areas of coastal plain are on the north shore of the Aegean in the hinterland of Thessaloníki and on the plain of Drama. The plains do contain areas of modern agriculture.

Figure 10.2   Greece and Neighboring States.

The Peloponnisos

The Peloponnisos is separated from the northern landmass of Greece by the Gulf of Corinth with, in the east, a narrow isthmus joining the Peloponnisos to the area around Athens. In classical times, small vessels were hauled across the isthmus from the Aegean into the Gulf of Corinth. The modern Corinth canal was opened in 1893, shortening the voyage from the Adriatic to Piraeus by over 200 miles (Photo 10.1). The city of Corinth was one of the most prosperous in ancient Greece, founding many colonies including Siracusa in Sicily. The Romans destroyed the city in 146 B.C. and rebuilt the settlement (44 B.C.) as a Roman colony. St. Paul established a Christian community in Corinth. In modern times the town has been subjected to devastating earthquakes in 1858 and 1928. Today, it is a town of modest size.

The mountains of the Peloponnisos generally trend north to south, with several peaks exceeding 6,000 feet. At higher levels there are remaining stands of coniferous trees, but the greatest part of the mountains have suffered erosion and goats retard the regeneration of any vegetation other than Mediterranean *maccia* scrub.

The coastal plains are cultivated, as are valleys running to the interior. One of the most extensive valleys is that of the Evrótas, which drains to the south coast of the Peloponnisos. The present-day town of Spárti is close to the ruins of the ancient city of Sparta.

Figure 10.3 Greece and The Islands.

The Peloponnisos, except at higher levels, has a Mediterranean climate. Winter is the rainy season. Frosts do not occur on the coastal plains but on the inland uplands, winters can be cold and bleak, particularly when a depression tracks over Greece. In summer the region is hot and dry with summer crops depending on irrigation.

The Greek Islands

There are thousands of islands large and small in the Ionian and Aegean Seas, which we refer to collectively as the Greek islands. Many of the islands are tiny and consist of bare rock. They are featured on tourist posters as white limestone islands, surrounded by the blue Mediterranean Sea, shimmering in the summer sun, cooled by the Etesian winds blowing from the northeast during the long summer, with a canvas sail windmill emphasizing the breeze. Many of the islands have only become a part of the modern Greek state in the last century. For example, the Dodecanese islands, including Rhodes, were Turkish, then under Italian control from 1912 to 1945 and made a formal part of Greece in 1948. A number of the islands were used by western powers as bases for trade with the Ottoman Empire and the Middle East before being incorporated into the expanding Greek state in the nineteenth and twentieth centuries.

The Ionian Islands lie close to the west coast of Greece in the Ionian Sea south of the Adriatic and the Strait of Otranto. The most northerly major island in the group is

Photo 10.1 Corinth Canal.

Source: Age fotostock/SUPERSTOCK.

Corfu (Kérkyra). Cephalonia (Kefallonía), Itháki, and Zákynthos lie in the south. Itháki was the home of Odysseus, when he was at home. The Ionian islands were under Venice from the fourteenth century until 1797 when the French took possession. At the end of the Napoleonic wars the islands came under British administration until ceded to Greece in 1864.

In the Aegean there are several major groups of Greek islands, including, from north to south, the Thracian Islands with Thásos, the Sporades, the eastern Aegean islands containing Lésvos and Khíos, the Cyclades with Náxos, and the Dodecanese in which the major island is Rhodes. Crete, at the south end of the Aegean, is the largest of all the islands, extending 160 miles from west to east.

The eastern Aegean islands and the Dodecanese form a string of Greek-speaking settlements off the coast of Turkey. In the medieval and early modern era the Genoese (1261–1566) used islands like Khíos for trading purposes. The Dodecanese were controlled by the crusading Order of St. John of Jerusalem (1309–1522) who used their base at Rhodes to attack Muslim shipping.

In the Dodecanese, the islands of Khíos, rising to 4,160 feet, and Rhodes (3,986 feet) rise high enough to induce plentiful winter precipitation to water fertile coastal plains growing the full range of Mediterranean crops. The Knights of St. John developed the city of Rhodes into a well-fortified naval base. Many of the buildings erected by the Order still stand and are part of the attraction of Rhodes as a tourist center. Upon coming to power in 1520, Süleyman the Magnificent determined to remove the Christian crusaders from his coast as the Order's navy was close to the sea route from Constantinople to Egypt, which the Ottomans had taken in 1516–17, giving them a place in the spice trade that flowed through the Red Sea. After a long siege lasting from June to December 1522, the Order was forced to accept terms and leave Rhodes with the fleet, armaments, records, and moveable wealth. A number of Rhodeans, dependent upon the Order for a living, also left the islands, many finding their way to Malta where the Order founded a new base in 1530.

Crete

The largest of the Aegean islands is Crete. Here Minoan civilization flourished until 1600 B.C. Crete became part of ancient Greece and was taken by Rome in 67 B.C. before coming under Byzantium, although subject to Arab attacks. Crete was gained by Venice in 1204, being known as Candia. After a long struggle the Turks got control of most of Crete in 1669. Greece tried unsuccessfully to take Crete from Turkey in 1897, after which allied powers administered the island until it joined Greece in the years before World War I.

In World War II German forces occupied the island and set up airbases to attack British positions in Egypt and Suez (Psychoundakis, 1998). The west-to-east trending mountain range is highest in the west (8,058 feet), although peaks in the east still attain 7,000 feet. The island produces the usual Mediterranean crops on the coastal plains and in the valleys, which penetrate into the central highlands. Now tourism is the major industry.

Climate

We think of Greece as a Mediterranean land but the Mediterranean climate is not present in much of the uplands. In the Píndos mountains and adjoining uplands, the winters are cold, the summers hot but not wholly dry. In the winter, cold air from central Europe and associated winds can produce harsh conditions (Figure 10.3).

The coastlands fronting onto the Ionian and Aegean seas do have a Mediterranean climate, with warm, sunny summers and mild, moist winters. There are contrasts. The coastal areas adjoining the Adriatic and Ionian seas receive 30–40 inches of rainfall a year. The east coastlands looking to the Aegean, including the region around Athens, receive less than 20 inches of precipitation a year and summers are generally dry. In the Adriatic during winter, the *bora* brings colder air down the sea to meet maritime tropical air. Fronts develop along the boundary between the cold and warm air and precipitation can be heavy where mountains and higher ground are close to the Adriatic coast. By contrast, the Greek lands facing the Aegean are in a rain shadow and receive less precipitation.

Greek Macedonia and Thrace experience continental influences. Winters are colder than Mediterranean norms, hard frosts are recorded at Thessaloníki, and rainfall does occur in summer months. In winter, cold winds come to the coast from the continental interior.

In the Píndos mountains is deciduous forest but in Mediterranean climate environments the Aleppo pine is an important part of the natural vegetation. The coniferous tree has the capacity to withstand summer heat and drought. Few stands of the pine remain. Much forest was cleared in antiquity and regeneration is retarded by grazing goats. Many hillsides are in *maccia*-type vegetation, with low-growing herbs and scrubs.

Agriculture

Just under 20 percent of the Greek workforce is in agriculture and produces less than 10 percent of the GDP. Nearly a third of Greek exports are agricultural products. The Greek figures for percent of workforce in agriculture, the percent of GDP derived from farming, and the dominance of agriculture in exports are high in EU comparisons. Before the 2004 enlargement Greece stood apart from all other EU countries.

In areas of Greece with a Mediterranean climate there are two cycles in the agricultural year, one in the fall, the second in spring and early summer. In September temperate grains and vegetables are sown, to be harvested in March, April, and May. Many crops, particularly grains, are grown on terraced hillside fields that lack irrigation, for the moderate rainfall is concentrated in the winter months.

By June, summer conditions with long, sunny days, little cloud, stable air, and drought are established. Farming is concentrated in areas where irrigation water is available. Spring and early summer see intense agricultural activity as seedlings are set out, carefully watered, and rapidly mature to produce high yields of tomatoes, squashes, and other vegetables. In summer, the terraced hillsides are grazed by goats and, lacking water, lie idle in the heat. As fall rains arrive, hillside fields are ploughed and sown with grains, some of coarse varieties that provide fodder for animals.

Fruit trees are an important part of Greek agriculture and include citrus (harvested early in the year), figs, almonds, peaches, nectarines, and grapes to be consumed fresh, dried into raisins, or fermented into wine.

Away from the coasts, in the Greek uplands, there are few extensive areas of agricultural land and because of hard frosts traditional crops, like citrus and olives, are only found in favorable locations. Along the north shore of the Aegean, in the hinterland of Thessaloníki and Kavála, there are agricultural plains growing grains and tobacco. This was a region in which Greeks coming from Turkey, after World War I, were resettled. Villages were laid out and settlers acquired farms of an economic size. Farming in the region is modern compared with traditional Greek agriculture.

In most Mediterranean lands, and Greece is no exception, agricultural land is being abandoned. Coastal land is absorbed into towns and resorts. The small, terraced hillside fields go out of cultivation as they are difficult and unprofitable to farm. There used to be seasonal movement from the Mediterranean lowlands to upland summer grazing. Some transhumance still takes place, but the activity is much reduced.

Population Distribution and Urban Places

Historically the distribution of rural population has been strongly influenced by the quality of soil and the availability of precipitation and irrigation water. On the west coast, adjoining the Ionian and Adriatic Sea, rainfall totals are relatively high and so are rural population densities, but there are few cities other than Pátra.

Around the Aegean Sea, the center of the Greek world, rural population densities are lower than in the west, but there are more cities (Figure 10.3), including Athens-Piraeus, Thessaloníki, and Lárisa, inland from the Aegean on the plain of Thessalía.

Athens

The Athens conurbation, including the port of Piraeus, Kallithéa, and other suburbs, contains more than three million people, over a quarter of the population of Greece (Table 10.1). The conurbation contains about half of all Greek industry, in addition to being the major commercial, banking, and governmental center. The modern city is marked by overcrowding, traffic congestion, and on still days, smog (*netos*). Athens has always suffered dust storms in the dry summers when the Etesian winds blow from the northeast. (Photo 10.2)

Athens displays many of the site and situation characteristics of Greek cities. For defensive purposes the city was established four miles inland, with the citadel set on top of a hill rising steeply to three to four hundred feet above the plain. The hilltop Acropolis was a feature of many Greek urban places. Beneath the Acropolis was the lower town and the intensively cultivated agricultural plain. At the coast was a port connecting the city with trade routes and the wider world. The Greeks, of course, founded numerous colonies, ports, towns, and cities around the Aegean, the Black Sea, and in southern Italy and Sicily, including Naples, Taranto, Messina, Catania, and Siracusa.

Thessaloníki (Salonika)

The second largest city in Greece is Thessaloníki, a major port and manufacturing center on the Gulf of Thermaïkós. The valley of the Varda, which drains to the Aegean, provides a route north to Skopje in Macedonia and over the mountains to the Danubian lands. Thessaloníki, founded in 315 B.C., later became an important Roman urban center on the routeway from the Adriatic to Constantinople. After the split of the Roman Empire into east and west entities, Thessaloníki was a part of Byzantinium. The modern city still possesses the medieval walls and White Tower of the time. The city was later occupied by the Saracens, Normans, Crusaders, and Venetians before coming under Turkish rule in 1430.

TABLE 10.1 Greek Urban Areas

Greater Athens with Piraeus	3,363,400
Thessaloníki	800,800
Pátra	185,700
Peristérion (suburb of Athens)	137,900
Iráklion	144,600
Lárisa	125,000
Vólos	118,000
Kallithéa (suburb of Athens)	110,000

Photo 10.2 Athens and the Acropolis with smog (*netos*).

Source: Aliki Saountzi/Aliki Image Library/Alamy Images.

Within the Ottoman Empire the city was multicultural. Under the Turks and Islamic law, Christians were allowed to practice their religion. After 1492 and their expulsion from Spain, large numbers of Jews arrived in the city, speaking a Judeo-Spanish tongue. In the nineteenth century with the coming of the railroad and increased European trade, many Macedonians, Serbs, and Bulgars came to work in the city.

Salonika, to be renamed Thessaloníki, was taken by Greece in 1912 and non-Greeks were removed, including many Bulgarians. After World War I, with the exchange of Greek and Turkish populations, and the expulsion of remaining Bulgars and Serbs, the city became populated predominately by Greeks for the first time in five hundred years. There was still a large Jewish population but in 1917 a fire destroyed the central section of the city containing Jewish neighborhoods. When the central streets were rebuilt, in a planned manner, the Jewish communities were not accommodated. In 1939 the Jewish population in the city still numbered 60,000 people. Adolf Eichmann oversaw the deportation of Jews to the death camps in 1943. Between 1913 and 1947 Thessaloníki went from being a multicultural city to one that was largely Greek (Mazower, 2005).

▶ THE IBERIAN PENINSULA—SPAIN AND PORTUGAL

From the fifteenth to the early nineteenth century Spain and Portugal possessed large overseas empires. After the Napoleonic wars, the Spanish mainland territories in Central and South America became independent and Brazil separated peacefully from Portugal. Spain lost the Philippines and her Caribbean territories in the Spanish American War. Portugal held Goa until 1961, when it was seized by India. The long-standing Portuguese colonies in Africa, Angola and Mozambique, became independent in the 1970s and Macao reverted to China in 1999. The Atlantic archipelagoes of the Azores and Madeira are autonomous parts of Portugal. The Canary Islands, off the Atlantic coast of Africa, are part of Spain, together with two provinces around the towns of Ceuta and Melilla in Morocco, North Africa.

Having lost New World empires, Spain and Portugal did not find that capital and entrepreneurial skills had been released to promote industrialization in the nineteenth century. Portugal lacked coal and iron ore. Spain had ore and coal in the north but the development of heavy industry was confined to a few areas, including Bilbao and Santander and Barcelona on the Mediterranean coast. Historically Barcelona had imported raw cotton from the Mediterranean region and produced cotton textiles. Slowly, in the nineteenth century, the new textile technology was adopted at Barcelona. In the adjoining Ebro valley many small plants relied on water power.

Economic growth was slow in Spain and Portugal during the second half of the nineteenth century, compared with northern Italy. Spain and Portugal entered the twentieth century as the backward countries of western Europe. Political instability, followed by fascist regimes, did not create the business environments that encouraged capital investment in modern manufacturing capacity, although fascist leaders promoted the growth of some industrial sectors. As Mexico has demonstrated there are advantages to being a relatively poor country, with cheap labor and a low cost of living, if you are next to affluent countries with inhabitants who have money to spend.

In the 1950s and 1960s France, Germany, Italy, and the Benelux countries did not want Franco and Salazar negotiating for EEC membership, but many Brits, Germans, and Scandinavians wanted a cheap vacation in the sun, on the coasts of the Iberian Peninsula. Package tour holidays, particularly to the Spanish Mediterranean coast, became popular. Spain was cheaper than Italy and the area around Barcelona, the Costa Brava, catered to British visitors. Restaurants served fish'n'chips in the English style and mass-market brands of British beer were on tap at every bar. Tour operators and the construction industry could not keep up with demand and each season some visitors arrived to stay in half-finished hotels. Life in traditional coastal fishing communities was overwhelmed by tourists, new hotels, bars, and restaurants (Lewis, 2006).

The Balearic islands, particularly Majorca and its capital Palma, were overdeveloped along with the Costa Brava. Other Iberian coastal regions, including the Costa del Sol, further south on the Mediterranean coast, and the Algarve, in southern Portugal, developed rapidly. Portugal was less involved in mass-market tourism than Spain, but in both countries the foreign exchange earnings and the economic stimulus from tourism were substantial. When countries shed their fascist governments in the 1970s, they were able to join EC in the 1980s, having developed some industries that could compete in Europe. Traditional agricultural products, including sherry, port, wines, brandy, fruits, and vegetables, benefited from easier access to European markets.

Cheaper labor in the Iberian Peninsula remained an advantage. Spain, with over forty million inhabitants, has a large internal market and began to manufacture the full range of consumer goods, including cars. On a lesser scale, Portugal, with a population of just over ten million, also attracted investment. In both countries there were financial institutions capable of expanding and meeting the capital and investment requirements of growing economies. In the case of Spain, the Bank of Santander operates on a European and Atlantic stage. Numerous multinational Spanish corporations have emerged, doing business throughout the EU, and in Latin America (Guillén, 2005).

Constitutionally, Spain and Portugal established democracy after the fall of fascism. In Spain, Franco arranged for the restoration of the monarchy upon his death. The succession to the crown skipped a generation, and much of the civil war baggage was left behind. The young Juan Carlos came to the throne in 1975 to guide the country to a constitutional monarchy, in which, as head of state, he signs into law legislation passed in the bicameral Cortes. In February 1981, Fascists attempted a coup, holding the lower house captive for two days. Juan Carlos was firm. Constitutional government was quickly restored. The issue of the Basque region remains and the ETA terrorist group staged bombings until recently (Beck, 2005). Since the bombing of commuter trains in Madrid in 2004, acts of terror are unlikely to advance campaigns for more regional autonomy, and ETA announced a cease fire in March 2006. The cease fire was welcomed by most Basques who understood that ETA activities had deterred tourists and retarded investment. However, hard-line Basque separatists still threaten terror, and revoked the cease fire in June 2007.

Physical Environments of the Iberian Peninsula

Climate

Climatically the Iberian Peninsula is a mini-continent surrounded by the sea on three sides, with the Pyrenees keeping out cold, northern, winter air. The peninsula is a large enough landmass to generate continental influences, warming rapidly in summer and cooling quickly in winter. The major external climate influences are the Azores High, the westerly winds blowing off the Atlantic, the Mediterranean climate regime in the east, and, to the south, North Africa and the Sahara.

The climate influences can be illustrated by studying basic climate data for cities on the Iberian Peninsula (Table 10.2).

The west coast of the Iberian Peninsula, represented by Lisbon and A Coruña, indicates how sensitive temperature and rainfall are to the impact of the westerly winds and the Azores High. If in summer the Azores High is weak, depressions and cooler air are more likely to be allowed onto the northwest coast of Iberia. At Coruña, on average, temperatures are lower, and rainfall higher, than at more southerly Lisbon.

Barcelona, on the Mediterranean coast, displays a mild winter, a warm summer, and an adequate rainfall in the winter months. Summer is generally dry. Further south on the Mediterranean coast, Cartagena is warmer and drier than Barcelona.

TABLE 10.2 Iberian Weather Data

	mm Jan°F	mm July°F	Annual rainfall total in inches
Lisbon	52°	72°	27
A Coruña	48°	66°	32
Cartagena	51°	75°	15
Barcelona	49°	74°	21
Madrid	41°	77°	17
Sevilla	51°	85°	20

Madrid, on the Meseta, displays continental influences. Winters are colder than at the Iberian coasts and frosts are common across the Meseta. Summers are hot, with daily highs in the 90s and all-time highs around 110°F. Rainfall is low, with most precipitation coming in spring and autumn. On the Meseta some years are too dry to produce a crop. Much land is pasture for grazing sheep. The climatologist Kendrew described the Meseta summer:

> *The summer drought is broken only by an occasional thunderstorm; the heat is intense, evaporation vigorous, and the vegetation is dried up. Failing irrigation, the landscape is semi-desert, brown and grey; dust is everywhere . . . the dry air hazy with the particles swept up by the strong wind. (1961, p. 372)*

In the south of the Iberian Peninsula at Sevilla, in the valley of the Guadalquivir, summers are hot, reflecting a more southerly latitude, the influence of the Iberian landmass, and the impact of North Africa. In summer when the *Leveche* wind blows from the southeast, across the sea from North Africa, plants wither quickly and humans are weakened by the intense heat.

Topography

We can divide the Iberian Peninsula into topographic regions, as shown in Figure 10.4. The fit between the historic regions of Spain and the physical environment is not good, suggesting that humans interpret and develop environments in different ways (Figure 10.5).

The Meseta   The core of Iberia is the Meseta, consisting largely of older, harder rocks of Paleozoic age. In the Alpine orogeny mountain ranges of sedimentary rocks were pushed up to the north and to the south of the Meseta. In the north, the Cantabrians and the Pyrenees (highest peak 11,168 feet), to the south the Adalucian mountains (the Betic Cordillera) including the Sierra Nevada. In the Sierra Nevada over twenty summits exceed 10,000 feet, the highest being Mulhacén (11,411 feet) south of Granada. In winter, when depressions track into the Mediterranean from the Atlantic, precipitation in the high Sierra Nevada comes as snow and temperatures, at altitude, can drop below 0°F.

In the mountain-building phase, the old rocks of the Meseta block were faulted, buckled, and tilted to the west. Sierras were created within the Meseta. The central Sierra, rising to nearly 8,000 feet, divides the Meseta into a northern Meseta tableland (2,200–2,700 feet) and a lower, southern Meseta tableland (1,700–2,000 feet). Additionally, the Iberian mountains, running from the Cantabrian mountains to the Mediterranean coast north of València, separated the Ebro valley from the Meseta, ensuring that the Ebro flowed southeast to the Mediterranean Sea.

Because of the tilting of the Meseta, the major rivers of the Iberian Peninsula, except the Ebro, flow from east to west and fall to the Atlantic, including the Douro, Tagus, Guadiana, and Guadalquivir (Figure 10.6). The Meseta is often represented as a plateau with extensive flat surfaces, as in La Mancha, but in many areas there are more steep hillsides and incised river valleys than might be expected. As we saw in the case of the Massif Central in France, regions of older rocks, peneplaned by the forces of erosion, were rejuvenated in the Alpine orogeny: surfaces were uplifted and streams began to downcut, vigorously eroding steep-sided valleys.

Figure 10.4 Iberian Peninsula: Topographic Regions.

Major River Basins of the Iberian Peninsula

The Duoro drains much of the north and west of the Iberian Peninsula, receiving tributaries from the southern flanks of the Pyrenees and Cantabrian mountains as the river flows west to the Atlantic at Oporto. (Figure 10.6) The Duoro is not navigable upstream but is used to generate hydroelectric power and to irrigate fields in the river valley, which is renowned for the country estates, *Quintas,* producing the fortified Port wine that is still a substantial export from Portugal. The vines are grown on steep terraced slopes in rocky conditions (Stanislawski, 1970). (Photo 10.3)

The Tagus rises east of Madrid and flows south before turning westward and running through Portugal to the twelve-mile-long estuary on which Lisbon is situated. The Tagus, (and tributaries), drains an extensive part of the west central Iberian Peninsula and is navigable through much of Portugal. Lisbon is sited on the north bank of the Tagus estuary. A bridge across the estuary goes south to Setúbal on the Sado estuary. Setúbal

Figure 10.5   Iberian Peninsula: Administrative Regions, of Spain.

contains a former royal residence and is an important shipping point for the traditional Portuguese exports of wines, sardines, citrus, and cork.

The Guadiana drains the south central Meseta and flows west before turning southward to fall into the Atlantic west of the resort and port of Huelva. The Guadalquivir rises on the south flank of the Sierra Morena and flows in a southwesterly direction to the Atlantic, north of Cádiz. The river has an Arabic name (the big river) and many landscape features in this part of Spain, Andalucía, reflect the fact that the region, the Kingdom of Granada, had an Arab ruler until 1492. The valley of the Guadalquivir is well cultivated and modern prosperous farms produce wheat, maize, onions, sunflowers, tobacco, olives, and high-value vegetables for sale in European markets. Córdoba with its palace is on the Guadalquivir. (Photo 10.4) Downstream, at the head of maritime navigation, is Sevilla, the city that controlled Spanish trade with the New World. The Archive of the Indies remains in the city, housed in one of many impressive buildings. Downstream of Sevilla, the Guadalquivir is tidal and runs through Las Marismas wetlands, used as a wildlife refuge and for rice cultivation. Near the Atlantic, on the east

Figure 10.6 Iberian Peninsula: Major Rivers.

bank, is the port of Sanlúcar de Barrameda where the convoys (*flotas*) left for the Indies in the colonial era.

The Ebro rises on the south flank of the Cantabrian mountains and flows southeast between the Pyrenees and the Iberian mountains to the Mediterranean, draining much of northeast Spain. The river is little used for navigation but along its course are hydroelectric power plants and irrigation works. Climatically the Ebro basin, being mostly inland, receives less precipitation, around 15 inches a year, compared with 20 inches at Barcelona. Summers are hotter than at the coast and the natural vegetation is steppe rather than Mediterranean.

The Coasts

In the north, the Cantabrian mountains, composed of limestones and some sandstones, drop sharply to the Bay of Biscay. Frequently, the upland meets the sea in precipitous cliffs. Only around Bilbao, Santander, and Gijón are there narrow coastal plains.

Photo 10.3   Portugal, Douro Valley, with vineyards.

Source: Marka/AgeFotostock.

In the northwest the Galician coast is composed of granites and metamorphic rocks that have been weathered into rounded forms but the coast remains rugged. In the east, Portugal is an extension of the Meseta but adjoining the Atlantic is a coastal plain interrupted by ranges of hills.

The south coast of Portugal is the Algarve with a rocky limestone coast in the west and a sandy shore in the east. Inland the mild climate and plentiful sources of underground water have allowed productive agriculture, including the cultivation of citrus fruits. In the modern era the Algarve coast has become a resort area with numerous golf courses that can be played for most of the year. Leaving the Algarve (*the west,* from an Arabic root) we cross the Rio Guadiana and the Portuguese–Spanish border and come to the Atlantic coast of Spain, which runs to the Strait of Gibraltar. The long, sandy stretch of coast, running from the mouth of the Rio Tinto to just south of the Guadalquivir estuary, is known as the Playa de Castilla and behind the long beach are the dunes of the Arenas Gordas. On or near the mouths of the rivers falling to the Atlantic are a number of historic ports, including Huelva and, on the now silted-up course of the lower Rio Tinto, Palos. On his first trans-Atlantic voyage, Columbus sailed from Palos and many of his key personnel came from the port. Just south of the Guadalquivir estuary, higher ground

Photo 10.4 La Alhambra and Sierra Nevada. Granada, Andalucía, Spain.

Source: AgeFotostock.

reaches the coast, forming a peninsula and the well-protected Bahía de Cádiz. In the immediate hinterland of the bay is Jerez de la Frontera, an important wine, sherry, and brandy producing city that exports via Cádiz. Cádiz is a Phoenician foundation (Gadir); the Romans called the port Gades. The port played a major role in the naval affairs of Spain's New World empire and became a base for the Spanish navy. It was from Cádiz that a combined French and Spanish fleet, under the command of Admiral Villeneuve, sailed in October 1805 to be destroyed by a smaller Royal Navy force, under Admiral Nelson, off Cabo Trafalgar. The Battle of Trafalgar, on October 5, 1805, ensured that Napoleon lacked the naval forces to invade Britain.

From Cádiz south to the Punta de Tarifa, the western end of the Andalusian mountains (Betic Cordillera) meets the coast in cliffs and rocky headlands. Passing through the Strait of Gibraltar we come to Spain's Mediterranean coast. Again the Betic Cordillera comes close to the sea and coastal plains are narrow, except where rivers, descending from the upland, broaden their valleys approaching the sea. Málaga, a Phoenician foundation and major port with over half a million inhabitants, is near the mouth of the Guadalhorce, the whole area, the Costa del Sol, being overbuilt to accommodate resorts. At Marbella, corrupt authorities have allowed thousands of villas, time shares, and apartments to evade the planning regulations. The town council was suspended by the central government in 2006. The inland city of Murcia is on a broader plain created by the Rio Segura and its tributary, the Guadalentin. The plain is one of the hottest regions in Europe in the summer but irrigation works allow a large production of fruits and vegetables.

North of Cap de la Nau is the Gulf de València, surrounded by a broad coastal plain. The region is watered by rivers and streams flowing from the eastern edge of the Meseta and the Iberian mountains. Around the Gulf de València annual rainfall is only about 15 inches but streams from the uplands, principally the Turia on which València stands, feed a system of irrigation canals that water the *huertas* around the city (Photo 10.5). The long growing season, the irrigation water, and the intense summer sunshine allow continuous cultivation of *huertas,* which yield crops in winter and summer. Carefully worked *huertas*

Photo 10.5 Aerial view of irrigated *huertas*. València province, Spain. Notice the small plots and intensive cultivation.

Source: AgeFotostock.

will yield three crops a year. A great range of temperate and subtropical crops are grown and extensive groves produce citrus fruits, including València oranges. (Photo 10.6)

Around the Gulf de València is a sandy coast, with dunes sometimes impounding water inland as in La Albufera. València is an industrial city, port, and fishing town. A few historic buildings remain in the city.

North of the Gulf de València, the coastal plain narrows until the broad lower valley of the Ebro and its delta are reached. The Ebro valley is productive where irrigation water is available. Where water is lacking, the flat lands in the valley are inhospitable, being hot in summer and, inland, cold in winter.

Barcelona is a Carthaginian foundation taken over by the Romans. The city became an important medieval port and shipbuilding center. Today, Barcelona is the leading industrial and commercial city in Spain besides being the major port (Photo 10.7). The fabric of the city reflects Roman, Moorish, Medieval, Renaissance, and Baroque influences and modern buildings include the work of the architect Gaudi. Along the coast, north of Barcelona, is the Costa Brava, running along the narrow plain to the Pyrenees.

The Distribution of Population and Urban Places

The Meseta, with marginal rainfall, hot summers, and frosts in winter, does not support intense farming or high densities of rural population. Over wide areas of the Meseta, rural population does not reach twenty persons per square mile and many communities are aging and declining. Unless small towns and villages are favorably located and able to attract retirees and visitors, the general trend, over recent decades, has been for population numbers to shrink as young people move away and older people die. Some small towns

Photo 10.6 Intense, irrigated agriculture is practiced on the coastlands of Mediterranean Spain. Here, polythene greenhouses, near Almería, protect and produce crops.

Source: David Pearson/Alamy Images.

Photo 10.7 Las Ramblas, Barcelona, Spain.

Source: Marka/AgeFotostock.

have brought migrants from eastern Europe but the general problem remains. Isolated places on the Meseta with limited economies do not have the job opportunities that exist in most coastal cities.

Population densities are higher on the margins of the Iberian Peninsula than in the interior. The Mediterranean coast, except where mountains come to the coast, is densely populated and densities have increased as a result of tourism and people migrating from the interior to work in expanding coastal places. Rural population densities are high in the valley of the Guadalquivir where agriculture is intensive and climate allows two crops a year. In the northeast, in the Ebro valley, rural densities are high where irrigation water is available. If water is lacking, farming is marginal and rural population sparse. (Photo 10.8)

Spain has a rich urban history and contains urban foundations made by the Phoenicians, the Carthaginians, the Greeks, the Romans, and the Moors, besides towns laid out as the *Reconquista* moved to the south, culminating in the fall of the Kingdom of Granada in 1492. In the imperial period, when Spain administered a global empire, many cities expanded and were redeveloped with gracious Renaissance and Baroque buildings. Cádiz, Sevilla, and other places looking to the Indies display in the urban fabric the riches of overseas trade (Figure 10.7).

Photo 10.8 Shepherd and sheep moving to new pastures. Navarra, northern Spain.

Source: Age fotostock/SUPERSTOCK.

Figure 10.7 Iberian Peninsula: Major Urban Places.

Madrid, the former Moorish fortress of Majrit, became the capital of Spain in 1561 on the basis of a central location, although it was not on, or close to, navigable water as are most European capitals. Political power resulted in the completion of many important buildings, including the Prado art gallery, within a gracious city. The late nineteenth century saw industrialization, an influx of migrants, and urban poverty close to the worst in Europe. In the Spanish Civil War (1936–39) Madrid held out against the Fascist forces of Franco but when the city fell, the General consolidated power at the capital. Madrid grew while Barcelona, the center of the Republican resistance and Catalan identity, was marginalized. There is no way to explain the existence of a conurbation of three million people in the center of the Meseta, a region of low rainfall and population density, except

TABLE 10.3 Major Urban Centers in Spain

Madrid	3,017,000
Barcelona	1,528,000
València	762,000
Sevilla	705,000
Zaragoza	621,000
Málaga	536,000
Bilbao	360,000
Las Palmas (Canary Islands)	371,000
Murcia	378,000
Palma de Mallorca (Balearic Islands)	359,000
Córdoba	315,000
Alacant	294,000
Vigo	288,000
Gijón	270,000

in terms of the decision of the Emperor Philip II in 1561 to make Madrid the capital of Spain and the Spanish Empire (Table 10.3).

Regionalism in Spain

Regional identities are strong in Spain. Recalling a few facts from Spanish history suggests some regional divides. The marriage of Ferdinand of Aragon (which included Catalonia) to Isabella of Castile, resulted in the eventual union of the Kingdoms in 1479. Of course the union did not turn Catalans into Castillians any more than the Scot, James I, ascending the throne of England in 1603, made the English into Scots.

The reconquest of Spain from the Moors culminated in the fall of the Kingdom of Granada in 1492. The southern region of Andalucía became part of the Spanish kingdom after centuries of Islamic control in which distinct patterns of speech, architecture, and agriculture evolved.

During Franco's fascist rule (1939–1975), the emphasis was on loyalty to the state and one Spain: "La España Una." This view of Spain as a unitary state was enforced, but it bred resentment in regions like Catalonia, which felt discriminated against as a center of the Republican movement, which Franco defeated in the Spanish Civil War (1936–39).

Four languages now have official status: Castillian, Catalan, Gallego (Galician), and Basque. In June 2006, after legislation in Madrid and a referendum in Catalonia, the Catalan "nation," via the parliament in Barcelona, received increased power over taxes, more control over immigration, and a judiciary independent of Madrid. Catalan became the official language of the region and schools are required to teach Catalan.

There are eighteen Spanish regions (Figure 10.5). It seems likely that Galicia, the Basque region, Andalucía, and the Balearic islands will want more autonomy than they presently enjoy. In 2006, ETA, the Basque terrorist organization, announced a cease fire, which was breached in 2007. It may be possible to accommodate Basque aspirations within the "Catalan model," but more autonomy would be seen by many Basques as a stepping stone to independence. Many of the "no" votes cast in the Catalonia referendum of June 2006 were made by Catalonian nationalists who felt the new constitutional arrangements did not go far enough toward independence. The Catalonian push for more autonomy is resented in other regions of the country.

If regional autonomy moves toward separatism, in theory, the breakaway units could be accommodated within EU but new countries would have to accept all existing European legislation. There are practical issues. It remains to be seen if additional emphasis on Catalan in Barcelona will make the city less attractive to transnational corporations wanting to trade into worldwide Spanish-speaking markets.

Portugal: Urban Places

According to the Population Reference Bureau, Portugal has just over half the population of 10.5 million living in urban places. This is a low percentage by western European standards. The country does not have many large cities, but Lisbon (3.8 million) and Oporto (1.6 million) contain a high proportion of the urban population.

Portugal still has relatively high rural population densities. Portuguese population densities, in general, tend to be higher at the coast than inland where less productive upland is encountered, except in river valleys. Population densities are high in the Douro and Tagus valleys. The Tagus valley is a routeway into Spain and population densities are relatively high along the river. South of the Tagus valley the Meseta extends into Portugal; agricultural productivity and rural population densities drop. The Algarve, as a result of year-round agricultural productivity and resort development, has high population densities but few major urban places.

Lisbon

The capital of Portugal certainly had a Roman presence, and before that was possibly a Phoenician/Carthaginian trading colony. The Moors controlled the city from 714 A.D. to 1147 A.D., when it was taken in the Reconquest of the Iberian Peninsula. Lisbon grew in importance in the 'age of discovery' and particularly after Portugal developed Brazil in the sixteenth, seventeenth, and eighteenth centuries. Brazil was a source of timber, dyes, large quantities of sugar, and, in the eighteenth century, gold and diamonds. Lisbon prospered on the colonial trade, for Portugal could not absorb all the imperial produce. The fabric of the city reflects the history of the place. The medieval streets and the cathedral are clustered on the hillside leading up to the castle St. George, with its Moorish and medieval fortifications. In the 1758 earthquake (*terremoto*) the old Alfama district, leading up to the Castelo de São Jorge, survived, although many buildings collapsed. In the lower quarter, along the harbor shore, a tsunami inundated the city, drowning thousands.

After the debris was cleared, the harbor shore was redeveloped in a lavish baroque style. At the waterfront an impressive Praça do Comércio was laid out, surrounded by major buildings. Running back from the square a rectilinear street plan was traced and the area built up with well-proportioned buildings a few stories high. Lisbon is by far the largest place in Portugal and the center of government, commerce, banking, industry, and culture. (Photo 10.9)

Photo 10.9   Praça do Comércio, Lisbon, Portugal.

Source: Pixtal/SUPERSTOCK.

▶ FURTHER READING

Beck, J.M. *Territory and Terror: Conflicting Nationalisms in the Basque Country*. Routledge: New York, 2005.

Birmingham, D. *A Concise History of Portugal*. 2nd edition. New York: Cambridge University Press, 2003.

Closa, C., and P. Heywood. *Spain and the European Union*. New York: Palgrave, 2004.

Guillén, M.F. *The Rise of Spanish Multinationals: European Business in the Global Economy*. Cambridge, UK: Cambridge University Press, 2005.

Hughes, R. *Barcelona*. New York: Alfred A. Knopf, 1992.

Kendrew, W.G. *The Climates of the Continents*. Oxford, UK: Clarendon Press, 1961.

Kulikowski, M. *Late Roman Spain and its Cities*. Baltimore: Johns Hopkins University Press, 2004.

Lewis, N. *Voices of the Old Sea*. New York: Carroll and Graf, 2006. (Originally published 1984).

Mazower, M. *Salonica, City of Ghosts: Christians, Muslims, and Jews, 1430–1950*. New York: Knopf, 2005.

McDonogh, G. *Global Iberia*. New York: Routledge, 2006.

Psychoundakis, G. *The Cretan Runner: The Story of German Occupation*. Harmondsworth, UK: Penguin, 1998.

Stanislawski, D. *Portugal's Other Kingdom: The Algave*. Austin: University of Texas Press, 1963. This well-illustrated book provides an account of the Algarve when many small traditional farms and crafts existed.

Stanislawski, D. *The Individuality of Portugal: A Study in Historical Political Geography*. Austin: University of Texas Press, 1959.

Stanislawski, D. *Landscapes of Bacchus: The Vine in Portugal*. Austin: University of Texas Press, 1970. An illustrated, detailed account of traditional vine cultivation in Portugal.

Tremlett, G. *Ghosts of Spain: Travels Through Spain and Its Silent Past*. New York: Walker and Company, 2007.

11

YES AND NO IN THE 1990s: AUSTRIA, SWITZERLAND, SWEDEN, NORWAY, ICELAND, AND FINLAND

▶ INTRODUCTION

The tearing down of the Berlin Wall (1989), the absorption of East Germany into West Germany, and the collapse of the Soviet Union altered European geopolitics. Estonia, Latvia, and Lithuania reappeared as independent states. Poland moved to embrace the west. Finland was no longer held to strict neutrality by the Soviet Union. Traditionally neutral Sweden stopped worrying about some imagined slight to the Soviets. Austria was not bound to the Soviet interpretation of the Austria State Treaty of 1955 when the country got a formal peace, and agreed to strict neutrality. The collapse of the Soviet Union meant a looser definition of neutrality that would allow Austria to join EU. Austria's neighbor, Switzerland, asked for public opinion on the merits of EU membership (1992) but Swiss voters did not like the idea. Norway reconsidered membership but the electorate again rejected the commitment, as in the seventies. Iceland and Lichtenstein, along with Norway and Switzerland, all members of EFTA, did not join EU.

In 1990 EFTA had seven members. By the end of the decade, membership was down to Switzerland, Norway, Iceland, and Lichtenstein, but EFTA had obtained its aims. In 1992 EFTA and EU signed the European Economic Area (EEA) agreement, which established free trade between the members of both organizations. EFTA countries now enjoy free trade with the EU but avoid the political commitment and Brussels regulations. When EU accepted ten new members in 2004, all had already signed on to the EEA agreement with EFTA.

▶ AUSTRIA

Austria is a landlocked central European country. The neighboring states are Germany, the Czech Republic, Slovakia, Hungary, Slovenia, Italy, Liechtenstein, and Switzerland. Most of Austria to the south and west is mountainous country in the Alps. The Danube, rising in southern Germany, flows through northern Austria to enter the extensive Danubian plains (Figure 11.1). Vienna is located in the area where the Danube exits the mountains and uplands of Central Europe and flows onto the Danubian plains. From prehistory the Danube, running to the Black Sea, has been a European routeway. Coming upstream, Vienna is situated at a point where the Danubian routeway constricts and is channeled into a narrower valley that leads to southern Germany. (Photo 11.1) (Photo 11.2)

Austria has other historically important routeways. The Brenner Pass (4,494 feet) across the Alps is a routeway from Verona in northern Italy, which connects with Innsbruck in Alpine Austria and, further north, Munich in Germany.

Emergence of Austria-Hungary

By the fourteenth century the Ottoman Turks were a great power on the lower Danube, with Serbia falling to them in 1389 at the battle of Kosovo. Constantinople, on the Bosporus waterway that controlled entry and exit to the Black Sea, fell to the Turks in 1453. In 1526, at Mohacs, the Hungarians were defeated. A few years later Budapest was occupied and the Turks unsuccessfully besieged Vienna in 1529. The Turks besieged the city again in 1683 and then the Austrians began to push the Turks back down the Danube. In its turn Austria became the major power in southeast Europe and the Balkans.

Figure 11.1 Austria.

Photo 11.1 The Danube flowing through the central uplands below Schloss Aggstein.

Source: SUPERSTOCK.

Photo 11.2 Vienna is situated where the Danube leaves the central European upland and enters the Danubian plains on the way to the Black Sea.

Source: Age fotostock/SUPERSTOCK.

Austria came to rule Slovenes, Hungarians, Slovaks, Romanians, Germans, and the mixed population of Transylvania in addition to the Czech lands, Lombardy in northern Italy, and, after the Napoleonic wars, part of partitioned Poland.

The northern Italians and the Hungarians were particularly restive. The Italians became part of a unified Italy in the 1860s. In 1867 via the *Ausgleich* (compromise), Hungary got its own parliament, with both Austria and the Kingdom of Hungary recognizing the emperor Franz Josef as head of state. Foreign policy and defense were conducted on a joint basis and the countries shared a common currency and tariff policy.

Several other national groups in the empire, including the Czechs in Austria and the Slovaks in Hungary, hoped that political power would be devolved and lead to parliaments in Prague and Bratislava. Archduke Ferdinand was in favor of the devolution of political power from the center to regional capitals but he was assassinated in Sarajevo in 1914, setting off the events that led to World War I.

As World War I dragged on there were food shortages, strikes, mutinies, and ethnic unrest. By the fall of 1918 the Empire of Austria and Hungary had crumbled even before the armistice was signed. Several of the ethnic entities, including the Czechs and the Slovaks, in the state of Czechoslovakia had independence recognized at Paris in the peace treaties.

Austria after World War I

Austria emerged from the Peace of St. Germain (September 10, 1919) as a downsized state with a large reparations bill and a clause in the treaty forbidding amalgamation with Germany. A similar clause in the Treaty of Versailles forbade Germany absorbing Austria.

In 1910 Vienna had a population of just over two million inhabitants, living in a city of fine buildings, well-planned thoroughfares, and a vibrant intellectual life with arts and music. However, the city had grown too rapidly as migrants had moved in from rural areas to find work. Educated Austrians came to serve in the imperial bureaucracy, the banks, and commercial houses of Vienna. Between 1850 and 1900 the city grew fivefold. Even by nineteenth-century standards Vienna was overcrowded, sewers and drinking water provisions were inadequate, and diseases, like tuberculosis, were common. World War I exacerbated the problems. When the empire collapsed, Austria was downsized in defeat and shorn of an imperial role. Vienna lost status, power, and functions.

At the end of the war, Vienna was short of food, unemployment was high, administration weak, and political stability lacking. By 1923 the population of Vienna had decreased to 1.8 million but 28 percent of all Austrians lived there. The Czech lands that had supplied raw materials, coal for heating, and industrial goods were now part of independent Czechoslovakia and there were frontiers, customs posts, tariffs, and a separate Czechoslovak currency to inhibit trade.

Before World War I Vienna was a primate city within the empire as a whole. Vienna was approximately two and a half times the size of Budapest (Table 11.1). After the war Vienna was over ten times the size of the next largest place in Austria (Table 11.2).

With the monarchy gone Austria became a republic in November 1918. Vienna was socialist, the remainder of the country conservative. Street fighting was common in Vienna, resulting in serious riots in the summer of 1927. The depression made things

TABLE 11.1 Urban Hierarchy of Austria-Hungary in 1910

Vienna	2,031,498
Budapest (H)	880,371
Prague (C)	223,741
Lemburg (P)	206,113
Trieste (I)	160,993
Kraków (P)	154,141
Graz	151,781
Brno (C)	125,737
Szeged (H)	118,328

C, to Czechoslovakia; H, to Hungary; I, to Italy; P, to Poland after World War I. Lemburg became the Polish city of Lwów (1919–1939); today, it is L'viv in the Ukraine.

TABLE 11.2 Urban Hierarchy of Austria in 1923

Vienna	1,865,780
Graz	152,706
Linz	102,081
Innsbruck	56,401
Salzburg	37,856

worse with the collapse of a major bank. Chancellor Dollfuss was murdered by a Nazi gang in July 1934, and the country came increasingly under German control. Hitler wanted the manpower, industrial capacity, and resource base of Austria, which possessed iron ore, oil, timber, and nonferrous metals besides lying on the Danubian routeway to Hungary, Yugoslavia, and Romania, all countries that had raw materials Germany needed to support a war economy. On March 12, 1938, German forces entered Austria without resistance. *Anschluss,* the union of Germany and Austria, came the next day. Hitler was greeted rapturously by crowds in Vienna. The archbishop of the city encouraged churches to ring their bells and fly Swastika flags. The Austrian army was absorbed into the *Wehrmacht.* (Photo 11.3)

After World War II, Austria portrayed itself as a victim of Nazi aggression, but many Austrians were pleased with the *Anschluss* and saw opportunities for advancement as part of Nazi Germany. Kurt Waldheim (b. 1918), Secretary-General of the UN (1971–1982) and president of Austria (1986–92), is a prominent example of Austrian involvement in Nazi war aims. Waldheim served as an intelligence officer with a German unit in Greece and Yugoslavia with a record of reprisals, which led to the unit commander being convicted and executed for war crimes. The existence of the far right freedom party in contemporary Austrian politics indicates that fascist concepts still attract voters in the country.

After World War II, Austria was divided into occupation zones, with the Red Army in Vienna. Food was short in 1945–46, but the country had the capacity to grow most of the subsistence crop requirement and by 1950 industrial production and living standards exceeded prewar levels, in spite of the Soviets extracting reparations and industrial equipment from their occupation zone.

Photo 11.3 Hitler arrives in a Mercedes in Vienna, March 1938. Notice the large crowd welcoming Hitler and swastikas hanging from buildings.

Source: Ulstein bild/The Granger Collection, New York.

The Soviet Union agreed to the reestablishment of an independent Austria with the provision that Austria be neutral and not have union with Germany. The Austrian State Treaty was signed in 1955. Realists noted that Soviet troops withdrew a short distance. Allied troops fell back much further from Vienna, which is close to the border with Hungary, and what was Czechoslovakia, both under Soviet control.

There was no question of neutral Austria joining the EEC in 1957, but it did join EFTA, the free trade organization, in 1960. With the end of the Soviet Union there was no one making the argument that joining EU was economic union with Germany and an infringement of the permanent neutrality to which Austria had agreed in 1955. Austria joined EU, along with Finland and Sweden in 1995.

Regions of Austria

The greatest part of the population of Austria lives in the two hundred-mile stretch of the Danube, which runs through the country. The Danube valley is flanked in the northwest by the Bohemian hills and to the south by the Alps.

Danube Valley

The Danube valley is relatively narrow in Austria compared with the great plains that open up downstream of Vienna (Photo 11.4). Although the Danube is navigable through Austria into southern Germany, higher ground often comes close to the river bank, constricting the valley and speeding water flow. Navigation is not easy upstream of Vienna and near the city the higher landscape of the Vienna woods, a spur of the Alps, increases the speed of the river.

Upstream of Vienna is the river town of Krems, where a scenic, but narrow, section of the Danube is a result of the Bohemian highlands extending down to the north bank of the river. The scenic section, the Wachau, stretches to the border with Germany. Linz, with a population of over 200,000, lies in the Wachau segment of the river. Linz, Roman Lentia, is Austria's third largest city and a major Danubian port.

Downstream of Vienna the Danube enters the western part of the Hungarian plain. Close to the borders with Slovakia and Hungary the Danube receives the Morava river,

Photo 11.4   Palace of Schönbrunn, Vienna. The palace is a legacy of Austria's imperial past.

Source: Mauritius images/Age Fotostock America, Inc.

flowing south from Moravia in the Czech republic. The Morava river is navigable on its lower course and its broad valley reminds us that, strategically, Czechoslovakia had become untenable after the *Anschluss;* the *Wehrmacht* had an open routeway, from the south, into the country. To the south of the Danube is Burgenland, a rich agricultural region that Austria obtained by treaty from Hungary in 1920.

The climate of the Danube valley is temperate but we are now in Central Europe where continental influences are experienced. Winters are colder and summers warmer than in western Europe. Maximum precipitation is experienced in July because thunderstorms are common in the warm summer. In northwestern Europe, on the other hand, most rainfall comes from depressions, which are more commonly experienced in winter months.

The Danube valley supports productive agriculture in a well-ordered rural landscape with villages and farms. Agricultural output includes grains, potatoes, sugar beets, vines, vegetables, dairy products, poultry, and meat for urban markets. Austria has more land than most European countries devoted to growing organic products, free of artificial fertilizers and pesticides.

Bohemian Upland

North of the Danube, on the stretch of the river from Krems to the Germany border, are the granitic uplands of southern Bohemia. The upland rises to 2,976 feet and is largely covered with woodland, contributing significantly to timber resources. Approximately 40 percent of Austria is wooded and much forest is conserved. Cut commercial timber is replaced with new planting.

The Alps

South of the Danube are the Alps, consisting of a series of segments running generally west to east. The first segment is the Alpine foreland, succeeded to the south by the Northern Limestone Alps rising to 8,251 feet. To the south again are the Central Alps dominated by igneous and metamorphic rocks, which attain 12,457 feet at the Gross-glockner in the Hohe Tauern. Along part of the border with Italy and Slovenia are the Southern Limestone Alps.

The Alpine region contains major north–south and east–west routeways, in the depressions between the easterly trending ranges. The Alpine environment contains land suitable for arable farming on lower-lying valley floors. There is coniferous timber on the hillsides and high summer pasture above the tree line. Livestock and dairy products dominate agricultural output in the Alps, which also generate large quantities of hydroelectric power and attract visitors in both summer, for the scenery, and winter, for the skiing.

The Economy

Like other countries in western and central Europe, well over half of GDP is generated by the service sector of the economy, particularly tourism. Agriculture produces less than 2 percent of GDP but manufacturing still contributes significantly, including shipbuilding on the Danube and iron and steel production, using electricity to produce high-grade steel and many engineering products of high quality. Austria receives over twenty million visitors a year. Innsbruck and Salzburg are highly attractive but Vienna is the largest tourist destination.

Austria has a population slightly in excess of eight million. The birth and death rate is nine per thousand Austrians. The population grows slowly by immigration. Fearing increased immigration, Austria will resist admitting Turkey to EU, proposing instead that Turkey be given a "privileged partnership" with the organization.

▶ SWITZERLAND

Switzerland does not have an extensive resource base: alpine grassland, timber, water power potential, some iron ore, a few nonferrous metal ores, a little coal, and only a small proportion of the land area is cultivatable. Further, looking at the map, Switzerland appears to occupy a landlocked, isolated position in central Europe. As we look more closely we see that two of Europe's major rivers rise in Switzerland (Figure 11.2). The Rhône flows westward into France before turning south to the Mediterranean Sea. The Rhine flows into and out of Lake Constance and then exits Switzerland downstream of the port of Basle. The Rhine is a major routeway linking central Europe to The Netherlands and the North Sea. The head of Rhine navigation is in Switzerland. (Photo 11.5)

To the south of the country the Alps appear to be a barrier but a number of passes provide routeways to France, Italy, and the Mediterranean lands. The passes include, from west to east, the St. Bernard leading to France (7,178 feet), the Simplon to Italy (6,581 feet), the St. Gotthard (6,916 feet) on the route to Milan, and the San Bernadino (6,775 feet).

On the routeways through the Alps, linking the environments of northern Europe with those of the Mediterranean, groups speaking French, German, Italian, and Romansch, as well as Protestants and Catholics, have found it in their interests to cooperate in the construction of a neutral country that promotes and safeguards trade, industry, and finance.

The neutral status of Switzerland was recognized by the European powers at the Congress of Vienna in 1815. Neutrality had been practiced long before the end of the

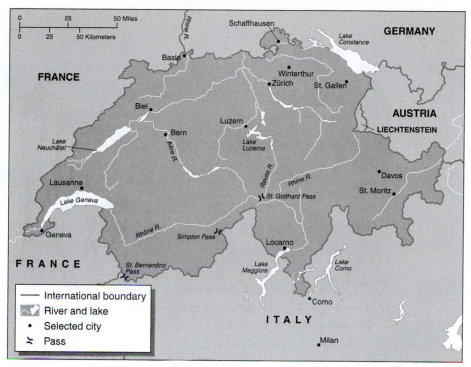

Figure 11.2   Switzerland.

Napoleonic wars, in which Switzerland was invaded, and the country had earlier stayed out of the destructive Thirty Year's War (1618–48). The Cantons of the Swiss federation have many powers. It was not until the middle of the nineteenth century that the government at Bern could coordinate foreign policy, standardize weights and measures, and establish a national postal system. Switzerland has a system of direct democracy in which important issues are referred to the voters, who were all male until 1971.

Holding Switzerland together and neutral has not been easy. In 1840 there was a civil war between the Catholic and Protestant cantons. Italian unification, the Franco-Prussian war, World War I, and World War II engaged the sympathies of French, German, and Italian speakers, particularly as neighboring countries were on different sides in conflicts.

Neutrality may keep a country out of wars but war disrupts trade. By 1915, Switzerland found that all the neighbors, except Liechtenstein, were at war. France and Italy to the west and south were in conflict with Germany and Austria to the north and east. When World War I started in the summer of 1914, the first Swiss casualty was the tourist industry. Hotels and resort towns suffered as visitors returned home. Few foreigners arrived for the winter sports later in the year. Many hotels just closed.

In 1914, Switzerland imported foodstuffs and raw materials and exported watches, machinery, textiles, and processed foods. Trade became difficult. By the end of the war food was short and prices high. In 1918 there was a general strike, broken by the army. The Swiss and the Dutch, at opposite ends of the Rhine routes, learned that profits from trade disappear when the routes are disrupted by war.

Photo 11.5   The Rhine in Basle. Notice the container loaded ship passing under the Middle bridge and the trams crossing the bridge.

Source: Keystone/Gaetan Bally/NewsCom.

After World War I, by a countrywide all-male vote, the country joined the League of Nations, which recognized Switzerland's neutral status and established headquarters at Geneva. Neutrality and the location of the League helped Switzerland play a role in international affairs. The Treaty of Locarno (1925) was signed in the Swiss lakeside resort town of the same name. The treaty, in which Britain and France made concessions to Germany in return for recognition of the postwar boundaries in western Europe, did not bring lasting peace. But before there was a second world war there was the Great Depression.

In the Depression, international trade halved. As a trading country, Switzerland suffered. Exports of Swiss products fell to a quarter of pre-Depression levels. Upmarket consumer products ceased to sell and industrialists avoided capital expenditure on Swiss-made machinery.

Switzerland faced increasing insecurity. In Germany Hitler came to power in 1933. In 1936, Mussolini declared the Rome–Berlin axis and the German Rhineland was remilitarized in contravention of Locarno. In 1938, Germany absorbed Austria (*Anschluss*) with a direct impact on Switzerland. Many Austrians, including members of Jewish communities, wanted to leave the country for fear of Fascists. The best exit route from Austria was through Switzerland and the Swiss, fearing an influx of refugees, began to tighten entry requirements, a process that continued until 1943 when the tide of war turned against Fascism. When France fell in June 1940, Italy joined the war to share the spoils and another flow of forced migrants developed who did not find entry easy, including the novelist James Joyce coming from France who had previously lived in Switzerland. Hundreds of thousands of refugees passed through Switzerland. Many refugees settled in the country or were put into camps. However, thousands were denied entry, including Jews fleeing Austria, Germany, and the Vichy regime in France.

After the invasion of the neutral Netherlands, Norway, and Denmark in 1940, Switzerland feared it would be occupied too. The armed forces prepared for retreat into the Alps to fight a long campaign. The tunnels through the Alps were set with explosive charges and this may have been a deterrent, for Hitler had to get coal and other resources to Mussolini.

Switzerland was more valuable to Germany as a functioning economy than an occupied territory. The Swiss and German economies were interlinked and starting in 1940, Switzerland was pressured into signing tough trade agreements by which the Swiss, on credit, supplied precision instruments and the other equipment used in the armament industry. It was either dilute neutrality or prepare for invasion. Germany removed gold reserves from the central banks of conquered countries and Swiss bankers handled the assets, allowing Germany to finance some imports.

Neutral Switzerland did not avoid hostilities. There were border incidents and the small Swiss airforce tried to protect national air space both against the Luftwaffe and British and American planes flying over Switzerland, on their way to Axis targets. In November 1940, Germany insisted that Switzerland black out cities at night, as British planes were using well-lit towns as navigation points. The lights went out and then the bombers mistook Swiss towns for Axis targets. The worst errors were in 1944 when U.S. planes bombed Schaffhausen, and other towns near the border with Germany. In 1949 the U.S. paid $62 million in compensation. Crippled allied bombers crash landed in Switzerland. Surviving crews were interned, as were escaped prisoners of war who reached Switzerland.

During the war, food and raw materials were short as imports declined. Good grazing land was ploughed up to grow crops, livestock herds were decreased, and Switzerland was able to home grow about 80 percent of basic foodstuffs. Some food-processing industries just ceased because it was nearly impossible for the chocolate industry, for example, to acquire cocoa beans.

After World War II, Switzerland, with the advantage of not being fought over, quickly rebounded economically. The 1948 Olympic Games were held in London. The Winter Games were at the Swiss resort of St. Moritz. Quickly the walkers and climbers returned in summer and the skiers or bobsledders filled the resorts in winter. In a Europe suffering scarcity, the Swiss could sell everything they made. By the 1960s and early 1970s labor was short and foreigners were admitted to the workforce. Today about a fifth of the population is foreign born.

Switzerland and International Organizations

Switzerland declined to join the UN at the beginning, but the voters agreed to membership in 2002. However, several UN agencies were long established in Switzerland, including the International Labour Organization, the International Bureau of Education, and the World Health Organization, all in Geneva. The Universal Postal Union is in Berne. Other international organizations with major offices in Switzerland include the World Trade Organization (Geneva), Bank of International Settlements (Basle), and, since foundation in 1863, the International Red Cross in Geneva.

In 1957 Switzerland did not sign the Treaty of Rome, establishing the EEC, for the six were merging aspects of sovereignty. Switzerland did become a founding member of the European Free Trade Association (EFTA) in 1960. EFTA was simply an agreement by member states to make trade between them easier by removing tariffs on manufactured goods. When in 1992 it was decided to have free trade between EFTA and EEC and create the European Economic Area (EEA), Swiss voters, now male and female, did not approve. Switzerland did not join EEA or approve an application to join EU, although a Switzerland–EU trade agreement was signed. By this time Swiss bankers, manufacturers, and multinational corporations were aware of the economic advantages of EU membership. Swiss voters saw that direct democracy and political organization by Canton would be diluted if many powers were exercised from Brussels.

Switzerland and Globalization

Switzerland, like The Netherlands, experienced the forces of globalization early. Trade goods flowed through the country and so did ideas and technological developments. The Swiss did not invent watches but a locksmith in the Jura region in 1679 succeeded in repairing the watch of a visitor and made a copy of the action. Soon watches were being produced in quantity and the episode illustrates an element in Swiss economic development. Inhabitants are good at taking existing equipment and improving it and producing it more efficiently. During the Napoleonic wars, with British textiles excluded from the continent by Napoleon's decrees, the Swiss textile industry expanded to meet demand. Soon the Swiss were manufacturing and selling across Europe textile machinery as other countries industrialized. Swiss corporations became transnational long before it was fashionable. Switzerland was a small market, and in order to grow, corporations opened subsidiaries in other countries or bought corporations and improved manufacturing processes. Nestlé, founded in 1867, the always present example, has taken over food-processing companies all over the world and created a network to distribute Nestlé brands. There are other transnational corporations based in Switzerland, including the pharmaceutical companies Ciba, Geigy, and Sandoz, now all merged into Novartis.

Merchant houses trading in goods frequently evolved into banks that loaned money and provided banking services to traders and manufacturers. Switzerland, along with The Netherlands and Britain, because of their international trade connections, developed large numbers of merchant banks and investment banks. Banking interests, financing international trade, provided services to a wide range of clients.

There were many reasons why neutral Switzerland was attractive to those who needed a secure, discrete place to deposit money to finance international trade operations or to build up wealth beyond the reach of despots, dictators, and tax gatherers. Although the famed nameless but numbered Swiss bank accounts could be abused by those avoiding taxes, or laundering money, transnational corporations needed to have currency reserves that were secure and did not fluctuate greatly in value. Switzerland and the Swiss franc provided this service to global business.

Regions of Switzerland

Switzerland can be divided into the Jura, the Mittelland, and the Alps. The greatest part of the population, the major cities, and the most productive agriculture are found in Mittelland (Figure 11.3).

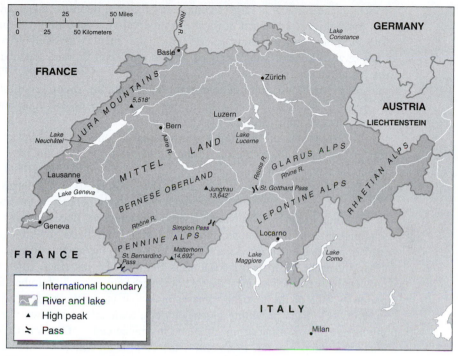

Figure 11.3 Switzerland regions.

The Jura

The Jura Mountains consist of limestone folds of Jurassic age, which trend northeast to southwest. Streams flow in the valleys between the folds, occasionally cutting from one valley to another by gaps known as *cluses*. In the limestone country, there are many dry valleys, sink holes, caves, and underground caverns. The range rises up to 5,518 feet at Mont Tendre, with a well-developed surface at about 4,000 feet. The higher parts of the region are exposed, contain little vegetation, and are of limited agricultural value. Agriculture, particularly dairying and cheese making, are well developed at lower levels and many parts of the Jura are forested with beeches and conifers.

Being exposed to the moist westerly winds, snowfall in the Jura is heavy and even today the region is not easy to cross in winter. The Jura was the first region in Switzerland to develop watchmaking and most of the small towns have a history of manufacturing. There are no large towns in the region. The northeast of the Jura looks toward Basle, the central area to Neuchâtel and other towns on the west shore of the lake of the same name. The international boundary with France runs southwest to northeast along the edge of the region and small French border towns provide some services. The Jura region, which became a canton in 1979, is French speaking and has displayed separatist tendencies (Jenkins, 1986).

Mittelland

The Mittelland is part of the Alpine foreland that extends into Germany and Austria. The Mittelland, central Switzerland, is sometimes referred to as a plateau to distinguish it from the neighboring Alps and Jura. There are not extensive flat areas in Mittelland, which, at the surface, is largely composed of glacial deposits, lakes, and valleys. In the Ice Age, glaciers advanced down the Alpine valleys, depositing till and end moraines. The moraines became dams behind which the many Mitteland lakes formed. As we have seen in Denmark and Germany glacial clays can become productive farmland and the Mittelland is intensively cultivated, particularly in river valleys. The crops include the grains and root crops of temperate lands and, on south-facing slopes, vineyards. Livestock raising and dairying are of great importance as milk is a raw material in the butter-, cheese-, and chocolate-making industries.

The southwest of Mittelland drains to the Rhône, Lake Geneva, and France. The rest of the region drains mostly to the Rhine, with the Aare drainage system being an important Rhine tributary draining much of the heart of the Mittelland.

The Mittelland is densely populated. The landscape is full of farms, villages, towns, and cities surrounded by productive fields. Sixty-eight percent of the Swiss live in urban areas, which is not high by western European standards. Because Switzerland has strong administration in the Cantons, and not a highly centralized government machine in the capital, regional cities are vibrant. Table 11.3 shows the towns and cities at the top of the urban hierarchy. In a central location is Bern, the capital since 1848, on a rocky ridge protected by an incised meander of the river Aare. Bern is number four in the urban hierarchy, one of the few European capitals that is not the biggest city in the country. In terms of centers providing global services, Switzerland has two world cities. Zürich, the largest city, is a major financial and insurance center. Geneva, second on the list, is home to a number of major organizations including the International Red Cross, the World Health Organization, and the International Labor Organization. The League of Nations was housed in the Palais des Nations, still used for international work today. The third largest city is Basle, close to the territory of France and Germany, just below the head of navigation on the Rhine. All the major Swiss cities have interesting sites and situations. Geneva sits on both sides of the Rhône as the narrow river leaves Lake Geneva. Zürich has a similar site on both sides of the Limmat, which flows out of Lake Zürich.

The climate of the Mittelland is continental, with cold winters and warm summers. At Bern the mean monthly temperature for January is 32°F. In July the mean monthly is 65°F but the mean conceals the fact that temperatures rise quickly in the day and fall rapidly when the sun sets. In a continental climate, there is more precipitation in summer

TABLE 11.3 Swiss Urban Populations

	City Population	Metropolitan Area
Zürich	336,800	943,400
Geneva	173,500	457,500
Basle	166,700	
Bern	122,700	319,100
Lausanne	114,500	288,100
Winterthur	88,000	119,700
St. Gallen	69,800	132,500
Luzern	57,000	181,400
Biel	48,800	84,200

than in winter. Thunderstorms in summer are common in Mittelland. Bern receives nearly forty inches of precipitation over the year.

The Alps

The Alps run from the Mediterranean coast through France, Italy, Switzerland, Germany, and Austria where the range ends at the Danubian plain. In Switzerland, the Alps make up more than half the country.

The Rhine and the Rhône rise in the Alps. The Rhine rises in the vicinity of the St. Gotthard pass and flows northeast before turning northward for Germany. The Rhône flows out of the Rhône glacier and drains to the southwest, passing through Lake Geneva before leaving Switzerland for France. At their headwaters, the Rhine and the Rhône are only 15 miles apart and the intervening land is drained by the river Reuss, which flows north through the mountains in a spectacular glaciated valley marked by high waterfalls, including the Reichenbach Falls where Arthur Conan Doyle tried to kill Sherlock Holmes in the novel *The Final Problem* (1893). Holmes survived to appear in the *Adventure of the Empty House* (1903).

At Reichenbach Falls, Conan Doyle tells us the Reuss, "swollen by the melting snow, plunges into a tremendous abyss" (Photo 11.6). Below the falls the valley broadens into the characteristic U shape. There is good agricultural land on the delta of the Reuss where it enters Lake Lucerne. The Reuss flows thorough the lake, past the city of Luzern, which occupies both banks of the river, and out onto the Mittelland to join the Aare, a tributary of the Rhine.

The valleys of the Rhône and Rhine divide the Alps into two major parts. North of the Rhône–Rhine line, in the west, is the Bernese Oberland, and in the east are the Glarus Alps. South of the line are the Pennine Alps, Lepontine Alps, and the Rhaetian Alps. Of course, at a more detailed scale, more subdivisions of the mountains would be identified. The segments of the Alps named above contain formidable peaks. The Bernese Oberland has the Jungfrau (13,642 feet) and the Eiger (13,025 feet). The Pennine Alps peak at the Matterhorn (14,692 feet). The Lepontine and Rhaetian Alps are not so high but have peaks rising up to around 10,000 feet.

Land Use in the Alps   Although the Alps appear to be full of bare rock surfaces and snow-covered peaks, there is agriculture in the valleys and grazing on the grassland above the tree line. Depending on altitude, valley floors can be used to grow grains, root crops, and hay. South-facing slopes have a longer growing season and are favored sites for farms, villages, and small towns. During the summer months the grass on the high *alpen* grows. Cattle are driven up through the trees on the valley sides (trees extend up to 5,000 feet on north-facing slopes and higher on those looking south) to the high pasture. Part of the farming family will live in a chalet, milk the animals, produce dairy products, and make hay. Others in the family tend the crops on the valley floor, grow vegetables, and cut hay on the streamside meadows. As cooler weather arrives at the end of summer and the grass stops growing on the *alpen,* the cattle are brought down to graze fields from which hay has been cut or crops harvested. As winter arrives the cattle are put into barns and fed the summer hay.

The system in which people and beasts move from one environmental niche to another on a seasonal basis is known as *transhumance* or *sömmerung* in German.

THE DEATH OF SHERLOCK HOLMES.

Photo 11.6 Sherlock Holmes apparently falling to his death, at Reichenbach falls.

Source: Alamy Images.

Transhumance represents the type of labor-intensive traditional farming that has been in decline. However, increased demand for traditional-tasting food, produced organically, has made some transhumance profitable. Nevertheless, many high pastures have been abandoned. There has been a decrease in the number of farms and increasing outmigration as people leave the highlands or move to town.

Switzerland has little coal and the traditional source of energy is water power. Over half of all Swiss electricity is generated by hydroelectric power stations, which are a common sight in the country and in Alpine valleys. Hydroelectric power plants cannot be built profitably high in the valleys because the greater the altitude, the longer the freeze season in which electricity is not produced.

The greatest contribution to Swiss GDP from the Alps is the income derived from visitors. In 1866 Edward Whymper and his party climbed the Matterhorn, accelerating English interest in walking and climbing. In the same decade physicians recognized the contagious disease of industrial cities, tuberculosis, was rare in the Alps. At first it was thought that cows might be a factor but fresh, clean air and sunshine kept the disease out. Davos became a town with sanatoria to treat the disease in well-to-do patients.

The number of visitors to Switzerland grew rapidly in the late nineteenth century. As in World War I, World War II nearly shut down the tourist industry, although, in the second war, refugees with the means found accommodation in resort towns.

The Future

Switzerland enjoys high living standards and has an industrial and service economy that adapts to changing circumstances. The country is not without controversy, particularly relating to the banking industry. The handling of Nazi gold in World War II haunts. Long dormant accounts of victims of the Holocaust became an issue as banks had not worked consistently to trace relatives. In 2001, leading banks completed a $1.25 billion payment in reparation, but the issue lingers. Although Swiss banks do not reveal information relating to clients and their accounts, the practice of anonymous, numbered accounts was abolished in 1991. Inevitably there have been suspicions that those associated with the Nazi regime were able to hide wealth in Switzerland during World War II.

The Swiss voted to join the UN in 2002, with 55 percent of the electorate in favor of membership. Joining EU remains an issue. In 1992, the electorate voted against joining the European Economic Area. In 2001, 78 percent of voters in a referendum were against EU membership. The government plans to reopen the issue.

► SWEDEN

Sweden, like Switzerland, had to deal with complicated issues of neutrality. When World War II started in September 1939, Sweden declared neutrality and hoped to keep northern Europe out of the war. The concept had a short life. On November 30, 1939, the Soviet Union attacked Finland to take territory. Sweden sent supplies and volunteers to Finland but stayed out of the war, refusing a British and French request to send troops to Finland via Sweden. The Swedes feared that the allies might occupy the Kiruna iron workings, which supplied half of Germany's imported ore. Sweden helped negotiate the peace between the USSR and Finland signed in March 1940. The following month Germany occupied Denmark and attacked Norway. Again Sweden stayed out of hostilities, but on June 18, 1940, with the Norwegians still resisting, allowed German troops and armaments across Sweden to Norway. France was collapsing and Sweden felt it had to comply or face a German invasion. In June 1941, Germany moved troops from Norway through Sweden to take part in the attack on the Soviet Union. Some Swedes were sympathetic to the German cause, wanting to see the Fatherland establish a superstate, as the Swedish geopolitician Kjellén had suggested.

Sweden supplied iron ore to Germany in Swedish ships. Like Switzerland, Sweden did not avoid hostilities and merchant vessels were lost to German confiscation, mines, and allied air attacks.

At the end of World War II the Soviet Union was in command of the south shore of the Baltic Sea all the way to the Jutland Peninsula. Sweden built up armed forces in World War II and continued to do so in the Cold War. The defense buildup, with orders for vehicles, aircraft, surface ships, submarines, and artillery, benefited Swedish ship builders and corporations like Saab and Volvo.

Resources and Regions

Physical Background

Sweden shares the Scandinavian peninsula with Norway (Figure 11.4). Along the peninsula, from northeast to southwest, runs a mountain mass of Caledonian age. In northern Sweden, the range forms the boundary between Norway and Sweden; further south the

Figure 11.4 Sweden.

main mass of the mountain range lies in Norway. For the most part Sweden faces the Baltic, Norway the Atlantic. Norway has a maritime climate; Sweden is more influenced by continental cold in winter. The Gulf of Bothnia, the northern area of the Baltic, freezes in winter. The Norwegian coast is ice free to North Cape (Nordkapp). Norway and Sweden were under the Scandinavian ice cap during the last glaciation. The ice scoured rock and soil from higher ground and deposited moraines and other material in coastal areas and toward the south.

Although lacking a broad resource base, Sweden does possess rich iron ore, workable deposits of lead, copper, zinc, and plentiful timber. Half the country is forested. There is no coal but the hydroelectric power potential has been widely exploited.

Climate

The Swedish climate gets colder from south to north. The mean monthly temperature for January at Göteborg, in the southwest, is around freezing. Moving northeast to Stockholm, away from Atlantic toward Baltic influences, the mean temperature for January is below freezing. At Haparanda, on the north coast of the Gulf of Bothnia, ice covered in winter, the coldest month has a mean temperature of $11°F$. Inland at Kiruna, around $68°N$, January and February have mean monthly temperatures around $10°F$, and readings frequently drop to minus ten or worse. The averages conceal extremes. A winter depression tracking over northern Scandinavia, bringing maritime air, can raise temperatures markedly at Kiruna. At Stockholm, when the winter high pressure system of cold Siberian air extends westward, across the frozen Gulf of Finland, the city can experience subzero temperatures.

In summer, the temperature differences are not so marked. Göteborg and Stockholm have mean monthly July temperatures in the low 60s. In the north, the means are in the mid-50s, reflecting the long hours of sunshine.

In the far north, daylight hours vary markedly between winter and summer. Stockholm gets less than eight hours of daylight in the shortest days. On midsummer night the sun sets for two hours. At Kiruna, in the north, the sun never sets in the endless June days. At midwinter there is only twilight.

Much of Sweden is in rain shadow on the western side of the mountains of Norway. The far south, with maritime influence, receives 25–30 inches of precipitation. Stockholm and the surrounding area, in the lee of the Scandinavian highlands, receives around 20 inches.

Snow is plentiful in winter and much of the country is covered for several months, although snow cover is less persistent in the south. Going north along the Baltic coast on the Gulf of Bothnia, ice is an increasing problem in winter. On average, Sundsvall is closed for about a month each year. Luleå, near the head of the Gulf of Bothnia, can be frozen from December to May. Luleå is the main port for the shipment of Kiruna iron ore. When the port freezes, ore goes by rail through the mountains to the ice-free Norwegian port of Narvik.

Regions of Sweden

Sweden can be divided into:

1. Norrland
2. The central lowlands, containing Göteborg in the west and Stockholm in the east

3. Småland, an igneous upland

4. Skåne, with a landscape of moraines similar to much of Denmark

Norrland

Norrland (Figure 11.4) is flanked in the west by the ancient igneous rocks of the Caledonian mountains and slopes to the Baltic Sea in the east, with a coastal plain composed of marine deposits. When the Scandinavian ice cap melted, the first impact was for the sea to rise and sediments were laid down in coastal areas. Then slowly, isostatic readjustment, which still continues, saw the land rise. As a result, much of Scandinavia has old shorelines above the present sea level and coastal areas covered in marine deposits that make productive farmland.

Norrland has extensive coniferous forests. The igneous rocks of the region contain iron ore and nonferrous minerals, which are mined. The forests provide timber, and the streams are harnessed for energy outside the winter months. Most settlements are on west–east flowing lakes and streams close to the coast where agriculture is possible and some fishing practiced.

In northern Norrland there are communities of Lapps (Samit, as they call themselves). The largest number of Samit live in Finland and are discussed later.

Central Lowlands

Traveling south from Norrland, at around the latitude of Gävle, we enter the Central Lowlands of Sweden. The region runs from the Baltic shore in the east to the Skagerrak in the west, which gives access to the Atlantic world. Central Sweden, containing the major cities of Uppsala, Stockholm, and Göteborg, is the historical and political core of the country. Like Norrland, central Sweden is underlain by predominantely igneous paleozoic rocks, but the landscape is markedly different, being covered by glacial depositional features. The climate is less cold in winter and warmer in summer. Forests are more varied, containing deciduous as well as coniferous species, and arable farmland is more extensive. The surface of central Sweden is marked by lakes, large and small. West to east across the region, the larger lakes are Vänern, Vättern, Hjälmaren, and Mälaren, with Stockholm situated on the channel that links Mälaren to the Baltic Sea. The lakes of central Sweden are connected by the Gota Canal, with 66 locks, that runs from Göteborg to Stockholm. (Photo 11.7)

Faults in the older rocks have created depressions, which are filled with lakes, including Vättern and Hjälmaren. Other lakes are impounded behind glacial moraines. Eskers, low ridges of fluvial deposits made by under ice meltwater streams, provide sinuous routeways of slightly higher ground and often form piers into and across lakes. Most soils have developed from glacial materials including moraines, outwash sands, and coastal marine deposits laid down in the postglacial rise in sea level, before isostatic adjustment raised shorelines and exposed the marine sediments.

Historically, the central lowlands have been productive agriculturally and industrially. Much Swedish industry had origins in numerous small iron works that used local ore deposits and charcoal from nearby forests to make iron, which became an important export. Small in scale, the iron industry was practiced in many communities with the early technology of iron and then steel making being widely spread.

Photo 11.7   Riddarholmen, Stockholm, Sweden.

Source: PhotoDisc/SUPERSTOCK.

Småland

South of the Central Lowland is Småland, into which Lake Vättern protrudes. Except on its coastal plain, Småland, which rises up to around 500 feet, consists of igneous and metamorphic rocks with thin soils. The minerals associated with igneous activity have been exploited and Jönköping is one of several manufacturing centers.

Skåne

Skåne is covered in moraines and other glacial deposits. The intensely cultivated landscape is reminiscent of Denmark (Photo 11.8). Historically, the region was a part of Denmark and is now linked, across The Sound, by the Malmö road bridge to Copenhagen. Skåne is milder and has a longer growing season than more northly regions of Sweden and is more open to maritime climate influences than central and northern Sweden.

The Economy

Sweden was a late industrializer, for in the first half of the nineteenth century the country was largely agricultural. Iron ore was mined and smelted with charcoal in many small operations. Timber was exploited and textiles were only beginning to evolve from cottage industry to mechanized production units.

In the second half of the nineteenth century came economic diversification. Lars Ericcson invented his telephone system, the ball-bearing industry using high-quality steel was established, the mass production of matches started, and Alfred Nobel invented dynamite, a product with a worldwide market. By the early twentieth century, Swedish industrialists were building ships, making steam turbines, cars, medical equipment, chemicals, and beginning to export modern, fashionable furniture made of timber from Sweden's forests. Demand for Swedish iron ore, steel, machinery, and armaments was heavy in World War I and Sweden benefited from the policy of neutrality, which kept the country out of the war.

Like all other European countries, Sweden suffered in the Great Depression. In the late 1930s, as in Switzerland, the state, workers, and employers came together to

Photo 11.8   There are extensive areas of good, arable farmland in Skåne, southern Sweden.

Source: Age fotostock/SUPERSTOCK.

create a social contract that would protect employees in economic downturns and create a framework in which wages could be negotiated without resorting to strikes. Sweden developed a welfare state. The government-built infrastructure, provided a social security network, and saw that workers were properly housed. The large corporations operated as capitalist institutions.

Planning Urban Environments

In 1947, Sweden passed a Building and Planning Act, which controlled housing density and gave municipalities the sole power to approve residential subdivisions. Stockholm got the power to design and build satellite towns providing dwellings, workplaces, and services. The towns contained many housing units in high-rise buildings and they were built along transportation corridors, creating what Peter Hall describes as "the Social Democratic landscape of the 1950s and 1960s." Since the 1980s, as in most western European countries, more urban expansion has been allowed and state agencies have been less involved in the design and creation of housing units for workers, leading Hall to suggest that a "placeless landscape" has emerged (1998, p. 877).

Population

Currently 9.1 million people live in Sweden. The birth rate per thousand of the population is 11, the death rate 10. By natural increase the population grows by 0.1 percent per annum. However, Sweden has admitted more migrants and refugees in recent years and

TABLE 11.4   Sweden: Urban Places over 100,000

Stockholm	758,148
Göteborg	474,921
Malmö	265,481
Uppsala	179,673
Linköping	135,066
Västerås	128,902
Örebro	125,520
Norrköping	123,303
Helsingborg	106,525
Jönköping	118,581
Umeå	106,525
Lund	100,402

the Population Reference Bureau estimates there will be 9.9 million Swedes by the year 2025. The total fertility rate of 1.8 is above the European average of 1.4. Sweden has generous child benefits, maternity leave, and childcare provision and this has helped to keep total fertility rates above those experienced in many western and southern European countries.

Summary

Since the late 1930s, Sweden has followed a middle way economically, with corporations being the engines of the economy and the state providing a range of benefits and a safety net within a well-run welfare system. Sweden has not adopted the Euro. The country retains the national currency, the Krona, and control over money supply and interest rates.

▶ NORWAY

In 1397 Denmark, Norway, and Sweden were brought into a union from which Sweden broke in 1523. Norway became a province of Denmark in 1536. At the end of the Napoleonic wars, by the Treaty of Kiel (1814), Norway became part of Sweden, with Denmark retaining the Faroe Islands, Greenland, and Iceland. Norway became independent of Sweden in 1905. (Figure 11.5)

Norway is one of the few countries in the economically developed world where a considerable proportion of the economy is based on the exploitation of natural resources. Given the limited extent of agricultural land, Norwegian farming is not export orientated, but the forests (about a quarter of the country is forested) produce timber, pulp, and paper for export. Norway has a fishing fleet of over 8,000 vessels, nearly half of which are inshore boats crewed by a few men. The larger trawlers produce a surplus catch and Norway is one of the world's largest exporters of fish. Increasingly fish, particularly salmon, are raised in fjord farms.

Norway possesses extensive water resources as the Caledonian mountains receive high precipitation from the westerly winds blowing off the North Atlantic. The streams falling from the high fjeld to the fjords are harnessed to generate hydroelectric power. The electricity is used to smelt metals, including iron, copper, zinc, lead, and aluminum.

Figure 11.5 Norway physical environmental regions.

Norway has North Sea oil and gas resources and is a major exporter, after Saudi Arabia and Russia. The chemical industry benefits from home-produced hydroelectricity and petroleum products. Norway is a significant part of the global shipping industry, with a large merchant marine including cargo vessels, tankers, and cruise liners.

Regions of Norway

Norwegians, at the national scale, perceive at least six regions in Norway at the national scale (Figure 11.6). Eastern Norway (*Østlandet*) focuses on Oslo fjord and the Skagerrak. Over the Hardanger Plateau—a granitic region of fjelds, national parks, and glaciers—is West Norway (*Vestlandet*). West Norway and its fjords face the Atlantic, with the port of Bergen, founded 1070 A.D., being the major settlement. Bergen was the capital of Norway in the twelfth and thirteenth centuries. Today, Bergen, the second largest city in Norway with a population of over 200,000, is a major center of fishing, shipbuilding, engineering, and oil and gas service companies.

Norwegians recognize the Trondheim (*Trøndelag*) region as distinctive. The port of Trondheim is on a major fjord into which several valleys run and there are extensive areas of agricultural land around the city. Trondheim, founded 997 A.D., today has a population of around 150,000, working at fishing, shipbuilding, timber industries, and engineering. In addition, services, including technical education, are important.

The three northern provinces of Norway—Nordland, Troms, and Finmark—constitute *Nord-Norge*. At the southern tip of Norway is *Sørlandet,* which looks to Kristiansand (65,000) as the regional center. The regions recognized by Norwegians are not solely based on physical environments; the character of places such as Bergen and Trondheim, and their hinterlands, distinguishes the regions in national consciousness. Differences are marked at relatively local scales, with settlements in one fjord having customs and dress that set the area apart from an adjoining fjord.

Physical Environmental Regions

Looking at the map we can distinguish three major environmental regions in Norway (Figure 11.5):

1. The relatively extensive area of lower land around Oslo Fjord, which flows into the Skagerrak.

2. The Caledonian Mountains that run from southwest to northeast through Norway. The Vikings referred to the mountain range as The Keel.

3. The west coast of Norway is a Fjord coast, with numerous offshore islands. Mariners going north sailed inside the islands to gain protection from the Atlantic. The searoute became known as The North Way, the name now applied to the whole country.

Oslo Fjord

The Oslo Fjord lowland largely corresponds to Østlandet. The region is drained by the largest river in Scandinavia, the Glåma, and its tributaries. The Glåma tributaries emerging from the Caledonian Mountains in the west flow through deep, glaciated dales to enter the

Figure 11.6 Regions of Norway.

lowland, with many valleys displaying finger lakes produced by glacial scouring. Much of the lowland surrounding the lakes is covered by glacial till and outwash material. As elsewhere in Scandinavia, and on the North European Plain, glacial clay has been made into arable farmland and pasture.

Much land in the lowland is in coniferous forest and the area around Lake Øyeren is a major forestry region. There are numerous pulp and paper mills along the Glåma, (navigable to Sarpsborg), as well as hydroelectric power plants and shipyards on the lower course.

The Oslo conurbation contains approximately 800,000 inhabitants, being much larger than the second city, Bergen, with just over 200,000 people. Other towns in the lowland region include Fredrikstad, south of Oslo near the border with Sweden and linked by ferries to Denmark. To the southwest of Oslo, on Norway's south coast, is Kristiansand, another town providing ferry links to Denmark.

Caledonian Mountains

The Caledonian Mountains are of Paleozoic age and have been eroded down to expose the igneous and metamorphic rocks formerly deep in the high mountain range. The range is now a plateau punctuated with peaks rising above the erosion surfaces. In the mountains, coniferous trees will grow up to about a thousand feet above sea level, higher in more favorable places, but soils are thin and on the exposed plateau surfaces vegetation is grass, sedges, mosses, ferns, and heather. This upland environment is known as the fjeld. In Britain, it would be referred to as moorland or fell, the latter word reflecting the Viking influence on topographic names in highland Britain. The fjeld is snow-covered in winter and in summer provides rough grazing for livestock.

The Norwegian mountains are highest in the south, Galdhøpiggen rising to over 8,000 feet with sufficient altitude to sustain glaciers. In general, the extensive Hardanger plateau stands at 3,000 feet. Further north the Finmark plateau averages a thousand feet above sea level. Nord Norge has longer winters and although lower, does have ice fields. (Photo 11.9)

There are no major towns in the upland but there are small communities along the routeways, particularly where they cross the Hardanger connecting Oslo to Bergen and the routes from Oslo north to Trondheim.

The Fjord Coast and Offshore Islands

Fjords are a product of the glacial era. Ice flowing down valleys from the Caledonian Mountains gorged out deep, U-shaped valleys. When the ice melted, the sea inundated the overdeepened valleys, creating the fjords that characterize the west coast of Norway.

At the head of fjords is an area of flat land covered by marine deposits that emerged at the end of the Ice Age. The fjord head flatland is usually the settlement site and provides land on which arable crops and vegetables can be grown. (Photo 11.10)

Traditionally, communities exploited the fisheries of the fjord and the open sea. Inland, often at the top of steep, bare rock surfaces, were coniferous forests and above them the *seter,* the high summer pasture. Fishing could be practiced for much of the year as the ice-free fjords and island coast gave some shelter against winter storms. In spring, arable land was tilled and sown. In summer, hay was made by hanging it up to dry. Cattle were driven to the *seter* and, in a way similar to the transhumance practiced in the Swiss and Austrian Alps, part of the family would move up with animals to milk

Photo 11.9 Svartisen Glacier. Arctic Circle. Norway.

Source: Age fotostock/SUPERSTOCK.

them and make dairy products. As they exploited the environmental niches, inhabitants became farmers, fishers, foresters, and boatbuilders. People had a range of skills, a high degree of self-reliance, and were protective of community resources. The isolation of the farming, fishing, and forestry communities was lessened when the state built a railroad north to Bodø and ferry services created links between many, but not all, fjords.

The traditional fjord lifestyle is in decline. Many small communities have lost inhabitants. Now nearly 80 percent of Norwegians live in towns. Making a living from small fjord head fields is difficult, even with generous farm subsidies. The seasonal migration to the summer *seter* with the dairy cows is not as widely practiced as it was. Norwegians do not wish to see the traditional lifestyle and landscape disappear; many enjoy fjeld walking, climbing, skiing, and fishing as recreation. Holiday cottages in fjord environments are popular. The isolated communities along the fjord coast provide a base for people wanting to enjoy the harsh but beautiful environment of the coast and the adjoining mountains and glaciers. The Norwegian government, using subsidies, attempts to keep small communities viable.

From south to north along the fjord coast the major towns are Stavanger, Bergen, Trondheim, Narvik, and Tromsø. Few of the offshore islands, with the exception of the extensive Lofoten Islands with their coal deposits, have many settlements. Overall population densities in the fjord coast and islands are low.

Norway extends from 58°N in the south, well into the Arctic circle to North Cape at 71° 10'N. The coast of Norway is ice free year round due to the North Atlantic drift.

Photo 11.10   Geirangerfjord. There are many small communities at the head of Norway's fjords. Tourist activities help to keep the small settlements viable.

Source: Mauritius Images/Age Fotostock America, Inc.

Urban Places and the Distribution of Population

With thirty seven Norwegians, on average, to each square mile of the country, Norway has the lowest population density of all European countries, with the exceptions of Iceland and Russia. Of course, large areas of Norway have only one or two people per square mile, and the higher environments are nearly empty for much of the year. Most Norwegians live in urban areas or close to towns and cities. As Norway's prosperity has grown in the last half century, more people have made careers in the cities. Fewer Norwegians have gone into farming, fishing, forestry, mining, and petroleum extraction, as the primary sector of the economy has become increasingly mechanized.

The top ten urban areas by number of inhabitants are listed in Table 11.5. No other town exceeds a population of 40,000 people.

Summary

Only about 3 percent of Norway is in cultivation. From the Viking era on, Norwegians have sailed outward from the fjords to exploit the resources of inshore waters, the North Sea, and the Baltic. Orient the map so that you look from Oslo, down Oslo fjord, to the Skagerrak and the Kattegat. Vikings based on Norway, Sweden, and Denmark crossed these waters easily. Orient the map to look out from Bergen onto the North Sea. You are looking into a sea that by the ninth century was becoming a Viking lake. It is true that in the early phases of contact the Vikings were raiders, as pagans particularly attracted to

TABLE 11.5 Norway Urban Populations

Oslo	773,498
Bergen	205,759
Stavanger/Sandnes	162,083
Trondheim	140,631
Fredrikstad/Sarpsborg	93,273
Drammen	86,732
Porsgrunn/Skien	83,409
Kristiansand	61,400
Tromsø	49,372
Tønsberg	43,346

soft targets like monasteries. The famous monastery of Lindesfarne on Holy Island, off the coast of Northumberland, England, was pillaged in 793 A.D. By 1000 A.D., trading places had been established as at Jorvik (York). Settlements were established in eastern England, the Scottish Islands, Dublin, Normandy, the Faroe Islands, Iceland, Greenland, and eventually Vinland in America. In the Atlantic island settlements, such as Iceland, Norwegians were in the majority, although Swedes and Danes were represented. Vikings sailed up the Baltic to make settlements and trading posts around Lake Ladoga in Russia, and went south, via Kiev, into Black Sea trade.

The history of sailing, trading, and settling in distant places is still reflected in the present. There are communities of people originally from Norway in the Dakotas and Minnesota. The number of well-trained Norwegian merchant seamen, the Norwegian shipping and cruise lines, together with the deep-water long-range fishing fleets, are all a reflection of a maritime past that encouraged people to learn sailing skills in fjords and look outward for opportunities. North Sea oil was an opportunity Norway was well placed to exploit, already having a shipbuilding industry that could build oil rigs, fleets of tankers, and a chemical industry based on hydroelectricity that could use oil as a feed, as well as an engineering sector capable of constructing pipelines and refineries.

Norway has one of the highest living standards in the world. On quality of life indices Norway is always near the top of the list. Like Sweden, Norway runs a welfare state providing generous benefits to Norwegians. State corporations run a number of industries including the railways. The state initially capitalized the oil exploration industry but that corporation, Statoil, has been privatized, although the government remains a large shareholder.

Approximately 20 percent of Norway's GDP is derived from oil and gas and nearly 50 percent of exports by value are oil and gas. Oil royalties go into a state fund, which smooths the budget and provides subsidies to farming and communities on isolated fjords. The government does not want small fjord communities to die, for they are a part of the national heritage. The northern fjords are a valuable resource that attract cruise boats sailing to see Arctic ice and the midnight sun.

Under EEA Norway enjoys free trade with EU countries. Norway has negotiated greater access for Norwegian fish products to EU markets in exchange for EU countries being allowed fishing quotas in Norwegian waters. A majority of Norwegians resist EU membership. The 1972 referendum rejected EEC membership by a narrow majority.

In the 1994 referendum 52.4 percent of the voters were against EU membership. Norway, with reliance on primary products, is wary of EU energy policies, EU fisheries policy, and the Common Agricultural Policy (CAP). Norway is a country that imports much food. No matter how much CAP subsidy money is received, Norway cannot become self-sufficient agriculturally. Further, under CAP rules, Norway would pay higher prices for imported foodstuffs. Better to buy on the global market where the prices of agricultural commodities are far lower than in EU. Norway is a land of small farms. Experience has shown that such units are unable to take advantage of many CAP programs and that subsidies go disproportionately to large farms. Norway subsidizes farming but under its own rules aimed at preserving rural communities. The Common Fisheries Policy, based on overoptimistic estimates of fish stocks, has been disastrous for many traditional fishing communities in EU countries. Stocks of cod have been depleted, quotas reduced, and vessels laid up. Norway, by avoiding pooling traditional fishing grounds with EU countries, has helped fishing communities, with smaller vessels, survive. Most Norwegians want to preserve the small farming and fishing communities even if they do not want to live on a remote fjord.

► ICELAND

Introduction

Impossible as it seems, think of Iceland as a sub-Arctic Hawaii! Both are located on mid-oceanic ridges where the ocean floor is pulling apart, allowing basaltic lava to flow to the surface. Iceland and Hawaii are largely composed of recent volcanic rocks and volcanoes continue to add material to the islands. As a result of volcanic eruptions in 1964, Iceland gained the offshore island of Surtsey. Iceland has many thermal springs and the Geysir spouts hot water two hundred feet into the air. It is from this site that we took the word *geyser*. (Photo 11.11)

Beginning around 870 A.D. Iceland was settled by Norwegian Vikings, although some Danes and Swedes were involved. Iceland became a base for the settlement of Greenland. In 1397 Denmark, Norway, and Sweden were unified under one crown. When the kingdom broke up in the sixteenth century, Iceland remained under Danish control. After World War I, Iceland had internal self-government but continued to recognize the Danish crown as the constitutional head of state. In 1944, Iceland became a republic.

Iceland was of strategic importance in World War II as Germany occupied Denmark and Norway and used the Norwegian fjords as bases for submarines and surface vessels like the battleship *Tirpitz*. To prevent Germany getting another base in the North Atlantic, Britain sent troops to Iceland in May 1940. The U.S. deployed forces to Greenland and, on July 7, 1941, took over the garrisoning of Iceland, allowing British troops to be used elsewhere. The deployment of American troops took place before the U.S. was in the war. President Franklin Roosevelt had a strong strategic sense and was aware that Hitler's ambitions would lead to a battle for the Atlantic searoutes, which reached its height in 1941–43.

In the Battle of the Atlantic, Iceland was an airbase from which antisubmarine planes could patrol. Iceland was also used as a stopover for allied planes crossing the Atlantic and as a meteorological station. After World War II, Iceland joined NATO and became a base to track Soviet submarines. Now, with the Soviet threat diminished, the U.S. keeps

Photo 11.11 Hot springs in Iceland. Note the geothermal power stations on the far side of the lake.

Source: Alamy Images.

less than a squadron of aircraft on Iceland and employment for Icelanders working on the bases has been cut.

After World War II, when commercial airliners lacked their present range, Iceland was used as a refueling stop. That function has now gone but Icelandair still stops at Reykjavik to serve Icelanders and bring tourists.

Iceland has a strategic position in terms of North Atlantic fishing resources. There are extensive fishing grounds around Iceland and the fjords provide shelter for fishing vessels. Trawlers from other countries operated around Iceland outside the three-mile limit. Iceland was so dependent upon fishing that in 1964 it decided to extend territorial waters to twelve nautical miles and to fifty miles in 1972. In 1975, non-Icelandic boats were pushed 200 miles offshore. These measures led to the "Cod Wars," with British fishery protection vessels trying to stop the Icelandic authorities enforcing the new limits. For a time, in 1975–76, Iceland broke diplomatic relations with Britain.

Today, the fishing limits around Iceland are largely respected, and the Law of Sea Convention in 1982 recognized the 200-mile exclusive economic zones (EEZs). The disputes died down relatively quickly because in practice it was dangerous to fish the waters around Iceland without access to the ports and harbors of the country. When deep depressions and strong storms, including ice storms, develop in the North Atlantic many boats would seek shelter in Iceland. That would lead to severe fines if the Icelandic authorities thought you had been fishing illegally.

Fisheries account, directly, for over 10 percent of Iceland's GDP and indirectly for much economic activity. Over recent decades the fishing industry has diversified. Icelandic companies are involved in fish processing, packaging, and distribution, having established plants in European and North American ports. A supermarket chain, Iceland, has grown out of the retailing of fish in European markets.

Physical Environment

The mid-Atlantic oceanic ridge, the boundary between the Eurasian and North American plates, runs across Iceland from southwest to northeast. Along this faulted axis are active volcanoes and there are earthquake zones at each end of the axis. Iceland largely consists, at the surface, of volcanic rock, glaciers, and glacial outwash material.

Iceland lacks coal and oil. Early in the twentieth century, streams were harnessed to generate hydroelectricity. Starting in the 1940s, the geothermal energy in hot springs was used to heat Reykjavik and greenhouses growing fresh vegetables. Cucumbers and tomatoes are grown near the Arctic circle. Deeper boreholes are exploiting hotter sources of water.

Iceland imports much oil for automobiles and the fishing fleet. Geothermal energy is used to split H_2O into hydrogen and oxygen, the hydrogen becoming fuel in public buses. The policy is to move to hydrogen-powered vehicles and the reduction of oil imports.

Ocean Currents and Air Masses

Iceland is in the North Atlantic Drift and sea surface temperatures are mild for the island's high latitude. To the north, the cold East Greenland current flows through the Denmark Strait between Iceland and Greenland. Sea surface temperatures, along Iceland's northern shore, are 10 degrees colder than along the south coast. Historically, the Greenland ice cap has been a source of pack ice in Icelandic waters, but the ice is less frequent as a result of global warming.

The major air masses influencing Iceland are the maritime polar air, originating over the far North Atlantic, and the maritime tropical air that flows from the Gulf of Mexico across the Atlantic to western Europe. Reykjavik, in the southwest corner of Iceland, has a January monthly mean temperature of approximately 34°F. The July mean is 52°F. Inland the topography rises rapidly to 2,000 feet. Temperatures are lower and snow cover sustained in the winter months on the higher ground. Reykjavik receives 30 inches of rainfall a year but on the uplands precipitation can exceed 150 inches.

The interaction of the maritime polar and maritime tropical air leads to the creation of depressions and storms along the front between the airmasses. So prevalent are low pressures and depressions over Iceland that climate maps identify an Icelandic low as a major feature of the Atlantic weather system. On the island, rainfall is plentiful, clouds persistent, clear days few, and sunshine hours low. Intense low pressures produce gale-force winds and with large storms Iceland can experience hurricane-force winds (winds in excess of 74 mph).

Topography

The topography of Iceland is dominated by lava flows and glacial features (Figure 11.7). The highest point in the country is 6,950 feet, but the average elevation is 2,000 feet above sea level. Glaciers are extensive. In the southeast the Vatnajökull, the largest glacier in Europe, rises to over 3,000 feet. The Hofsjökull and the Langjökull are other major masses of ice. Around the margins of the ice sheets, smaller glaciers can be seen eroding valley sides and meltwater streams flow from the ice, carrying, sorting, and depositing silts, sands, and gravels. Glacial lakes are common. (Photo 11.12)

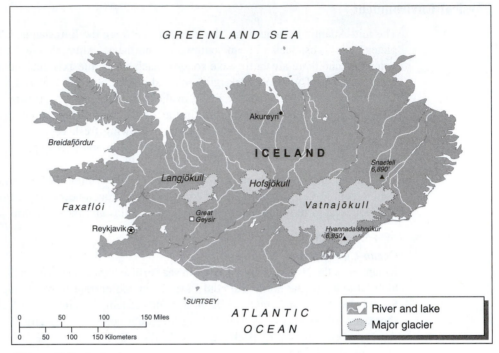

Figure 11.7   Iceland.

Settlement

Much of Iceland is too high, icy, and exposed to be used for human settlement. At lower levels are many lakes, marshes, bogs, and other ill-drained areas. Today, settlements of any size are at the coast in sheltered bays or fjords.

If we had visited Iceland 150 years ago, the settlement pattern would have consisted of Reykjavik, a few small fishing communities at the coast, and thousands of small farms on better-drained ground. Farming families raised potatoes and vegetables and grazed cattle and sheep. Many farmers supplemented the diet by fishing inshore waters in small boats.

The modern fishing industry is a product of the European industrial revolution and the industrial cities that emerged with the growth of manufacturing. Industrial societies built large trawlers to fish distant waters, with powered equipment and steel hawsers to haul in large nets and their catch.

Trawlers from British ports started to fish Icelandic waters in the last decades of the nineteenth century. By World War I, Icelanders were acquiring trawlers and selling catches at fishing ports in Europe, particularly the UK. The British national dish had a home-grown potato on the plate, but the fish in the fish and chips was likely to be from Icelandic waters, whether caught by a Scottish, English, or Icelandic vessel.

The rise of large-scale fishing made it possible for young men to take waged work on vessels. This encouraged rural-to-urban migration and the decline of small farms. The towns in Iceland have coastal locations and are active in the fish processing industry.

Photo 11.12 Vatnajökull (Europe's largest glacier). Iceland. Cloudiness and precipitation are characteristic of Iceland's climate. At high levels the precipitation comes as snow which maintains the glacier.

Source: Age fotostock/SUPERSTOCK.

Now the country is 93 percent urban. The largest place is Reykjavik (111,517) in the southwest, facing into the Faxaflói, the Atlantic and the westerly winds.

Summary

Iceland is populated largely by Icelanders who speak the official language, Icelandic. It is one of the relatively few European examples where there is ethnic homogeneity. The total population is 0.3 million but the total fertility rate at 2.1 is as high as any in Europe.

The country has diversified from fishing, although the industry remains central to the economy. Iceland joined EFTA in 1970 and, as a result of EEA, has free trade with both EFTA and EU countries. Iceland has agreed to allow some fishing quotas for EU vessels in its waters, but is unlikely to become an EU member because of the Common Fisheries Policy. Joining EU would mean sharing Iceland's fish resources with Europe. The Common Fisheries Policy (CFP), revised in 1983, was based on optimistic estimates of fish stocks and the CFP has failed to establish sustainable exploitation of fisheries. The CFP assigns quotas and controls fishing, but abuses are widespread and enforcement, largely on a national basis, flawed. In 2005 France was fined the equivalent of $20 million for catching undersized fish in EU waters. The French government paid the fine. It is doubted that French fisherman will take much notice.

Iceland, along with Japan, is against the ban on whale fishing. In 2006, harpooning of whales for commercial purposes restarted. As Iceland attracts ecotourists and whale watchers, the killing was quickly reflected in a declining number of visitors.

► FINLAND

During the sixteenth and seventeenth centuries, Sweden gained control of much of the territory we now recognize as Finland, Estonia, Latvia, and Lithuania. With the rise of Russia, under Peter the Great, Sweden lost territory on the southern and eastern shore of the Baltic Sea. Starting in 1703, Peter laid out St. Petersburg at the eastern end of the Gulf of Finland.

As a result of a campaign in 1808–09, Sweden lost Finland (treaty of Hamina, 1809), which came under Russian rule as a Grand Duchy, with the Tsar, who occupied the dukedom, being represented by a Governor-General. The Finns did retain many governmental, legal, and cultural institutions but by the early twentieth century Russia was attempting to Russify Finland. In World War I, with the Russian Empire defeated, and in the midst of a revolution, the Finns declared independence in 1917. The newly independent country then entered a civil war in January 1918 between the Whites and the Reds, the latter receiving support from the Bolsheviks.

The boundary of newly independent Finland and the Soviet Union was established by the Treaty of Tartu in 1920. The Finnish boundary was close to Leningrad, as St. Petersburg was renamed in 1924, and the Soviets were not comfortable with Finnish proximity to the strategic city.

Finland had a boundary dispute with Sweden over the Åland Islands. The inhabitants of the strategic islands, at the mouth of the Gulf of Bothnia, spoke Swedish and wanted to join Sweden after World War I. The League of Nations (1921) ruled that, by previous treaties, the demilitarized islands should remain an autonomous part of Finland. Sweden accepted the ruling of the League. Swedish influence is still strong in Finland. Many settlements were founded by Swedes and have both Swedish and Finnish names. Nearly 300,000 Finns speak Swedish as their first language and many more can converse in Swedish. Both Finnish and Swedish are recognized as official languages.

After the signing of the misnamed Non Aggression pact between the Soviet Union and Germany in 1939, the Soviets could be aggressive with neighboring states. In November 1939 the USSR demanded that the Finns withdraw the frontier from proximity to Leningrad, give up territory, and allow Soviet bases in Finland. The Finns refused; the Red Army attacked. The Soviet army looked incompetent in the Winter War (1939–40), but had the numbers to pressure Finland into a settlement. In March 1940, Finland signed the Treaty of Moscow and gave up the territory Stalin demanded.

In June 1941, Germany attacked the Soviet Union in Operation Barbarossa. Finnish forces retook the lost lands and other areas containing Finns but at the end of World War II yielded again, giving up more territory including the Petsamo district in the north, with the ice-free Arctic port, nickel deposits, and connection to the North Atlantic. The Soviet Union got a land boundary with Norway.

In addition, the Soviets got naval facilities on the Gulf of Finland, west of Helsinki, reparations payments, and a policy of strict neutrality from Finland. With the loss of Karelia and other territory, over 400,000 Finns were forced to leave their homes and relocate within the new boundaries of Finland. To accommodate farming families, some existing Finnish farms were subdivided and many "cold farms" were cut from the coniferous forests, with farmers having to clear the land before a crop could be planted and harvested. A disadvantage of "cold farm" creation was that the land was more valuable as a forest resource and, over time, the new farms were unlikely to provide a good standard of living for settlers. Many who left Karelia and other territory taken by the Soviets settled in cities where work was available. There is no widespread support in Finland to regain the lost territories from Russia now that the Soviet Union has collapsed, although some would like to live on old family lands. However, the Finns have a far higher standard of living than the Russians and the costs of creating modern infrastructure in the former territories would be large, to say nothing of the problems of absorbing the Russian population (Paasi, 2002).

Finnish recovery after World War II was relatively rapid. The country hosted the Helsinki Olympic games in 1952, got Soviet troops out in 1955, and as western Europe expanded economically, the demand for Finnish timber, wood products, pulp, and paper grew. The Soviet Union restricted Finland's role in international affairs but in 1961 it became an associate member of EFTA and entered into a trade agreement with the EEC in 1974. Membership of EU became possible after the collapse of the Soviet Union and was achieved in 1995.

The Physical Environment

Geologically, Finland is part of the Fenno-Scandinavian shield, composed of ancient crystalline rocks. However, the greatest part of the country is covered by glacial deposits and glacial lakes (Figure 11.8). In the north, parts of the underlying crystalline rock have been exposed by eroding glaciers. Central and coastal Finland display the features of glacial deposition. In the south are bands of moraines, running west to east, behind which are numerous glacial lakes.

The natural vegetation of three quarters of Finland is coniferous forest, with some birch trees. In the southwest, with a less severe climate, deciduous species are found in addition to conifers. The conifers grow into the Arctic Circle, but in the far north the growing season is about three months, too short to support tree growth. The vegetation is of a tundra type. The word *tundra* is derived from the Sami (Lapp) language. Tundra vegetation consists of low-growing shrubs, grasses, sedges, mosses, and lichens. The tundra environment is frozen for most of the year, thawing at the surface for three months in the summer. As in Norway, Sweden, and parts of Russia, the Lapps have traditionally followed the reindeer herds but now many have settled to a less nomadic existence. (Photo 11.13)

Climate

Finland lies to the east of Norway and Sweden. Continental climate influences are more marked and the impact of Atlantic climatic forces less evident. The French geographer Jean Gottmann, then at the University of Oxford, described the Finnish climate:

The whole country gives an impression of dampness. Practically every place is frozen from December to March; in July, Helsinki on the south coast averages 62°f, but severe frosts are

Figure 11.8 Finland.

possible to June. . . . The white carpet of snow covers the landscape for 120 days every year around Helsinki and 210 days in the north. The Baltic Sea freezes in winter along the Finnish coasts. (Gottmann, 1969, p. 561)

Urban Centers and Population Distribution

Nearly a fifth of all Finns live in, or close to, the Helsinki metropolitan area (Table 11.6). Population density decreases to the north away from Helsinki and the Gulf of Finland. In the northern upland, the most important town has less than 40,000 people. There are innumerable small rural settlements strung along moraines and eskers, which run beside the lakes and marshes of central Finland. In spite of the dominance of the urban places along the Gulf of Finland, nearly 40 percent of Finns still live in rural areas.

Photo 11.13 Sami reindeer herder lassoes a reindeer in a corral North Norway. Traditionally Sami (Lapps) followed the reindeer herds. Now the animals are often kept in corrals and the Sami have become sedentary.

Source: Alamy Images.

Economy

The area of cultivatable land covers only 5 percent of the country, with the principal crops being barley, oats, wheat, and rye together with potatoes, which do well on damp clay soil. Considerable areas are in pasture land and meadows, which produce hay as a winter feed for livestock.

The prosperity of Finland was initially based on the forests, timber products, wood pulp, and newsprint. The rise of national, mass-circulation newspapers in Europe increased the demand for newsprint. All the Baltic countries benefited from the demand

TABLE 11.6 Major Towns and Cities in Finland

Helsinki (Helsingfors)	555,474
Espoo (Esbo)	213,271
Tampere (Tammerfors)	195,468
Vantaa (Vanda)	178,471
Turku (Åbo)	172,561
Oulu (Uleåborg)	120,753
Lahti	96,921

(Swedish names in parens)

for paper. At first, Finland produced wood pulp, the export of raw materials providing the income to import the machinery for paper production. Industry grew as imported machinery mechanized the production of textiles, footwear, and other basic products. Shipbuilding, engineering, and electronic products appeared as the economy diversified. Today, Nokia is one of the world's largest producers of communication equipment.

▶ SUMMARY

The total population of Finland is 5.3 million people, with the birth rate (11) exceeding the death rate (9) per thousand of the population. Living standards are high. The country ranks well in terms of efficiency, is attractive to outside investors, and Finnish corporations have invested in the Baltics, particularly Estonia, where relatively cheap labor allows manufacturing of electronic equipment close to home.

▶ FURTHER READING

"Geography in Switzerland." *Geographical Helvetica;* 58(2), (2003).

Gottmann, J. *A Geography of Europe*. 4th edition. New York: Holt, Rinehart, and Winston, 1969.

Hall, P. *Cities in Civilization*. New York: Pantheon Books, 1998. Contains chapters on Vienna and Stockholm.

Jenkins, J.R.G. *Jura Separatism in Switzerland*. New York: Oxford University Press, 1986.

Kaplan, D.H., and J. Häkli (eds.). *Boundaries and Place: European Borderlands in Geographical Context* Lanham, MD: Rowman and Littlefield, 2002.

New, M. *Switzerland Unwrapped: Exposing the Myths*. London: Taurus, 1997.

Paasi, A. "Place Boundaries and the Construction of Finnish Territory," in David H. Kaplan and Jouni Häkli (eds.), *Boundaries and Place: European Borderlands in Geographical Context*. Lanham, MD: Rowman and Littlefield, 2002.

Sawyer, S. (ed.). *The Oxford Illustrated History of the Vikings*. New York: Oxford University Press, 1997.

Thorhallsson, B. (ed.). *Iceland and European Integration: On the Edge*. London: Routlege, 2004.

Valsoon, T. *Planning in Iceland: From Settlement to Present Times*. Reykjavík: University of Iceland Press, 2003.

Whited, T.L., et al., *Northern Europe: An Environmental History*. Santa Barbara, CA: ABC-Clio, 2005.

IV

ENLARGEMENT IN THE TWENTY-FIRST CENTURY

CHAPTER

12

EASTWARD EXPANSION 2004: THE BALTICS, POLAND, THE CZECH REPUBLIC, SLOVAKIA, HUNGARY, AND SLOVENIA

► INTRODUCTION

In the twentieth century eastern Europe did not experience economic and political stability. Prior to World War I, the countries of the region were under the imperial rule of Austria-Hungary, Germany, or Russia. Most people worked the land as laborers, peasants, or small farmers. Large landowners with great estates controlled local affairs and, in general, tried to keep traditional societies traditional.

World War I broke the imperial powers. The Baltics, Poland, Czechoslovakia, Hungary, and Yugoslavia became independent states. There were attempts to redistribute land but most large estates survived, as young countries attempted to establish democratic institutions, central banks, an independent currency, and national identity. Failure came quickly. By 1927, Poland was under Pilsudski's strong-man rule. The Great Depression crippled nascent democracy in every eastern European country except Czechoslovakia, which was dismembered by Germany in 1938–39.

The August, 23, 1939 non aggression pact, between the Soviet Union and Nazi Germany, allowed the totalitarian powers to pillage eastern Europe prior to turning on each other. Days after signing the pact Germany attacked Poland on September 1, 1939. The Soviets took eastern Poland later in the month. Then the Soviet Union occupied the Baltics and took Bessarabia, now part of Moldova, from Romania in June 1940.

When Germany attacked the USSR in June 1941, all eastern Europe came under Nazi control and economies were harnessed to Hitler's war machine. A number of eastern European countries, including Hungary, Romania, and Croatia, allied with Germany against the Soviet Union.

In 1945, German defeat brought Soviet occupation of eastern Europe. Socialist governments came to power and implemented land reforms, creating small farms for peasants by confiscating larger land holdings. Under Soviet pressure, the socialist governments were replaced by hard-line communists, who took orders from Moscow. In Stalin's view it was not good enough to be a Communist, you had to be a Moscow-oriented Communist. The early postwar socialist governments were disposed of by show trials, firing squads, and deportation. State farms and the collectivization of agriculture followed as Stalinist policies were implemented.

Economic policy for eastern Europe was oriented to the Soviet Union. Comecon (1949), created after Stalin had stopped eastern European countries joining the Marshall Plan, allocated resources within a Soviet empire centered on Moscow.

All the countries of eastern Europe, in the Soviet "Near Abroad," developed centrally planned economies. Most industries, including mining, iron and steel, chemicals, railroading, heavy engineering, and vehicle manufacture were controlled by state organizations, programmed to meet output targets, in the context of economic development plans compiled by central authorities in communist states. Communist ideology demanded work for workers, full employment, and the output needed to fulfill central planning goals. Plants became more concerned with output and employment than efficiency. In western terms many factories were overstaffed, lacked investment in new technology, and were careless of environmental impacts.

In the 1950s and 1960s there were uprisings against Soviet rule—East Germany (1953), Hungary (1956), Czechoslovakia (1968)—and unrest in Poland much of the time. Regimes became more tolerant of small farms and personal plots on collectives and state farms. By the 1970s, small businesses were allowed in Hungary and in most of eastern Europe small-scale entrepreneurs were able to operate even if they were employed at state enterprises during the day. Consumer goods, although limited in supply, became available.

The commanding heights of economies, the basic heavy industries, remained in the control of the state. When the Soviet empire collapsed in the early 1990s the protected, planned, overpeopled industries were not prepared for life in markets disciplined by competition. In 1991, the GDP of Hungary and Czechoslovakia, the more industrialized countries, fell by over 20 percent. The economies of Poland, Romania, and Bulgaria declined by over 10 percent.

The U.S., via NATO enlargement, engineered by President Bill Clinton's Czech-born Secretary of State, Madeleine Albright (née Korbel), reoriented the strategic and military stance of eastern Europe. Economic reorientation was more complex but if there was to be a Europe stretching from Portugal to Poland, as envisioned by Coudenhove-Kalergi in the 1920s, post-Soviet eastern Europe had to be brought into EU. The fear was that unless the region was incorporated, living standards would stagnate and societies would fall under nationalist dictators, as happened in the 1930s. During the 1990s, Yugoslavia was providing a model of what could go wrong in a post-Communist state.

Economic transition for eastern Europe was not easy. Many uncompetitive plants closed and unemployment rose. In some countries, like Bulgaria and Romania (Chapter 13), the old communists were prepared to sacrifice leaders but not centrally controlled economies. Other countries, including the Czech Republic, Hungary, and Poland, privatized state-run industries and encouraged foreign direct investment to bring in capital and modern technology. Policy was aimed at EU entry, which involved increased competition together with the opportunity to sell products into an enlarged market. The countries of eastern Europe did have some advantages, including an industrial labor force that could, with retraining, make the transition to modern manufacturing. Multinational companies saw opportunities for investment, encouraged by the knowledge that the region would enter EU. Volkswagen, for example, took over Czech Skoda cars in 1994.

In May 2004, the Baltics, Poland, the Czech Republic, Slovakia, Slovenia, and Hungary were admitted to EU. Bulgaria and Romania joined in 2007.

▶ THE BALTIC STATES: ESTONIA, LATVIA, LITHUANIA

Between the early sixteenth and early eighteenth centuries, the territory now occupied by Estonia, Latvia, and Lithuania was under Swedish control. Sweden had built up a south Baltic empire running from Finland to the North German plain. Sweden came into conflict with the rising power of Russia under Peter the Great, and lost the territory on which St. Petersburg (f. 1703) was built and gave up the Baltics in the Great Northern War. The war ran from 1700 until concluded by the Treaty of Nystad in 1721. The Baltics were valuable to Russia because the Gulf of Finland, around St. Petersburg, freezes in the winter and the Baltic ports of Estonia, Latvia, and Lithuania are more open.

In World War I Germany attacked Russia, occupying the Baltic lands. With German approval, Lithuania declared independence on February 18, 1918, Estonia on February 24, and Latvia on November 18, 1918. All had to repel Soviet troops sent to suppress independence. As Poland reestablished independence, after World War I, it took the city of Vilnius, and surrounding territory, from Lithuania.

During the 1920s and 1930s the Baltics failed to establish democratic institutions and came under authoritarian regimes—Lithuania in 1926, Estonia and Latvia in 1934. Authoritarian takeover was common in the new states of eastern Europe, and in established states like Germany, Spain, Portugal, Italy, and Greece.

Late in 1939, the Soviet Union demanded the right to create bases in the Baltics and in 1940 incorporated the three states into the USSR. Those thought to be resisting Soviet policy were eliminated or deported.

In the summer of 1941, the northern thrust of Germany's operation Barbarossa, aimed at Leningrad, came through the Baltics. As the Red Army retreated, towns and villages were destroyed and inhabitants killed. Many in the region associated Germany with the independence gained in 1918 and welcomed troops. However, in World War I, German policy had been to break the Russian hold on the non-Russian people of eastern Europe. Had Germany won World War II the policy would have been to push non-Germans out of the Baltics to make room for Nordic settlers.

When the Soviets reoccupied the Baltics in 1944 there were more executions, deportations, and some sustained resistance. Because of the strategic importance of the Baltic shore, the USSR invested heavily in the region, resulting in an influx of Russians, many of whom remained in the post-Soviet era. Twenty-five percent of the population of Estonia is Russian and in Latvia nearly one-third of the population is Russian.

Physical Environment

All the lands around the Baltic were glaciated with the eastern and southern shores of the sea today displaying the topography of glacial deposition rather than the scouring and erosive force of ice, as seen in Norway and northern Sweden.

In the glacial deposits, moraines, eskers, and outwash fans, an integrated drainage system has not yet developed, as is the case in parts of glaciated Wisconsin. The landscapes of the Baltics display many lakes, marshes, and extensive peat bogs. Large areas of the Baltic states are occupied by coniferous forests and there is much poorly drained grazing land.

Beneath the glacial deposits are the ancient rocks of the Baltic Shield, but little solid geology shows at the surface. Higher ground rarely rises above a few hundred feet. In the southeast of Estonia, the Nannja Upland, at Suur Manamägi, attains just over 1,000 feet, the highest point in the Baltic states. In Latvia the Vidzeme Upland rises to 1,017 feet at Mt. Gaizins and in the southeast of the country, an upland, rising to over 800 feet, stretches into Lithuania where it is known as the Baltic Highland, but all three Baltic countries are predominately low lying, ill drained, and forest covered with arable farming only on better-drained land (Figure 12.1).

The resource endowment is not large, consisting of timber, sands, gravels, limestones to make cement, peat, and some phosphates. In northern Estonia Paleozoic oil shales are exploited to generate most of the country's electricity. Two-thirds of Latvia's power is hydroelectric. Lithuania exploits peat resources, but most electricity is produced by one nuclear power station, which, by agreement with EU, is to be phased out. All three Baltic states have small oil and gas fields, none of which produces enough fuel to satisfy local needs.

Resources and Economic Activity

The Baltics moved rapidly to liberalize economic activity as the Soviet Union faded, becoming associate members of EU in the mid-1990s. In 1997 Estonia, Latvia, and Lithuania agreed to remove tariffs on goods moving between them, giving impetus to free trade. The Baltic countries have relatively small economies that are not well diversified. Agriculture, although providing a small percentage of GDP, still employs considerable numbers of people because modernization and mechanization have been slow. In Lithuania, in 2004, agriculture, forestry, and fishing contributed 5.7 percent of GDP but employed 17.3 percent of the workforce. In all countries farming is dominated by livestock production to produce meat and dairy products. Arable farming produces grains, potatoes, sugar beet, fruits, vegetables, and fodder for animals. All three countries have fleets equipped for fishing in the Baltic and the Atlantic.

Timber and timber products, including pulp, chipboard, cut timber, and furniture, exploit the coniferous forests and the less extensive stands of deciduous hardwoods. Manufacturing is predominately food processing, textiles, basic machinery, and consumer goods. EU entry has spurred economic activity as Finnish and Swedish companies have invested in production facilities to produce telephone and electrical equipment, attracted by the present low labor costs in Baltic countries. Tourism has been stimulated as the old centers of cities like Tallinn are attractive. Relations with Russia may ease because the Baltics will be applying EU entry regulations when admitting Russians and the bigger power will have to argue with Brussels if it does not like the fact that Russians can no longer enter the Baltic states whenever they want to. Further, the minorities of Russians, Belarussians, and Ukrainians in the Baltics will be protected by EU human rights standards.

Figure 12.1 The Baltics.

Population: Numbers, Distribution, and Urban Places

In Estonia (1.3 million), Latvia (2.3 million), and Lithuania (3.4 million) death rates exceed birth rates and at the present time population numbers are slowly shrinking. Since World War II there has been concentration of people in major urban areas, accompanied by a declining and aging rural population. Joining EU has provided economic stimulus and attracted investment but allowed people to move to more prosperous economies in western Europe.

TABLE 12.1 Estonia: Major Urban Areas

Tallinn	420,000
Tartu	110,000
Narva	70,000
Pärnu	50,000

Estonia

The density of rural population reflects the distribution of better-drained agricultural land. In Estonia, considerable areas are in lakes and marshes including Lake Peipus and the ill-drained wetlands in the hinterland of the Baltic port of Pärnu. The major Estonian towns are ports, with the exception of Tartu on the inland north–south routeway linking settlements on the Gulf of Finland to Latvia in the south.

Tallinn the capital, on the Gulf of Finland, contains nearly a third of the population. The historic old city is attracting tourists and Baltic cruise boats use the port in the summer months. Rapid economic growth has altered the skyline of the city, as modern hotels and office blocks are built (Photo 12.1). The economy of Estonia has expanded rapidly since the country joined EU. Estonia intends to adopt the Euro as currency in the near future.

Photo 12.1 At Tallinn, Estonia, high rise buildings overlook medieval streets.

Source: Jon Arnold/Danita Delimont.

TABLE 12.2   Latvia: Major Urban Areas

Riga	850,000
Daugavpils	120,000
Liepāja	100,000
Jelgava	72,000
Jūrmala	60,000
Ventspils	50,000

TABLE 12.3   Lithuania: Major Urban Areas

Vilnius	600,000
Kaunas	400,000
Klaipéda (Memel)	200,000
Šiauliai	140,000

Latvia

Riga, the capital, is a major port on the Gulf of Riga. Daugavpils is on the important overland routeway from St. Petersburg to Lithuania, Kalliningrad, and Poland. Liepāja, Jūrmala, and Ventspils are Baltic ports. Jelgava is on the routeway from Riga to Lithuania, Kalliningrad, and Poland.

Riga is the largest city in the Baltics. Historically, it has been on an important port, particularly when Latvia was part of the Russian Empire and, later, the Soviet Union. Riga is normally ice free and serves as a Baltic port year round, as opposed to St. Petersburg (Leningrad), which closes for part of the winter. The Soviet Union invested in both commercial and military installations at Riga; as a result, the city, which contains over one-third of Latvia's total population, has a large number of Russian-speaking inhabitants. Approximately one-third of all Latvians are of Russian origin. The minority wants schools that teach in the Russian language. Latvian law requires that the majority, but not all, courses be taught in Latvian.

Lithuania

Lithuania has the largest population of the Baltic states. There are higher densities of rural population in the southern part of the country, due to better-drained soils and, being further south, a longer growing season.

Vilnius, the capital, and Kaunas are in the east of the country on the higher ground that carries the routeways from Russia, through Latvia, to Poland. The major port is Klaipéda (formerly called Memel) on the Baltic. Immediately after World War I Memel was in east Prussia. The Lithuanians, needing a Baltic port, seized the city in 1923. It was returned to Germany in March 1939 but became part of Lithuania as Klaipéda after World War II, when the Soviets readjusted boundaries and deported the German inhabitants of the city.

Summary

Over recent years the Baltic states have experienced rapid economic growth and joining EU accelerated the trend as the countries became more attractive to foreign investment.

▶ POLAND

Introduction

By the end of the Napoleonic wars, the Polish state ceased to exist. Poles were divided into those who lived in the Russian Empire, the Poles in Germany, and the Poles under Austrian rule. At the end of World War I as the Empires of Russia and Austria-Hungary collapsed and Germany was defeated, Poland reappeared on the map. The new Poland was put together from three different administrative systems—German, Russian, and Austrian. There were no national institutions, so they had to be created. For example, a central bank and a national currency were not in place until 1924. Political parties had to be developed and in the 1920s a hundred parties were contesting elections and a large number were represented in the Diet and the Senate, established under the 1921 constitution.

Poland illustrated all the problems of boundary making for new states in the postwar era. U.S. President Woodrow Wilson spoke of self-determination and the Europeans talked of the nationality principle, meaning that boundaries were to be drawn around national groups to create new states. The American geographer Isaiah Bowman, Wilson's chief territorial advisor in Paris (1918–1919), experienced the difficulties of this approach for the national and ethnic groups were not neatly divided on the ground but clumped into clusters of villages and towns that intermingled (Smith, 2003).

The leaders of the aspiring new national states had strategic and economic concerns. Where possible they wanted an outlet to the sea, defensible borders, and economic assets like coalfields. It was impossible on the North European plain to give Poland defensible boundaries, but it did get the formerly German coalfield in upper Silesia. The Vistula, flowing to the Baltic, on which Warsaw stands, was in the midst of an undoubtedly Polish region. On the lower course of the river were German and Polish communities, with the Poles having a narrow, numerical majority. The Baltic coast port, Danzig, was undoubtedly German. The peacemakers created a Polish corridor to the sea and placed Danzig (Gdańsk) under the League of Nations to give Poland a Baltic outlet.

The peacemakers at Paris in 1919 could decide which German and Austria territories were to become part of the new Poland but had no control over Poland's eastern border with the emerging Soviet Union. Poland and the Soviets went to war. Eventually the Poles won and included parts of Lithuania, the Ukraine, and Belarus in their state. When Poland and the Soviet Union signed the Treaty of Riga in 1921, Poland was two-thirds Polish and included within its boundaries Germans, Lithuanians, Ukrainians, Belarussians, Ruthenians, Hungarians, and a large Jewish population. Germany and the Soviet Union felt they had been robbed of territory that under the nationality principle should have remained under their control.

Nor was Poland satisfied with its territory, for in the aftermath of Munich (1938), when Germany dismembered the Czech lands, Poland took territory (Teschen) from Czechoslovakia, as did Hungary.

The Nazi–Soviet Non-Aggression Pact

The Munich Agreement (1938) dismembered Czechoslovakia and led to the Soviet Union doing business with Nazi Germany. The Soviets saw Munich as an attempt, by France and Britain, to turn Hitler to the east. When Hitler suggested a deal to partition

Poland, Stalin was receptive. The notorious Ribbentrop–Molotov Non-Aggression Pact on August 23, 1939, divided Poland and eastern Europe up between Hitler and Stalin, a few days before the *Wehrmacht* attacked Poland on September 1, 1939. As Poland was *blitzkrieged,* the Red Army came from the east to claim the Soviet sphere on September 17, 1939.

Poland left the map in September 1939, but the Poles did not disappear. Polish warships joined forces with the Royal Navy. Polish pilots made their way to Britain, joined the Royal Air Force, and in the Battle of Britian (1940) helped to make a German crossing of the English Channel impossible. When Germany attacked the Soviet Union, Polish soldiers in Soviet prisoner of war camps were allowed to find their way out of the USSR, via Iran, to join the western allies. Polish units were prominent in the Italian campaign and the Normandy invasion. Battle honors include driving German forces out of the rubble of the monastery at Monte Cassino in southern Italy. Before the war, Poland had started work on German codes and the Enigma machine. The intelligence came to Bletchly in southern England where the codes were broken.

A Polish government in exile, led by General Sikorski, was established in London to represent Polish interests and act as a conduit for information coming from occupied Poland. In 1943, news of the Katyn Massacre (1940) of Polish prisoners of war by the Soviets reached London. Sikorski wanted an International Red Cross investigation into the killings. Stalin broke diplomatic relations with the government in exile. Britain and the U.S. would not confront the USSR over the issue, fearing Stalin might make a separate peace with Germany.

In April 1943, Warsaw's Jewish ghetto, knowing it was to be eliminated, fought German units for a month, losing 60,000 members. In 1944 the Nazis and the Soviets, although at war, cooperated once more against Poland. On August 1, 1944, Polish nationalists launched the Warsaw uprising against German occupation. The Red Army, advancing from the east, came to a halt. German units destroyed Warsaw as they killed resistance fighters (Davies, 2004). Stalin did not want Polish nationalists to liberate Warsaw (Broekmeyer, 2004). The Soviets had assembled their own Polish government (the Lublin government), which was placed in power when the Red Army did occupy Warsaw. Members of the Polish government in exile, returning from London to Warsaw, were arrested and charged with treason. Officers, after serving in the west, returned home to help reestablish the Polish armed forces. Some were tried for political crimes, as everyone with Western connections came under suspicion. Wisely a Polish government in exile was maintained in London, for Poland was not liberated at the end of World War II. There was no help from the West as the Stalinists took over. The postwar U.S. ambassador to Poland titled his account of his Warsaw term *I Saw Poland Betrayed* (Lane, 1948).

In 1939 Warsaw had a million inhabitants. In 1946 some 300,000 people huddled in the ruined city. Half of the prewar population was dead, others had fled. There were no whole buildings in the central city, all the Vistula bridges were destroyed, and Warsaw was a pile of rubble. The city has been rebuilt and some of the central areas recreated in traditional form.

After World War II, the Soviets restructured the territory of Poland, moving it to the west (Figure 12.2). Lithuania had Vilnius restored to it. East Prussia became Polish, except for the territory around Königsburg, the former capital of East Prussia, which, as Kaliningrad, became part of the Russian Soviet Socialist Republic. The Germans of East Prussia, along with those from the Polish corridor and Danzig, were deported westward,

Figure 12.2 Poland.

as were those living in Pomerania and Lower Silesia. The western border of Poland now ran along the line of the river Oder (Odra) and its southern tributary, the Neisse. The human disruption was appalling. Poles were forced out of the expanded Soviet Socialist republics of Lithuania, Belarus, and the Ukraine. German communities were given short notice to pack what they could carry and marched to East Germany. Many died along the way, for by the end of the war German communities were composed of women, children, and old men unfit for military service.

Poland saw the apparatus of a totalitarian terror state installed, along with centralized economic planning, a large army, and politicians who wanted Poland serving in a Soviet Empire.

Poland was never content under a Communist regime and Soviet control. Free elections came in 1991. Lech Walesa, who had led the Solidarity movement from the shipyards in Gdańsk (the former Danzig), was elected to the presidency, and a moving reconciliation took place. Walesa invited the Polish government in exile home and inherited the office of President not from the discredited Communist regime but from the last leaders of the Polish government in London. The leaders returned the insignia of Poland and the manuscript of Poland's 1935 constitution, which the exiled government had taken out of the country and struggled to uphold. Poland joined NATO in 1999 and the EU in 2004.

Physical Environment

Physically we can divide Poland into three major regions (Figure 12.3):

1. Northern Poland
2. Central Poland
3. Southern Mountainous region

Northern Poland

Glaciation   In Norway and much of Sweden, landscapes display the features of glacial erosion. On the south shore of the Baltic, from the Jutland Peninsula in the west, to south Finland in the east, landscape features have been created by glacial deposition. From Jutland, along the North German plain and through Poland, are a series of large

Northern Poland – Ice covered in the last glaciation which deposited clays, sands, and gravels. Drainage is poorly integrated, with many lakes and marshes.

Central Poland – South of the ice in the last glaciation. Received deposits of wind blown loess. Better drained than the north, the region carries major west to east routeways.

Mountainous South – The southern border of Poland is made up of the Sudeten and Carpathian mountains, separated by the Moravian Gap. Routeways run through the mountains connecting the Danubian and Baltic lands.

Figure 12.3   Regions of Poland.

end moraines that lay along the front of the great glacier coming from Scandinavia. The end moraines consist of clays and rocks pushed along by the glacier. Associated fluvial deposits were laid down by melt-water streams coming from glaciers in retreat.

The soils resulting from glacial deposition are highly variable. The moraines being pushed by the glaciers were hundreds of feet high and left a hummocky, poorly drained, landscape as streams have not had time to form integrated drainage networks. There are many lakes, marshes, and bogs, often with small streams linking several ill-drained depressions in a system of internal drainage from which water cannot escape to the sea. These features are well displayed in the Masuria region of northeast Poland. In northern Poland, much land is used for grazing. Arable farming is limited to better-drained areas.

Another contrast between Norway and Sweden and the south Baltic shore results from the fact that the greatest weight of the Scandinavian glacier lay on the more northerly lands. When the ice melted, and the weight lifted, the northerly lands slowly rose, compensating for the sea level rise resulting from ice melt. Norway and Sweden have emergent coasts. On the south Baltic shore the sea level rise submerged coastal areas.

The Scandinavian glacier had an impact on the drainage patterns on the northern European plain. As the ice retreated, and the climate warmed, the rivers flowing from the Alps and Carpathians toward the Baltic found the course to the sea blocked by moraines or ice. Look at the courses of the Vistula, the Elbe, and the Rhine; all have been elbowed to the west by glacial obstruction in the lower courses. Eventually, the Vistula found a route to the Baltic through the moraine field, as water piled up until it overflowed the moraines and cut a channel to the sea.

The Baltic coast of Poland is marked by sand dunes, spits, and lagoons. The port of Szczecin (former Stettin) is over twenty miles inland on the drowned estuary of the river Oder (Odra). The other sizeable Baltic ports are Gdynia and Gdańsk on the Gulf of Danzig. Gdańsk was an important medieval city state and a member of the Hanseatic League, with a legislature. Gdynia was a fishing settlement, which, after World War I and Polish independence, was built up into a major port to divert cargo from Danzig, a German city but in the Polish tariff system.

Central Poland

Central Poland was glaciated but not in the last glacial phase. There has been more time for the forces of weathering and erosion to smooth moraines and the drainage is better integrated. Central Poland was south of the Scandinavian glacier in the last ice advance. Central Poland was a cold, windswept waste, with the wind picking up fine particles of silt to be deposited as loess. The fertile European loess belt stretches from the North Sea across Europe well into Russia. In Poland, the loess belt provides the resource base for areas of productive farming. Warsaw, the capital, is located in the east-central part of the region, on the west bank of the navigable Vistula.

Southern Mountainous Region

In Poland's southern mountains, which include the Sudeten Mountains in the west and the Carpathians in the east, there is more evidence of glacial activity. The mountains were high enough for valley glaciers to form. The valley glaciers gouged the typical U-shaped valleys and at the ice front, moraines and outwash material were deposited.

A visitor to Warsaw has difficulty thinking of Poland as mountainous country but in the south the Sudeten Mountains and the Carpathians rise to 8,711 feet. Many Poles vacation in the mountains, which were a favorite region of John Paul II who was Archbishop of Kraków before he became Pope. (Photo 12.2)

Rivers

Poland's two major rivers rise in the southern mountains. The Oder (Odra), and some of its tributaries, rise in the Sudeten Mountains and in the Moravian gap that divides the Sudetans from the Carpathians. The Oder (Odra) flows north and then northwest to Wroclaw and it is navigable from that city to the sea.

The Vistula rises in the Carpathians and flows through Kraków to Warsaw. In March, ice flows down the river, often piling up and causing flooding. Thirty miles from the Baltic, the Vistula forms distributaries before falling into the Gulf of Gdańsk. The delta lands have drainage works and much land has been reclaimed for agriculture.

Distribution of Population and Economic Activity

In northern Poland the major concentrations of population are found in and around the ports on the Baltic shore (Figure 12.4). Szczecin was the German city of Stettin until the end of World War II when it was incorporated into Poland and the German population expelled. Stettin handled cargo going to and from Berlin and the cities were linked by canal in 1914. Trade between Berlin and Szczecin was disrupted in the postwar era. Now with Germany and Poland in EU Szczecin, and its downstream outport Swinoujście, will handle more German cargo.

Photo 12.2   In southern Poland, the Carpathian mountains dominate the landscape.

Source: Age fotostock/SUPERSTOCK.

Figure 12.4   Poland: Urban Places.

The largest concentration of population on the north coast is found in the Gdánsk–Gdynia conurbation, which includes the Gulf of Danzig resort town of Sopot. Because agriculture in northern Poland is not intensive, rural population densities are moderate and the market towns serving agricultural hinterlands are usually modest in size.

Central Poland, with more agricultural potential than the north, contains a number of important cities, including Poznan, Łódź, and Warsaw. Poznan and Warsaw are on a major east–west routeway across the northern European plain. Warsaw was bombed, shelled, and occupied by German forces in 1939. Destruction was completed between 1943 and January 1945 when the Red Army finally took the city. The old town, with its marketplace and cathedral, was rebuilt after the war. (Photos 12.3 and 12.4)

The major cities in the south of Poland are situated on a routeway that runs close to the Sudeten Mountains and the Carpathians. Kraków on the Vistula is an important center of Polish culture, with a university dating to 1364 and many fine medieval buildings. Kraków is also a major industrial center, with the nearby Nowa Huta built in the Communist era.

Nowa Huta, literally new steel plant, was part of a planned Communist industrial complex started in 1949 that included the Lenin steel works, a cement factory, and a

TABLE 12.4   Top Ten Polish Cities in Rank Order

Warsaw	1,620,000
Łódź	810,000
Kraków	740,000
Wrocław	640,000
Poznań	580,000
Gdańsk	460,000
Szczecin	420,000
Bydgoszcz	390,000
Lublin	360,000
Katowice	350,000

tobacco plant. The steel works was built on a site near Kraków, which did not possess iron ore or coal and the raw materials had to be transported in. Workers were housed in large blocks. No churches were built for a population that rapidly exceeded 100,000 inhabitants. A Stalinist landscape of heavy industry, with tower blocks housing workers, appeared, omitting the religious buildings and symbols of traditional cultural landscapes. The state was creating an atheist landscape close to Kraków, a center of Polish Catholic culture.

The Archbishop of Kraków, future Pope John Paul II, appointed a priest to build a church. Rome sent stone from St. Peter's tomb with instructions to build a church around the rock. The church opened in 1977 and many others followed, illustrating how religion can be a focal point of resistance when groups feel their culture is threatened, distorted, or destroyed.

In or adjoining the valley of the Odra are Poland's important Upper Silesian and Lower Silesian coalfields and associated industrial cities including Katowice and Wrocław. Katowice is a major mining and heavy manufacturing center. Wrocław, a historic river port on the Odra, was badly damaged in World War II. After the war, the formerly German city of Breslau was part of the German lands transferred to Poland. The German population of Breslau was expelled and the urban area was peopled with Poles and renamed Wrocław. The city was repaired and built into a major manufacturing center in the communist era.

In addition to the Silesian coalfields, Poland has deposits of lignite and brown coal. There are small oil and gas fields on the flanks of the Carpathians and some nonferrous minerals in the southern mountains.

Summary

In joining the western European world to which Poland has always thought it belonged, the country has undergone transformations. The Solidarity movement based on Gdańsk shipyards helped to bring the Communist regime down and won electoral success in the 1990s, only to fade as other political parties built bases elsewhere in the country. The Gdańsk shipyards could not compete once the state subsidies, characteristic of Communist economies, ceased. Many other, older, heavy industries have had difficulty being competitive, but in some areas including transportation equipment, the country is producing products that do sell into EU markets. The Polish agricultural sector is benefiting from

Photo 12.3 Fortified gate, Warsaw, Poland.

Source: Age fotostock/SUPERSTOCK.

the Common Agricultural Policy payments to farmers, although the impact is greatest with the larger farming units.

Warsaw is being transformed as new shopping malls are built on the outskirts and in the central area of the city. Roads are a problem. There are plans for major west-to-east and north-to-south highways, but they do not presently exist. There is no Warsaw beltway and far too many vehicles crowd the city center.

▶ THE CZECH REPUBLIC

Evolution of the Czech State

The ancient Kingdom of Bohemia was absorbed into the Hapsburg Empire after 1527 and came under the control of Vienna. In the early nineteenth century the Czech lands, which had many craft and cottage industries, began to adopt modern manufacturing practices. The country had coal and capital. The capital came from enterprising landowners who invested in industry. There were several coalfields including Kladno, west of Prague; Ostrava in the east, close to Poland; and Chomutov and Most in the northwest, close to the Ore mountains. Ores of iron and nonferrous metals were widely available in the mountains. Forests provided timber for fuel and construction.

By the beginning of the twentieth century the Czech lands contained substantial industrial regions and accounted for a major part of the manufacturing activity in Austria-Hungary. Industries included textiles, iron and steel, railroad equipment, furniture,

Photo 12.4   Warsaw was reduced to rubble in 1945. Central squares were restored.

Source: Age fotostock/SUPERSTOCK.

footwear, heavy engineering, glassware, and ceramics. At Pilsen (Plzeň) the giant Skoda works produced armaments and manufactured motor vehicles.

Plzeň was a major brewing center but every town had a brewery including Budweis (České Budějovice), home of Budějovický Budvar. Budějovice is an important communication center, south of Prague, near the border with Austria. Within Austria-Hungary, Czech living standards were relatively high. Czechs hoped for a parliament at Prague and the type of autonomy Hungary had achieved within the Dual Monarchy.

The leading figure in the creation of Czechoslovakia was Tomáš Masaryk (1850–1937), a professor of philosophy and a member of the Austrian parliament. Masaryk did not subscribe to pan-Slavism, the idea that Russia would play the lead in freeing Slav peoples from the control of Austria-Hungary. To Masaryk, the Russians were non-Europeans and inherently tyrannical. Masaryk, married to an American, looked to friends and countrymen in France, Britain, and the U.S., where several states, including Texas and Nebraska, had large communities of immigrant Czechs.

When World War I started, Masaryk left home and took a teaching post at Kings College in the University of London, working with R.W. Seton-Watson on an influential journal, *The New Europe* (1916–1921). Masaryk's articles educated many on the dangers of Russian-led pan-Slavism and the depth of German ambitions in *Mitteleuropa*. Masaryk's warnings against both Germany and Russia, together with his view that eastern Europe needed support from the West, are ideas that parallel those of Halford Mackinder. Masaryk and Mackinder probably met, for both were in Seton-Watson's circle and Mackinder also taught at London University, at the London School of Economics.

TABLE 12.5 Ethnic Groups in Czechoslovakia, 1920

Czech	7,000,000
Slovaks	2,000,000
Germans	3,100,000
Hungarians	745,000
Ruthenians	461,000
Jews	180,000
Poles	76,000

Mazaryk visited the U.S. toward the end of the war, and in Pittsburgh did the deal with the Slovaks leading to the creation of Czechoslovakia and its recognition by the U.S., in 1918. Not everyone welcomed the arrangement. Many Slovaks thought, correctly, that their interests would be secondary to those of the more numerous and wealthier Czechs. The state of Czechoslovakia was not a nation-state. (Table 12.5)

The placing of Czech and Slovaks in one state was partly to counteract the strong German presence in the Czech lands adjoining Germany. Another difficulty was that Slovakia had focused economically on Hungary and Budapest. The Czechs were oriented to Austria and Vienna. The west-to-east routeways linking the Czechs and Slovaks were not good and while Bohemia and Moravia had industrial cities, Slovakia was largely agricultural.

Czechoslovakia came into existence on October 28, 1918, and was formerly recognized by the Treaty of St. Germain on September 10, 1919. Professor Masaryk became President Masaryk (1918–1935). There was little difficulty in organizing the institutions of democracy and the country was active in international affairs. Foreign minister Beneš (1884–1948) made an alliance with France and joined the "Little Entente" with Yugoslavia and Romania (1921). Relations with Poland, that coveted Teshen, and Hungary, that wanted Ruthenia, were tense.

During the Great Depression Czechoslovakia suffered a major loss of exports as world trade collapsed. A larger problem was the emergence of the Nazi regime in Germany. The Sudeten Mountains and the Ore Mountains, adjoining Germany, had been in the Kingdom of Bohemia; they had never been a part of Germany, although many Germans had settled there. Although Germany had rejected President Wilson's offer in the fourteen points speech of a negotiated peace with boundaries arrived at by self-determination, after the war German geopoliticians, like General Haushofer, still argued that the Sudetenland should be in Germany.

In the spring of 1938 Hitler announced he would absorb Austria and take Czech territory that contained German speakers. The *Anschluss* with Austria took place on March 12, 1938. Austria became part of Germany. The Austrian army was absorbed into the German army. Germany now neighbored the Czech lands in the north, west, and south. Potentially, German forces could penetrate the Morava river to cut Czechoslovakia in half.

In September 1938 Hitler demanded the Czech lands in which there were German populations. There was to be war. France, with a treaty requiring it to aid Czechoslovakia, would not act without British armed support. Thinking he could appease Hitler, the British prime minister, Neville Chamberlain, who had never been on a plane, flew to Munich to attend a conference convened by Benito Mussolini. Hitler, Mussolini, Chamberlain, and the French prime minister Daladier were in the conference room. The Czechs were not. At the end of the meeting the Czechs, led by Beneš, were told that Britain and France would give no support. Hand over the territories with German speakers to Germany.

Prime Minister Neville Chamberlain flew back to Britain to be greeted as a peace maker. Roosevelt telegraphed his congratulations; "Good man," declared the President. No major British newspaper denounced the Munich settlement. Churchill told the House of Commons, on October 5, 1938, that Britain had "sustained a defeat without a war." His voice was unwelcome.

In March 1939, German forces occupied the remaining Czech lands and Slovakia became autonomous. President Roosevelt commented, "The hopes that the world had last September, [at Munich] that the Germany policy was limited . . . to bringing contiguous German people into the Reich and only German people . . ." had gone. Apparently there was no limit to German ambition (March 31, 1939). Britain and France now gave Poland, the next German target, a territorial guarantee, having relinquished the resources, armament capacity, armed forces, and strategic position of Czechoslovakia. In 1939 the Czech industrial sector became part of the German war machine. The country had the capacity to manufacture 40,000 vehicles a year along with tanks and artillery. Huge quantities of armaments were hauled out of Czechoslovakia into Germany to equip the expanding *Wehrmacht*. Czech tanks were used against the Poles in 1939, Western Europe in 1940, and on the Russian front in 1941. The Czech manufacturing sector was put to German war work and expanded. Many Czechs were forced to work in Germany.

A substantial number of Czechs, understanding after Munich that the Nazis were coming, got out of the country, moving to France, Britain, and the U.S. Czech pilots flew in the Battle of Britain (1940) and helped make invasion across the English Channel impossible. Many professional men and women from Czechoslovakia contributed to the western war effort. Among those who left Prague was the Czech diplomat Korbel, who took his family to Britain, where daughter Madeleine was educated, until the Korbels, along with the Czechoslovak government in exile led by Eduard Beneš, returned to Czechoslovakia in 1945.

After the war the Czechoslovak territory taken by Germany, Poland, and Hungary— was restored. The Germans and some Hungarians were deported. The Soviet Union took Ruthenia in the east of the country, which was attached to the Ukraine S.S.R.

During and after the war, Beneš tried to build good relations with the Soviet Union, but progressively, in the postwar years, Moscow-orientated communists got control of the government. President Beneš resigned in June 1948 and died a few weeks later. His foreign minister, Jan Masaryk, son of President Masaryk, was found dead beneath an open window at the foreign ministry. Few believed the official explanation of suicide. People like Korbel, who had spent the war in the west, were increasingly under suspicion. Korbel obtained a posting to the UN and settled in the United States where later, as Professor Korbel, he taught political science at the University of Denver. Back in Czechoslovakia the Communist party purged itself. The party secretary Rudolf Slansky, who had spent World War II in Moscow, was charged, along with thirteen other leading Czechoslovak communists, with Trotskyite–Titoist–Zionist activities. Eleven of the fourteen were executed. Slansky and several others were Jewish. Stalin did not like the new state of Israel's pro-Western stance.

The "Prague Spring" of 1968 was an attempt by the Czech communist party to liberalize the regime. At first the liberalization was allowed by the Soviet Union and adjoining Warsaw Pact countries. French filmmakers recorded the relaxed atmosphere in the streets of Prague and "hippies" held signs declaring "Make love not war" while smoking peace pipes. Hard-line communists in the Warsaw Pact feared that liberalization would spread to their countries and undermine regimes. On August 20 and 21 Soviet troops, together with forces from Hungary, Poland, East Germany, and Bulgaria, invaded Czechoslovakia. Alexander Dubcek, the Czech leader, ordered only passive resistance to the tanks. The French filmmakers recorded the arrival of the Warsaw Pact troops in their documentary *Oratorio for Prague*. By October, liberalization was diluted and under the Moscow Treaty 100,000 Soviet troops remained in Czechoslovakia. (Photo 12.5)

When Soviet control of eastern Europe crumbled, Czech industries found the translation back to a capitalist system achievable as many products were of high-quality, although production was overmanned by western European standards.

Free of Soviet control, the Czechs and Slovaks created separate states on January 1, 1993. When the issue of NATO expansion was considered by the Clinton administration, the Secretary of State for Foreign Affairs was Czech-speaking Madeleine Albright (née

Photo 12.5 Prague residents surround Soviet tanks in Prague on August 21, 1968 as the Soviet-led invasion by the Warsaw Pact armies crushed the so-called Prague Spring reform in former Czechoslovakia.

Source: Libor Hajsky/AFP/Getty Images.

Korbel) who held to the views of Masaryk and her father: the Russians should be kept out of Europe and the west should support the independence of eastern Europe. The Czech Republic joined NATO in 1999 and EU in 2004.

The Physical Environment

The core of the Czech Republic is the plain of Bohemia (Figure 12.5). The Bohemian plateau rises above the plain on all sides but in the north the river Elbe, which drains the plain, cuts through the plateau and flows into Germany. Embracing the plateau and plain are mountain ranges: in the southwest, the Bohemian forest (*Bohmer Wald*); to the northwest, the Ore Mountains; and in the northeast, the Sudeten Mountains. In the east, the Bohemian plateau merges into the plain of Moravia, another major topographic region within the Czech Republic. The plain of Moravia is drained to the south by the Morava, a tributary of the Danube. The valley of the Morava is an important passageway from the Danubian plains into Poland, the north German plain, and the Baltic shore.

Plain of Bohemia

The agricultural land in the plain of Bohemia is generally fertile, having had wind-blown loess deposited upon it during the Ice Age. The crops are the usual temperate mix for central Europe with its warm summers and cold winters—grains, root crops including potatoes, sugar beets, and hops for the famous Czech brewers.

Prague is on the navigable river Vltava, a tributary of the Elbe. On the east bank of the river, on higher ground, stands the historic core around the fortress of Hradčany,

Figure 12.5   Regions of Czech Republic.

with the St. Vitus cathedral and the old city below. To the west of Prague is the town of Kladno and the coalfield of the same name, which is a center of steel making and engineering. Prague, a city of fine baroque buildings, is at the center of an industrial region. (Photo 12.6)

Bohemian Plateau

The plateau, being higher than the plain, has a less warm summer and a colder winter. Crop mix reflects the climatic change, with rye and hardier grains being more prominent in cultivated areas on the plateau. The rivers are incised into the plateau, creating valleys with steep sides and narrow floors. Woodland is extensive and much land is used for grazing and livestock raising. Towns tend to be small service centers providing for the needs of surrounding agricultural districts. North of Plzeň, igneous rocks are exposed and around their margins are warm mineral springs, which are exploited at resorts such as Karlovy Vary (Karslbad) and Mariánské Lázně (Marienbad).

In the west, the major city on the Bohemia plateau region is Plzeň where industrialization was originally based on nearby sources of coal and iron ore. Besides steel, the city is famous for the brew Pilsen. Pilsen (Plzeň), the city, is the home of the Skoda

Photo 12.6 Prague. Gateway from the Charles Bridge, over the Vltava river, leading to the castle.

Source: Age fotostock/SUPERSTOCK.

engineering complex. Skoda was a major producer of armaments in the Austria-Hungary empire and a maker of vehicles in the interwar years. Today, Skoda cars are part of the Volkswagen group. In the south of the plateau the most important city is České Budějovice (Budweis, ger.), on the routeway from Prague to Linz on the Danube. The city is a historic fortress town and industrial and brewing center. Brewers still resent the use of their name on Anheuser-Busch products.

Moravia

Lying between the eastern edge of the Bohemian plateau and the Carpathians of Slovakia is the valley of the Morava river, flowing to the Danube. Besides being a routeway from the Danubian lands to the north European plain, the extensive valley is well cultivated. On the river Sázava, a tributary of the Morava, stands Brno, the second largest Czech city and another important manufacturing center.

In the north of Moravia there is an extension from Poland of the Silesian coalfield into the Czech republic. The coal was the basis of a region of heavy industry, with a number of towns having steel works and engineering factories. The regional center is the city of Ostrava.

The Mountains

The plain and plateau of Bohemia are bounded on three sides by mountains. The boundary with Poland, in the northeast, is the Sudeten Mountains. To the northwest, the boundary with Germany is formed by the Ore Mountains, which rise up to 4,080 feet. The Elbe cuts through the mountains downstream of Ústí nad Labem to reach Dresden in Germany. The boundary with Germany to the southwest is along the *Bohmer Wald* (Bohemian Forest). As the name Ore Mountains indicates, the ranges contain metals along with timber, water power potential, and some coal and lignite. The resources were attractive to craftsmen from Germany who settled to smelt metals and start small manufacturing plants. Thus the mountains came to house a population in which a majority spoke German. Hitler exploited this to demand that areas with a majority of German speakers be given to Germany in September 1938 at the infamous Munich conference. At the end of the war, the ethnic Germans were expelled and forced into Germany. Although the expulsion was not carried out as brutally as in Poland, many German families were forced to leave lands they regarded as home. The deportations remain an issue in both Czech and German domestic politics.

Population

Like many countries in eastern Europe the Czech Republic has a slightly higher death rate than birth rate per thousand of the population and the total fertility rate is down to 1.2. Prague (Photo 3.5) continues to attract migrants and is a prosperous, elegant tourist destination.

Prague

The Vltava river rises in the south of the Czech Republic and flows northward across Bohemia and joins the Elbe before cutting through the mountains to the north European plain and Dresden (Figure 12.6). The Elbe flows in a generally northwesterly direction

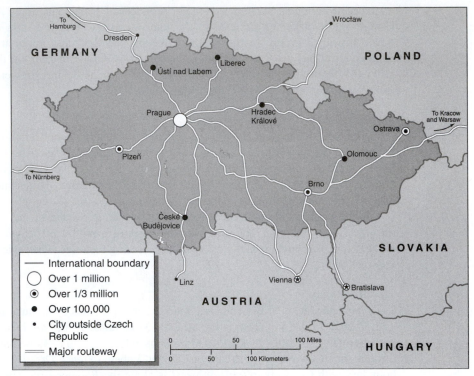

Figure 12.6 Czech Republic: Urban Places.

across the plain to Hamburg and the North Sea, linking Bohemia with one of Europe's major river routeways.

Prague is near the head of natural navigation on the Vltava at a point where fording (*Prah* = ford) was possible. Close to the fording place a hill provided a defensive site for a castle (*hrad*), occupied since the ninth century. On the opposite bank of the Vltava a trading settlement developed. The river was bridged in the mid-twelfth century and what we now call the Old Town (*Stare Mesto*) had attained borough status, a market, and fortifications by 1240. (Photo 12.7)

Charles IV, King of Bohemia and Holy Roman Emperor, had his court at Prague from 1346 to 1378. Charles was a builder, laying out New Town (*Nove Mesto*), extending the walls to enclose the expanding city, establishing Charles University in 1348, and having the Charles bridge erected in 1357 to provide a more convenient link between the *Hradčany* (Prague Castle) and the *Stare Mesto,* the Old Town, across the Vltava river. Over time the *Hradčany* became a complex of buildings, including St. Vitus Cathedral, palaces, and armament facilities, creating a Kremlin-like compound.

Bohemia and Prague became an early center of Protestantism under the teaching of Jan Hus (d. 1415). The last king of Bohemia died fighting with the Hungarians, against the Turks, at the battle of Mohács in 1526. Bohemia went to the Hapsburg, Ferdinand, Archduke of Austria. The Czech lands would be a part of the Austrian realm until October 1918.

Photo 12.7 Old Town Square, Old Town Hall and Tyn Church. Prague, Czech Republic.

Source: Age fotostock/SUPERSTOCK.

The Austrians attempted to reimpose Roman Catholicism, leading to persecution, civil unrest, and a decline of Prague as a trade center. In the seventeenth century, the Thirty Years War (1618–1648) disrupted trade across much of central Europe but in the second half of the century the Austrians insisted that internal tariffs within the realm be removed, creating a larger economic space and reviving trade.

Prague flourished in the late seventeenth century and in the eighteenth century. The revival can be seen in the urban fabric of *Hradčany,* the Old Town, and the New Town. The streets are lined with fine baroque buildings. There are few surviving medieval houses in central Prague, indicating that prosperity allowed the redevelopement, or refacading, of older structures to display the elegance of eighteenth-century architecture.

The population of Prague grew in the eighteenth century. Early in the nineteenth century Bohemia began to industrialize. Prague, and its region, had coal at Kladno and water power from the Vltava and its tributaries. Along with cities like Plzeň and Brno, Prague became an early center of industrialization in central Europe. (Table 12.6)

TABLE 12.6 Major Czech Cities: Population

Prague	1,178,000
Brno	379,000
Ostrava	379,000
Plzeň	166,000
Olomouc	104,000
Liberec	100,000
Hradec Králové	100,000
České Budějovice	100,000
Ústí nad Labem	97,000

In 1878 the walls of Prague were demolished to create green space, new roads, and room for buildings like the National Museum on Václavské Náměsti (*Náměsti* = square), which was an expression of quickening Czech nationalism. Although Czechs sat in the Austrian parliament, they wanted more autonomy and an assembly at Prague.

The city continued to modernize and by the 1890s an electric tram system was serving the city, which continued to grow as suburbs sprouted with the extension of the tram lines.

Prague benefited from becoming the capital of newly independent Czechoslovakia after World War I. The city was home to a parliament, the national bank, the machinery of government, and the ministry of external affairs. Suburbs expanded and the city built housing projects to provide accommodations for the growing workforce.

The Czech lands were taken over by Germany in the spring of 1939. There was no battle for the city, which was largely undamaged. There was an uprising against German occupation in 1945, as the Red Army approached. The *Wehrmacht* withdrew. Prague, overlooked by higher ground, is not a good place to make a stand in artillery warfare.

Postwar, a socialist government was replaced by hard-line Communists in 1948. The growth of Prague was rigorously planned, but fortunately the city avoided the restructuring of the center to provide space for a Stalinist square, as happened in numerous other European cities.

By the mid-1950s, five-story blocks of prefabricated panel housing for workers were being erected. To avoid urban sprawl, the blocks were built at a relatively high density along transportation lines. When the metro was built in the 1960s and 1970s the stations, serving residential areas, became small service centers with shops providing basic goods. Housing blocks erected in the 1970s were taller, the units smaller, with whole, factory-made bathrooms and kitchens being dropped into place as the block went up. The taller buildings, with more units, increased housing density and allowed more people to live close to metro stops.

In the 1990s the housing blocks were privatized into co-ops, with a committee taking responsibility for maintaining stairwells, roofs, and surrounding space. The metro stations had high-rise office blocks and hotels built nearby, to exploit access to the center, and shopping malls grew at metro stops.

Summary

Since 1993 when the Czech Republic and Slovakia became separate states, the Czechs have adjusted to a free market economy and the export opportunities opened up by EU membership. Although a new member of EU, the Czechs have independent views. Having freed themselves of Moscow control, they often find the Brussels bureaucracy tiresome. In 2005, the Czech President Vaclav Klaus called for the scrapping of EU and its replacement by a Free Trade Organization of European states.

▶ SLOVAKIA

Slovakia is a landlocked country bounded on the north by Poland, in the northwest by the Czech Republic, in the west by Austria, to the south by Hungary, and in the east by the Ukraine.

Physical Environment

Slovakia can be divided into two major physical regions (Figure 12.7):

1. The Carpathian Mountains in the north
2. The Danubian Plains in the south.

The Carpathians

The Carpathian Mountains are of Alpine age and consist of a series of folded ranges that run across Slovakia in a southwest-to-northeast direction and rise up to 8,199 feet in the High Tatra Mountains. When the young mountains were folded there was igneous activity and veins of copper and zinc were injected into the sedimentary rocks, in addition to high-quality igneous iron ore. A number of mineral-rich springs are a product of the mountain-building era and warm sulphurous springs, as at Piešťany Spa, provide treatment for rheumatic diseases.

The Carpathians are well wooded and the forests are exploited for timber and pulp. On the valley floors hardy grains and root crops are grown, but the greatest part of the agricultural land is in pasture used for livestock raising, particularly sheep. In the Carpathians, at altitude, winters are long and cold.

The Carpathian region of Slovakia is drained by rivers that flow between the mountain folds before turning south to join the Danube. The rivers generate hydroelectric power but have little navigational use. In the west, the prominent streams are the Váh, and the Nitra. In the east, the major rivers are the Torysa and the Ondava. Where the rivers leave the uplands, market towns developed at crossing points on the streams where

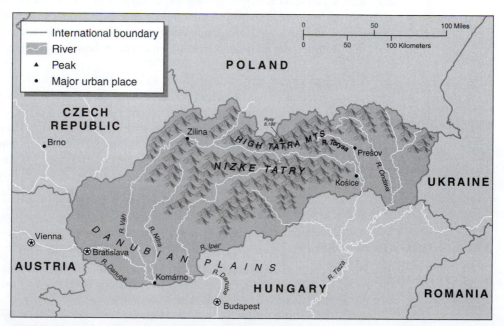

Figure 12.7 Slovakia: Physical Regions and Urban Places.

the produce of the Carpathian uplands could be sold on to the larger population centers on the Danubian plains.

The Danubian Plains

On the Danubian plains summers are hot and thundery. Winters can be cold on the plains, but the Carpathians keep out much of the colder air from northern central Europe. The Danubian plains, consisting of alluvium and loess, are predominately in arable crops—grains, root crops, vineyards, and orchards. The Carpathian livestock region and the arable Danubian plains complement each other in terms of agricultural production.

Distribution of Population and Urban Places

In the Carpathians most people live in the valleys where small market towns exist in addition to farms and villages (Figure 12.7). In the larger valleys—for example, the Váh—hydroelectric power plants provide electricity for nearby industries. The town of Zilina, on the Váh, has a range of manufacturing activities including chemicals, cement, textiles, and wood products using Carpathian timber.

Carpathian population densities are lower than on the Danubian plains in the south of the country. In general, population densities are higher near the river Danube, reflecting the productivity of the soil and the employment that exists in towns along the Danubian routeway. The capital of Slovakia, Bratislava, is on the Danube approximately forty miles downstream of Vienna in Austria. The Bratislava metropolitan area, with half a million people, is a port, a manufacturing center, a market for agricultural produce, and, being on the Danube close to the borders with Austria and Hungary, a communication node.

Downstream of Bratislava the rivers Váh and Nitra, flowing south from the Carpathians, make a confluence, before joining the Danube. Here on the north bank of the river is the Slovak town of Komárno. Across the Danube is the Hungarian town of Komorom. Historically, the two ports and manufacturing centers have complemented each other. Now that Hungary and Slovakia are in EU, interaction between the river towns is easier. (Photo 12.8)

The second largest city in Slovakia is Košice, with approximately 250,000 inhabitants, at the junction of the Danubian plains and the Carpathian Mountains. Košice is an ancient fortress, market town, and religious center on the river Torysa, a tributary of the Tisza, which flows out of Slovakia to travel north to south through Hungary to join the Danube in Serbia. Today, Koščise is a manufacturing and food-processing center serving eastern Slovakia. Upstream of Košice, on the Torysa, is Prešov, with nearly 100,000 inhabitants. The city is another historic fortress and religious and agricultural center, today possessing modern manufacturing in addition to the traditional brewing, distilling, food processing, and woodworking.

The Emergence of Slovakia

The Slovaks came under Hungarian control in the tenth century and remained there until the Treaty of Trianon (1920) separated Slovakia from Hungary and made the region a part of the new state of Czechoslovakia.

At the battle of Mohács (1526), the Hungarians were defeated by the Turks and, shortly, the Hungarian capital, Budapest, was ruled by the Turks. The Hungarian nobility

Photo 12.8 Komárno, Slovakia. Moving down the Danube, towards the Black Sea, different architectural influences appear.

Source: All Photo/Age Fotostock America, Inc.

were forced to operate from Slovakia where they held estates on the Danubian plains and in the Carpathians. For the most part the Carpathians were beyond Turkish control but, in the east of the upland, Košice and Prešov came under the Turks from time to time.

In the period when Turkey controlled Budapest, Slovakia was the base for Hungarian resistance, until Budapest was taken from the Turks by the Austrians in 1686. Bratislava, in Slovakia, remained an important administrative center, with the Hungarian Diet meeting in the city until 1848. The period in which the Hungarians operated from Slovakia had an ethnic impact. Today, about 11 percent of the population of Slovakia is Hungarian and as the Hungarians are found mostly on the Danubian plains of Slovakia, in places they constitute a considerable minority.

When Hungary negotiated the *Ausgleich* (compromise) with Austria in 1867 rather than signaling a move toward Slovakian autonomy, the Hungarians tightened control and tried to subdue the Slovaks, which helped crystallize Slovakian national consciousness. With the defeat of Austria-Hungary in World War I, Czechoslovakia emerged as an independent state. While the Czechs and the Slovaks had similar languages, the fit between the two ethnic groups was not good. The Czech lands looked to Vienna; the Slovaks were linked economically to Budapest. The Czechs had a more diversified economy than the agrarian Slovak lands. There were seven million Czechs and only two million Slovaks. Czech issues got more attention than Slovak issues.

Parts of Slovakia went to Hungary after Munich in 1938. As Czechoslovakia disappeared in the spring of 1939, Slovakia became autonomous, but under German control. In August 1939, German troops were stationed in Slovakia and the country allied with Germany. At the end of the war Czechoslovakia was reassembled minus, in the east, Ruthenia, which became part of the Ukraine. In 1993 the Czech Republic and Slovakia separated to become independent states. Both joined EU in May 2004. The countries are NATO members.

Summary

Birth and death rates per thousand Slovaks are about equal. The total fertility rate is 1.2 and population numbers will shrink slowly in coming decades, at present rates. Many of the heavy and armament industries of the Communist era have struggled to modernize, but with relatively low labor costs Slovakia has attracted investment from western Europe.

Slovakia has experienced economic growth in recent years as low wage rates and low taxes have attracted investment from western Europe. However, heavy industries and the armament industry of the Communist era have contracted. The realignment of manufacturing activity has resulted in an unemployment rate of over 15 percent of the workforce and although living standards have risen for some, many feel they are worse off now than under communism.

The discontents of those who have not enjoyed benefits under a free market economy, and EU entry, was reflected in the June 2006 election. The Christian Democrat party, associated with the free market reforms and EU entry, lost power to the left of center Smer party, which will lead a coalition government more sensitive to unemployment and poverty.

▶ HUNGARY

The Magyars were nomadic horsemen who arrived on the Danubian plains, probably from western Siberia, beginning in 896 A.D. The Magyars spoke a Ural-Altaic language, which was the foundation of modern Hungarian, a non-Indo-European language. When the Magyars arrived, the Danubian plains were an open grassland well suited to the lifestyle of the horsemen from the east. The former nomads adopted settled agriculture but continued to raise livestock, including horses. In 1000 A.D. the Hungarian Kingdom was acknowledged and Stephan I received a crown from the Pope, who recognized Hungarian autonomy within the Holy Roman Empire. A powerful kingdom developed, which gained control over Croatia (1091), Slovakia, and Transylvania.

In the fourteenth century the Turks pushed up the Danube valley, defeating the Serbs at the Field of the Blackbirds in 1389. In 1526, at the battle of Mohács, the Hungarians were defeated and lost most of their Danubian lands, including Budapest, to the Turks. Hungary was left in control of Slovakia, part of Transylvania, western Transdanubia, and territory stretching down to Zagreb in Croatia.

The Austrians eventually stopped the Turkish advance up the Danubian plains. The Turks were expelled by Austria from the Hungarian capital of Budapest in 1686 and by the Treaty of Karlowitz (1699) withdrew down the Danube, leaving the Austrians in control of a large region. The Austrians promoted colonization of the Danubian plains and many German-speaking settlers came to farm the Danubian lands. Hungarian nationalism remained, reinforced by the distinctive language that was not Germanic or Slavic. Hungary was under Austrian political control but in turn Hungary controlled Croatia, Slovakia, and Transylvania. Hungary was a multiethnic country within a multiethnic empire.

In 1848, all across Europe, there were antimonarchist demonstrations, nationalist uprisings, and demands for constitutional change. Uprisings occurred in Paris, Berlin, northern Italy (against Austrian rule), Prague, Vienna, and Hungary. The Hungarian revolutionary Kossuth did succeed, for a time, in gaining constitutional change, but when he rejected the monarchy, the Austrians, with Russian help, put down the revolt. Kossuth was a Hungarian nationalist and had no sympathy for the aspirations of the Croats and Slovaks who wanted less Hungarian control. Kossuth was forced into exile via Turkey and spent his time promoting revolutionary views in France, Britain, and the U.S. He was never allowed back into Hungary.

However, the struggles of 1848–49 were not without benefits. In 1867 the *Ausgleich* (compromise) resulted in the creation of the Dual Monarchy. Hungary would have a separate Parliament, acknowledge the crown as head of state, and have shared policies with Austria on foreign affairs, defense, tariffs, and a common currency.

The creation of Austria-Hungary resulted in the development of a common market. Operating in the enlarged market area, protected by a common external tariff, and possessing a common currency, manufacturers in Budapest prospered. Industrialization was rapid and Budapest grew in population numbers and as a center of transportation. (Photo 12.9)

Although there were advocates, in Austria and Hungary, of extending the *Ausgleich* model and creating parliaments for the Czechs, Slovaks, and Croats, the political development did not take place. The dissatisfaction of national groups was the cause of the breakup of the Austria-Hungary empire, which had ceased to function by the fall of 1918.

By the Treaty of Trianon (1920) the old Hungary was dismembered. Territory was detached and given to Poland. Romania got Transylvania, together with many Hungarians and Germans. Slovakia became part of Czechoslovakia, severing the economic

Photo 12.9 Budapest on the Danube.

Source: SUPERSTOCK.

links between Hungary and the former northern territory. Croatia became a part of the multiethnic state of Yugoslavia.

As at Vienna, Budapest lost many functions as it became the capital of a downsized state. Economic hardship was widespread, political stability fragile. In 1919, Béla Kun, a prisoner of war returning from Russia, where he had been indoctrinated with Communism, set up a regime in Budapest. Admiral Horthy, the last Commander-in-Chief of the Austria-Hungary navy, using the southern city of Szeged as a political base, was able to dislodge the Béla Kun regime and set up an authoritarian government with himself as Regent, effectively displacing the king.

Horthy, who held power for 24 years, wanted revision of the Treaty of Trianon and the return of territory detached from Hungary after World War I. This policy led to Hungary siding with Germany, which wanted revision of the Treaty of Versailles. After Munich (1938), Hungary gained some territory, containing Hungarians, from Czechoslovakia. Later, part of Transylvania with Hungarian communities was detached from Romania. In April 1941 Hungarian units helped occupy Yugoslavia along with German forces and declared war on the Soviet Union on June 27, 1941. Early in 1943 Hungary lost an entire army on the Russian front along the river Don. Hungary put out peace feelers to the Soviet Union, resulting in a German invasion of Hungary in March 1944, which accelerated the deportation to death of Jewish communities, despite the efforts of the Swedish diplomat Raoul Wallenburg, who issued Swedish passports to many Jews who did escape. The Red Army entered Hungary in September 1944. The last German troops left in April 1945, having blown up the bridges linking Buda and Pest across the Danube. Admiral Horthy (1868–1957) went into exile in Portugal. Wallenburg was arrested by the Soviets and died in captivity. Elections were held in 1945, the Communists got 17 percent of the vote, became part of the governing coalition, took control of the Ministry of the Interior and, with Moscow's help, established a hard-line Communist government controlling the apparatus of a totalitarian state, complete with secret police and detention without trial. A Soviet-style constitution was adopted in 1949. Collectivized agriculture and a five-year development plan were introduced in 1950.

Hungary was under the most repressive regime in eastern Europe. There were show trials. Cardinal Mindszenty, head of the Roman Catholic church in Hungary, was sentenced to prison in 1949. Lázló Rajk, a long-serving member of Hungary's Communist party and foreign minister, was charged with "Titoism" (Marshal Tito and Yugoslavia had broken with Moscow) and executed. The way the party could turn on its own members is presciently portrayed in Arthur Koestler's *Darkness at Noon* (1940). Koestler, Hungarian by birth, wrote the book after being imprisoned during the Spanish Civil War.

In Hungary, with repression rampant, there was an extraordinary efflorescence of creative talent. All the Communist countries of eastern Europe devoted resources to developing state-sponsored sports, for every "worker's paradise" needed a winning team. Hungary spent on a soccer team that played the game in a new way "by changing places and coming from behind." Players would interchange positions and make unexpected late runs in front of goal to create scoring opportunities. In 1954 a bewildered English team, using man-to-man coverage, lost by 7 goals to 1 in Budapest. After the 1956 Hungarian revolution many of the team left the country and played for leading western European clubs.

Photo 12.10 November 5, 1956. Hungarians did overthrow their communist regime but Soviet forces intervened and crushed the Hungarian Revolution. Many Hungarians fled the country.

Source: AFP/Getty Images/NewsCom.

The 1956 uprising started as a student demonstration that the authorities tried to repress, only to see factories go on strike with workers and professionals filling the streets. The totalitarian regime lost control of Budapest, secret policemen were hung from lampposts, and the Hungarian army could not reclaim the city. The revolt lasted two weeks and was put down by Soviet tanks coming through the hail of Molotov cocktails. Thousands of Hungarians were killed; many fled to the west. Reprisals and executions followed. (Photo 12.10)

The regime did loosen its grip in the 1960s, and by the 1970s economic control was being decentralized and small businesses allowed. Hungary moved toward a mixed economy in which there were large state-run enterprises together with innumerable small, privately run businesses. Hungary was able to make the transition from a centrally planned economy to a capitalist system more easily than some countries of eastern Europe. The state has divested itself of most of the land and assets it controlled in the Communist era.

Physical Environment

Topography
Topographically Hungary can be divided into four major regions (Figure 12.8):

1. The Little Alföld (Kisalföld)
2. The Great Alföld (Nagyalföld)

Figure 12.8 Regions of Hungary.

3. The mountains of Hercynian age running from the Bakony Mountains in the southwest to the uplands bordering Slovakia. The Bakony region contains extensive deposits of bauxite. The mountains separate the Little Alföld from the Great Alföld.

4. South of the Bakony mountains, in the west of the country, is Transdanubia, containing Lake Balaton. Transdanubia is higher and displays more relief than the Danubian plains. (Photo 12.11)

The Little Alföld and the Great Alföld are part of the Danubian plains. The lowland consists of alluvium with a covering of loess in places. There are ill-drained, sandy areas along the Danube, but generally the plains are fertile, with rivers providing irrigation water during the hot summer. The plains are heavily cultivated with grains (including corn), root crops, and vineyards.

Climate
Hungary has a continental climate with hot summers and cold winters. Thunderstorms are common in the summer months and more precipitation arrives in spring and summer than winter, with May, June, and October being the rainiest months. Annual precipitation is around 25 inches on the Great Alföld. At Budapest, in the north of the country, the

Photo 12.11   Lake Balaton is a popular resort area in Hungary. Tihany, by the lake, with Abbey Church.

Source: Age Fotostock America, Inc.

mean monthly temperature for January is at freezing (32°F). July has a mean monthly temperature of 71°F.

The forested Bakony Mountains, rising up to 2,339 feet, are colder than the plains, but not high enough to support permanent snowfields. To the south of the Bakony range the large, but relatively shallow, Lake Balaton can partially freeze in winter. There are excellent vineyards on the south-facing slope of the Bakony Mountains but the upland, and the lake, have been overexploited for the tourist trade and need rehabilitation.

Population Distribution and Urban Places

Hungary has a total population of approximately ten million people of which around one in five live in the capital of Budapest (Photo 12.12 and Table 12.7). The Danubian plains have population densities of several hundred people per square mile. Densities are lower in the mountains and in the Transdanubian region.

Budapest was originally sited on a hill where the Danube emerges onto the plains, having passed through the gap between the Bakony Mountains and the northern mountains of Hungary (Figure 12.9).

The population of Hungary is presently in decline. The birth rate per thousand Hungarians is nine. The death rate is thirteen. Like all of eastern Europe the total fertility rate is low at 1.3. Population numbers are shrinking by nearly half a percent each year. However, economic growth is around 4 percent per annum and the country is adjusting to EU membership.

Photo 12.12   Parliament on the Danube. Budapest, Hungary. The Hungarian version of neo-gothic displays similarities to the Parliament at Westminster.

Source: Kevin Galvin/SUPERSTOCK.

► SLOVENIA

Slovenia was part of Austria-Hungary until the Treaty of St. Germain (1920) severed the region from the empire to join Yugoslavia. In World War II Slovenia was divided between Italy, Germany, and Hungary. At the end of the war it was reattached to the Federal State of Yugoslavia with a communist government under Marshall Tito. As Yugoslavia began to break up, Slovenia was able to detach itself, in 1991, with only limited fighting and little damage to an economy that had already developed links to Austria and Italy.

TABLE 12.7   Hungary: Major Urban Centers

Budapest	1,900,000
Debrecen	220,000
Miskolc	180,000
Szeged	170,000
Pécs	160,000
Györ	130,000
Nyíregyháza	120,000
Székesfehérvár	100,000
Kecskemét	100,000

Figure 12.9 Hungary: Urban Places.

Physical Environment

The northern boundary of Slovenia, with Austria, is formed by the sharp peaks of the Karawanken Mountains. To the west, the Julian Alps, rising at Mount Triglav to 9,396 feet, run close to the boundary with Italy. At the lower southeastern end of the Julian Alps, limestone Karst country predominates. The region is marked by bare limestone surfaces. Settlements are found in *poljes* (collapsed limestone caverns), which possess soils and water at the lower levels. (Photo 12.13)

The greatest part of Slovenia has a ridge and valley topography, which is drained toward the east by tributaries of the Danube. In the north, the Drava is the main drainage system. The Sava drains the central region. Neither river is navigable in Slovenia.

Slovenia does have a short Adriatic coastline, south of the Italian city of Trieste. The Slovene Adriatic coast has a modified Mediterranean climate receiving rainfall in both winter and early summer. The temperature regime is more typically Mediterranean with hot summers and mild winters, although when the *bora* blows from the north, Slovenia suffers.

Inland from the Adriatic, continental influences are increasingly felt. At Ljubljana, (Figure 13.4) the capital at an altitude of 960 feet above sea level, the mean monthly temperature for January is well below freezing and snowfall is frequent. Summers are warm rather than hot, with altitude modifying the heat.

Photo 12.13 In the Dinaric Alps of Slovenia, are extensive areas of Karst limestone which lack surface drainage and vegetation.

Source: Marka/Age Fotostock America, Inc.

Over half of Slovenia is covered in forests. Timber, wood products, paper, and associated materials are an important part of the economy. There are few minerals other than the plentiful limestone for making cement. Brown coal is used in thermal power stations; there is one nuclear plant and hydroelectric plants on the rivers.

The Economy

Slovenia still has around 10 percent of the economically active workforce in farming. A number of traditional industries including building materials, furniture making, and wood products depend on the extensive forests and limestone regions of the country. Modern manufacturing includes consumer goods and cars. Trade with EU countries grew rapidly before Slovenia entered the organization and has continued to expand since 2004. The country is a competitive producer of consumer goods and relatively cheap labor has attracted investment. The economy grows at nearly 4 percent per annum and Slovenia adopted the Euro in 2007.

Distribution of Population

In the mountains, uplands, and limestone Karst regions, population densities are low. The remainder of the country is characterized by valleys where agriculture is practical in low-lying areas but, up slope, livestock raising rapidly becomes the dominant form

TABLE 12.8   Demographic Data

	Total population in millions	Birth rate per 1000	Death rate per 1000	TFR	Natural increase	Pop. estimate 2025 (in millions)
Estonia	1.3	11	13	1.5	−0.2	1.2
Latvia	2.3	9	14	1.3	−0.5	2.2
Lithuania	3.4	9	13	1.3	−0.4	3.1
Poland	38.1	10	10	1.3	−0.1	36.7
Czech Republic	10.3	10	11	1.3	0.0	10.2
Slovakia	5.4	10	10	1.3	0.0	5.2
Hungary	10.1	10	13	1.3	−0.3	9.6
Slovenia	2.0	9	9	1.2	0.0	2.0

Source: Population Reference Bureau. Data for 2006.

of farming. In the valleys are villages and small towns serving the needs of rural areas. Fifty percent of Slovenians still live in rural areas and there are few large towns.

Ljubljana, the capital, on the Ljubljanica river, close to its confluence with the Sava, has a population of 250,000 people. The next largest place, Maribor, has hardly 100,000 inhabitants. Slovenia has a population of around two million people. The birth rate is nine per thousand Slovenes and the death rate nine. The population is slightly shrinking in number and aging.

▶ SUMMARY

It is too early to declare whether or not EU enlargement into eastern Europe will bring economic success to the new members. Several general trends are apparent. The older, heavier eastern European industries have difficulty competing with more efficient producers in the west. Cheaper labor in the east has attracted direct investment into the territory of new members, but unemployment rates remain high. In Poland and Slovakia over 15 percent of the workforce is out of work. Aggregate farm income has increased as new markets for produce have opened in the west and EU agricultural subsidy payments are being phased in to eastern European farmers. The trends will obscure much restructuring as older industries close and new plants open. In farming, the larger units will benefit from the Common Agricultural Policy. Tiny farms of a few hectares are too small to benefit from EU programs, but may remain in operation as traditional subsistence units. In Poland, with an extensive agricultural sector, it is projected that the bigger farms will increase in size, subsistence units will increase in number, and medium-size farms will be split up, the bulk of the land going into enlarging neighboring farms, with the former cultivator retaining a parcel of land for subsistence while taking waged work in town.

The countries involved in the eastern European enlargement, with some exceptions, have death rates higher than birth rates (Table 12.8). The exceptions are Slovakia, Slovenia, and Poland, where birth and death rates are equal. As a result of negative natural increase of population the number of inhabitants in eastern Europe will shrink in the next twenty years, if present trends continue.

The situation may be worse than the projections on the natural increase account. As travel has become easier, as a result of the collapse of controlling Communist states and EU membership, many from eastern Europe have moved to western Europe and North America. Outmigrants tend to be young adults who are at the stage of life where they are about to form families. Total fertility rates may decline further as a result of the migration of young people who choose to settle and raise families in western Europe or North America.

► FURTHER READING

EASTERN EUROPE

Carter, F.W., and D. Turnock. *Envrionmental Problems of East Central Europe*. 2nd edition. New York: Routledge, 2002.

Crampton, R., and B. Crampton. *Atlas of Eastern Europe in the Twentieth Century*. New York: Routledge, 1996.

Crowley, D., and S. Reid (eds.). *Socialist Spaces: Sites of Everyday Life in the Eastern Bloc*. New York: Berg, 2002.

Dingsdale, A. *Mapping Modernities: Geographies of Central and Eastern Europe, 1900–2000*. New York: Routledge, 2002.

Palmer, A. *The Baltic: A New History of the Region and Its People*. Woodstock, NY: Overlook Press, 2006.

Pavlínek, P., and J. Pickles. *Environmental Transitions: Transformation and Ecological Defense in Central and Eastern Europe*. London: Routledge, 2000.

Rugg, D.S. *Eastern Europe*. London: Longman, 1985.

Traistaru, I., P. Nijkamp, and L. Resmini (eds.). *The Emerging Economic Geography in EU Accession Countries*. Aldershot, UK: Ashgate, 2003.

Turnock, D. (ed.). *East Central Europe and the Former Soviet Union: Environment and Society*. London: Arnold, 2001.

Turnock, D. *The Human Geography of East Central Europe*. New York: Routledge, 2003.

THE BALTICS

Pettai, V., and J. Zielonka (eds.). *The Road to European Union: Estonia, Latvia, and Lithuania*. Manchester and New York: Manchester University Press, 2003.

Smith, D.J. (ed.). *The Balkan States and Their Region: New Europe or Old?* New York: Rodopi, 2005.

Tammaru, T., and H. Kulu. "Ethnic Minorities of Estonia: Changing Size, Location, and Composition." *Eurasian Geography and Economics*, 44(2) (2003): 105–120.

CZECH REPUBLIC

Margolious, I. *Reflections of Prague: Journeys through the 20th Century*. New York: John Wiley, 2006.

Sayer, D. *The Coasts of Bohemia: A Czech History*. Princeton, NJ: Princeton University Press, 1998.

HUNGARY

Buckwalter, D. Highways and Regional Realignment in an Economic Frontier: The Case of Hungary. *Eurasian Geography and Economics*, 44(2) (2003): 121–143.

Foote, K.E., A. Tóth, and A. Ávray. "Hungary after 1989: Inscribing a New Past on Place." *Geographical Review* 90(3) (2000): 301–334.

POLAND

Broekmeyer, M. *Stalin, the Russians, and Their War 1941–1945*. Translated by R. Buck. Madison: University of Wisconsin Press, 2004.

Crowley, D. *Warsaw*. London: Reaktion Books, 2003.

Davies, N. *Heart of Europe: A Short History of Poland*. Oxford, UK: Oxford University Press, 1984.

Davies, N. *God's Playground: A History of Poland*. 2 vols. New York: Columbia University Press, 1981.

Davies, N. *Rising '44: The Battle for Warsaw*. New York: Viking, 2004.

Lane, A.B. *I Saw Poland Betrayed*. Indianapolis, IN: Bobbs-Merrill, 1948.

Michener, J. *Poland*. New York: Random House, 1983.

Smith, N. *American Empire: Roosevelt's Geographer and the Prelude to Globalization*. Berkeley: University of California Press, 2003.

SLOVAKIA

Kirschbaum, S. *A History of Slovakia: The Stuggle for Survival*. New York: St. Martin's Press, 1995.

SLOVENIA

Gow, J. *Slovenia and the Slovenes: A Small State and the New Europe*. London: Hurst, 2000.

Mrak, M., R. Matija, and C. Silva-Jauregui (eds.). *Slovenia: From Yugoslavia to the European Union*. Washington, DC: World Bank, 2004.

13

EASTWARD EXPANSION 2007: THE BALKANS

▶ INTRODUCTION: THE BALKANS

Protruding into the Mediterranean Sea are three major peninsulas: the Iberian peninsula, the Italian peninsula, and the Balkan peninsula.

The Balkan peninsula is a mountainous region bounded in the north by the Danube, in the east by the Black Sea, in the south by the Aegean, and in the west by the Adriatic. The name *Balkans* is derived from a Turkish word meaning "wooded mountains." Included in the Balkans are Romania, Bulgaria, European Turkey, Albania, and the new states emerging from Yugoslavia. Some states stretch beyond the Balkan peninsula. The breakup of Yugoslavia re-Balkanized the Balkans as Serbia, Croatia, Bosnia-Herzegovina, Macedonia, and Montenegro emerged. The term *Balkanize* refers to the breaking up of a region into small, usually hostile states. Balkanization is not yet complete, for Kosovo (90 percent Albanian) may break from Serbia. Greece is on the Balkan peninsula and gained territory in the Balkan wars, which preceded World War I. Today, Greece, an EU member, has relatively stable boundaries, although there are potential territorial disputes with Turkey, Macedonia, and Bulgaria.

Nineteenth-century European statesmen addressed the Eastern question, How was stability to be brought to the Balkans, as Turkish power declined, without generating great power competition? Several Balkan wars, prior to the outbreak of World War I, failed to settle issues. In 1917, under the Corfu Pact, the Serbs, Croats, Slovenes, and Montenegrans agreed to establish a Kingdom under the Serbian crown. The new state, recognized in 1918, included Bosnia-Herzegovina. The name Yugoslavia was adopted in 1929.

During World War II Yugoslavia fractured and came under Axis control. The Croats allied with Germany. Elsewhere numerous guerrilla groups operated. Eventually, Marshall Tito's communist partisans became the dominant force, creating a federal Yugoslavia, after the war. At first the regime was linked to Moscow but broke away in 1948. Marshall Tito became prominent in the nonaligned world and, at home, his powerful leadership held the federal state together and promoted economic development. Tito died in 1980. Failure to manage economic policy and control inflation led to the breakdown of the federal system. Slovenia and Croatia declared independence in 1991. Slovenia, with EU help, rapidly left the former Yugoslavia. Bitter wars between and within Croatia, Serbia, and Bosnia-Herzegovina resulted in breakup, a process that may not be complete. Now European leaders are proposing to settle Balkan problems by absorbing the region into EU. Romania and Bulgaria have negotiated entry. Croatia and Serbia want entry.

Romania and Bulgaria joined EU in 2007. Croatia and Serbia want EU entry, but will be delayed until nationals indicted as war criminals are delivered to the Hague and the International Court of Justice for trial. Bosnia-Herzegovina, Macedonia, Montenegro, and Albania are not currently in entry talks with EU.

▶ ROMANIA

The Physical Environment

Romania can be divided into three major physical regions (Figure 13.1):

1. The Danubian lands in the west, the south (Valachia), and the east, including the Dobruja
2. The mountains of Carpathia, Transylvania, and the Bihor Massif
3. Within the mountains of **2** lies the basin of Transylvania

Figure 13.1 Romania.

The Danubian Lands

The western Danubian lands are drained by streams that flow from the western flank of the Bihor Massif to the river Danube. Large alluvial fans extend out from the Massif onto the Danubian plain and the Banat: a rich agricultural region occupied to the west by one of Romania's neighbors, Serbia. The Banat produces grains, tobacco, sugar beets, cattle, fruits, and vegetables. A bituminous coalfield forms the basis of an iron and steel industry at Reşiţa, a mining and heavy industrial center.

At the south end of the Banat the river Danube, flowing east from Belgrade in Serbia, enters Romania, and cuts through a gap between the Balkan mountains and the Transylvanian Alps in a gorge that is narrowest and deepest at the Iron Gate (Photo 13.1). The Iron Gate gorge used to hinder navigation on the Danube, but many improvements culminating in a Yugoslav–Romanian hydroelectric dam (1971), which regulated flow on the river, now allows oceangoing vessels to sail up the Danube from the Black Sea to Belgrade.

Below the Iron Gate gorge, the Danubian plains widen again, with the river forming the boundary between Romania and Bulgaria. The southern Danubian plains of Romania slope from the south side of the Transylvania Alps to the river. Numerous streams flow out of the mountains down to the Danube across the gently falling plain composed of alluvium, and in places, a covering of loess. Along the Danube there are areas of wetland. The plains are farmed to produce grains including corn but much of the land is too dry for crop farming without irrigation, and there are considerable areas of pasture for cattle and sheep. The Danubian plains become drier toward the Black Sea where grassland (steppe) is the natural vegetation, the area being an extension of the steppelands found to the north of the Black Sea.

In the center of the southern Danubian plains (the region is known as Valachia) is the capital city of Bucharest, a spacious city that became a monument to socialist planning and a center of heavy industry (Photo 13.2). North of Bucharest, on the foreland of

Photo 13.1   The Iron Gate on the river Danube close to the Serbia-Romania border has been a navigation hazard.

Source: Robert Harding Picture Library/Alamy Images.

Photo 13.2 The monarchy of Romania did not survive the war and the communist take over, but the Royal Palace in Bucharest remains.

Source: Silvio Fiore/SUPERSTOCK.

the Transylvanian Alps, is the oilfield around the city of Ploiesti. Romania was a major source of oil for the German military in World War II and Ploiesti was the scene of a famous raid by U.S. Liberators, flying from bases in Libya in August 1943. Nearly a third of the bomber force was lost and five Congressional Medals of Honor won, in an effort to destroy oil installations essential to the Nazi war machine.

Near the Bulgarian port of Silistra the Danube turns north and no longer forms the boundary between Bulgaria and Romania. The river flows north along the west side of the Dobruja region. At the Romanian river port of Galaţi the river turns east toward the Black Sea and breaks into distributaries, which are extending the Danube delta into the Black Sea, as mud, sands, and silts are deposited. The delta region is a major wetland and wildlife refuge, in addition to a fishing ground (Photo 13.3). The core of the Dobruja is a block of limestone rising up several hundred feet. The surface does not support streams and the dry land is used largely for livestock grazing. The Black Sea coast of the Dobruja, south of the Danube delta, is marked by lagoons and the port of Constanţa (Table 13.1). The settlement has Greek origins but in the twentieth century it was developed as a major port handling Romanian trade and goods coming down Danubian road and rail links. A pipeline from Ploiesti exports oil.

The Mountains of Carpathia, Transylvania, and the Bihor Massif

The mountains of Carpathia, the Transylvanian Alps, and the Bihor Massif enclose the basin of Transylvania. The Carpathians, arcing south from Poland through the Ukraine into Romania, are a broad range rising up to 7,562 feet, at Vârful Pietrosu before trending west as the Transylvanian Alps. The Transylvanian Alps are not so broad as the Carpathians but rise higher to 8,346 feet at Vărful Moldoveanu. The Carpathians and the Transylvanian Alps are largely composed of sedimentary rocks, with igneous intrusions injecting mineral veins during the mountain-building era. The Bihor Massif is mainly

Photo 13.3 The Danube has an extensive delta where the river falls into the Black Sea. The delta lands are important wildlife habitats and traditional fishing grounds.

Source: Jon Arnold Images/SUPERSTOCK.

igneous and metamorphic rocks. Population densities are low in all the mountain regions that enclose, but do not cut off, the basin of Transylvania.

The Basin of Transylvania

The basin is a rolling upland drained to the west by tributaries of the Tisza and to the south by the river Olt, which cuts through the Transylvanian Alps to flow south to join the Danube. The altitude of the basin lowers temperatures and shortens growing seasons. Winters are long and snow covered. The enclosing mountains keep out some of the cold winter air from the north and east, but in the basin, surrounded by mountains, cold air drains to valley floors and lowers temperatures.

TABLE 13.1 Romania: Top Twelve Urban Places by Number of Inhabitants

Bucharest	1,922,000
Iaşi	322,000
Cluj-Napoca	318,000
Timişoara	318,000
Constanţa	310,000
Craiova	303,000
Galaţi	299,000
Braşov	284,000
Ploiesti	323,000
Brăila	217,000
Oradea	206,000
Bacău	176,000

Figure 13.2 Romania: Urban Centers.

Economic Development

Until a few decades ago, Romania was a country where most people made a living in the primary sector of the economy as farmers, foresters, and fishermen along the Black Sea shore, and as miners and quarrymen. Agricultural yields were low but some of the mineral deposits, including petroleum, coal, iron ore, bauxite, and nonferrous metals, were rich. Within the Soviet Empire Romania was seen as a producer and exporter of raw materials. Romania objected to being a mine and a quarry for eastern Europe and was able to develop basic heavy industries including metal smelting but to the present Romania is a country with a basic, undiversified economy. EU structural adjustment funds will ease the decline of older industries. As agriculture is forced to modernize, rural-to-urban migration will increase and urban places will grow in size (Figure 13.2).

Entering EU

Romania illustrates the changes that have to be made as a country becomes an EU member. All relevant laws have to be harmonized with EU practice and environmental regulations must be modernized and enforced. The civil service must have competitive

Photo 13.4   Romanian farming is largely horse-drawn and utilizes lots of hands wielding pitch forks and other hand tools.

Source: Hemis/Age Fotostock America, Inc.

entry rather than by the patronage practiced in the past. Bribery and kickbacks have to be eliminated, contracts openly bid on, and corruption stopped. The judiciary has to be independent. Minorities have to be given equal status and Romania has numerous Hungarians in the west, close to the border with Hungary. Romania has the largest Roma (gypsy) population, over a million, of any European country.

Much of Romania was part of the Roman Empire and the national language is in the Romance group. Seventy percent of Romanians supported EU entry, seeing it has an historic return to Europe, where the country belongs, after centuries of Turkish and then Russian domination.

All Romanians, whether for or against EU, understand that the country will be transformed by entry. A horse and cart, country living style, will not compete in Europe (Photo 13.4). The traditional methods of slaughtering animals, for meat and byproducts on the farm, will go by EU edict. The farming, mining, and manufacturing sectors will be transformed as modern methods replace the labor-heavy systems of the traditional past and the Communist era. The majority wants to accept the economic dislocation, hoping to bring Romania into the modern world (Sullivan, 2006).

▶ BULGARIA

The Bulgars came under Turkish rule in the late fourteenth century and stayed there until near the end of the nineteenth century. When the Bulgars rebelled in 1876 the Turks repressed the uprising with atrocities that provoked great power intervention. Bulgaria became autonomous within the Ottoman Empire and declared itself independent in 1908.

Hoping to improve access to the Aegean Sea, Bulgaria joined Germany and the Central Powers in World War I, losing territory in the peace settlements. In World

War II Bulgaria again joined Germany, declaring war on Britain in March 1941, but avoided a formal declaration of war against the Soviet Union, which did not save the country from Red Army invasion and then Communist takeover in 1944–46. Although political and economic reform started in the early 1990s, progress has been slow. This is not surprising as the country had few democratic political institutions to develop.

The history of Bulgaria could be characterized as a state of war from 1876 to 1918. During the interwar years (1918–1939) the monarchy became a dictatorship, political parties were banned, and the economy collapsed. World War II resulted in a Communist, one-party state, with a centrally controlled economy. From a western perspective, Bulgaria only began to create democratic political institutions in the last 15 years and there is little experience of an open, competitive economic system.

Topography

From north to south Bulgaria can be divided into four landform regions (Figure 13.3):

1. The Danubian Lowlands
2. The Balkan Mountains—the Stara Planina
3. The Sofia Basin, the Maritsa River, and the Black Sea Shore
4. The Rhodope Mountains

The Danubian Lowlands

The Danubian lands in the north of the country are composed of limestone, covered in places with loess. In general, settlement is confined to valleys where water is available.

Figure 13.3 Bulgaria.

Winters are cold and summers are hot and well suited to the cultivation of corn, if irrigation water is supplied.

The Balkan Mountains—The Stara Planina

The Stara Planina is a continuation of the Carpathian Mountains, coming south into Bulgaria before trending to the east and the Black Sea. The range rises to 7,794 feet and consists of sedimentary rocks with injections of igneous material. The range is well forested, lightly peopled, possesses mineral resources, and is much used for hiking and camping vacations.

The Sofia Basin, the Maritsa River, and the Black Sea Shore

The Sofia Basin, in which Bulgaria's capital stands, is drained by the river Iskŭr, which flows north to the Danubian lowlands and the river Danube. The dammed Iskŭr provides drinking water and power to Sofia and irrigation water to farmland around the capital. (Photo 13.5)

The Maritsa river drains much of the rolling lands lying between the Stara Planina in the north and the Rhodope Mountains to the south. The river flows in a generally west-to-east direction but after making a confluence with the major tributary, the Tundzha, the Maritsa flows south, becoming the border between Greece and Turkey before falling into the Aegean Sea at Enez in Turkey. The Maritsa is used for generating power, irrigating farmland, and the valley constitutes the main east-to-east routeway through Bulgaria.

The climate of the central valley has a less cold winter than the Danubian lowlands in the north, as the Stara Planina keeps out much cold air. In summer, the greater altitude of the region modifies temperatures.

The major town within the Maritsa valley is Plovdiv, the second largest place in Bulgaria, with a population of over 300,000 people. Plovdiv is an ancient city founded by the

Photo 13.5   Iskŭr flows past Sophia, capital of Bulgaria, before cutting a gorge through the Stara Planina mountains to join the river Danube. Cherepish Monastery in the foreground.

Source: Image Register/Alamy Images.

Thracians and later used by the Greeks, Romans, Byzantines, Turks, and Russians. The Roman amphitheater is still used for cultural events (Photo 13.6). Today, in addition to performing the traditional market functions for the surrounding fertile agricultural region, the city is a center of food processing, brewing, and manufacturing. Plovdiv is a transportation center on the routeway from Istanbul to Sofia. Most towns in the central valley originated as markets usually close to routeways leading down from the uplands where produce from the valley, and livestock from the hills, could be conveniently brought to market.

The Maritsa river does not flow through the east end of the central lowland, which is drained by a number of streams falling to the Black Sea. The major ports on Bulgaria's Black Sea Coast are, in the north Varna, a center of shipbuilding, manufacturing, and food processing. A Black Sea ferry links Varna with Odesa in the Ukraine. On the coast around Varna, which has an international airport, are numerous Black Sea resorts that serve tourists from many parts of Europe. Before Bulgaria joined EU, with the legal safeguards that provides for investors, vacationers from western Europe were buying second homes, possibly unaware that winter temperatures along the Black Sea shore are more central European than western European.

Toward the south of Bulgaria's Black Sea coast is the polluted port of Burgas, full of fishing vessels, fish processing plants, oil refining, and manufacturing activity.

Rhodope Mountains

The Rhodope Mountains are composed of igneous rocks and rise up to 9,596 feet at the Musola peak; there are extensive areas of plateau at 6,000 feet above sea level. In contrast to the Stara Planina, there are few passes through the Rhodope Mountains, which

Photo 13.6 With over a third of a million people Plovdiv, in the Maritsa valley, is a center of manufacturing and food processing. On the ancient routeway from western Europe to Istanbul the city has many historic associations. The Roman amphitheater still stages events.

Source: Age fotostock/SUPERSTOCK.

are isolated and snow covered in winter. The mountains are forested on the lower slopes and above the tree line are summer pastures.

Population Distribution and Urban Centers

Sofia

The capital of Bulgaria, Sofia, lies west of the Maritsa drainage basin and south of the Stara Planina on a plateau drained by the Iskŭr river. From Sofia there are routeways, north and northwest, through the mountains into the Danube valley. There has been a city in the area, between conquests, for over 2,000 years. Sofia is many times larger than second-ranked Plovdiv, in the valley of the Maritsa. (Table 13.2)

Bulgaria had little heavy industry until Five Year plans were implemented by the Communist regime after World War II. Close to Sofia, at Kremikovtsi, there are deposits of coal and ores and these become the basis of steel making and metal smelting. Sofia has a large range of manufacturing, including heavy engineering products, motor vehicles, food processing, and consumer goods.

The fabric of the city is a mix of socialist planning, with wide streets and large squares, together with historic buildings including ancient cathedrals and mosques. Most Bulgarians are members of the Eastern Orthodox Church, but there are substantial numbers of Roman Catholics and Muslims. (Photo 13.7)

Overall, Bulgaria's major cities, ports, and higher population densities are found in and around the Sofia Basin, the Maritsa river valley, and the Black Sea shore.

Photo 13.7 Bulgaria has a substantial, long standing, Muslim population. Banya Bashi Mosque, Sofia.

Source: Hemis/Age Fotostock America, Inc.

TABLE 13.2   Bulgaria: Major Urban Places

Sofia	1,096,000
Plovdiv	341,000
Varna	315,000
Burgas	193,000
Ruse	162,000
Stara Zagora	144,000
Pleven	122,000
Sliven	101,000
Dobrich	100,000

The Danubian lowlands do possess substantial towns and cities including Pleven and Ruse, a port and industrial center on the Danube. In general, Bulgaria's river towns cannot compete with the Black Sea ports because of the long route the Danube takes skirting around the Dobruga in Romania to reach the sea.

Summary

How well Bulgaria will perform in the more competitive economic environment of the EU is problematic. Relatively cheap labor may attract investment, but many industries inherited from the Communist era will have to streamline, downsize, and shed labor to compete.

▶ CROATIA

Croatia came under Hungarian rule in 1091 and remained there, with boundary variations, until December 1, 1918, when the region became part of the Kingdom of Serbs, Croats, and Slovenes, renamed Yugoslavia in 1929. Yugoslavia was invaded by Germany in the spring of 1941 and split up. Croatia allied itself with Germany, rejoining the reconstituted Yugoslavia, under Marshall Tito, after the war. In the early 1990s there was vicious fighting and ethnic cleansing as the Serbs and Croats fought for control of the Slavonija region.

Topographical Regions

Croatia can be divided into (Figure 13.4):

1. The Danubian lands, including Slavonija
2. The Dinaric Alps
3. The Adriatic coastlands

The Danubian Lands

The Danubian lands are drained by the east-flowing Drava in the north and the Sava in the south. The rivers join the Danube beyond Croatia. Eastern Croatia is composed of the Slavonija region. Zagreb, the capital, lies west of Slavonija on the Sava, above the head of navigation. Zagreb, a modern manufacturing and commercial city, owes its historic importance to a westward position that placed it beyond Turkish control when

Figure 13.4   Slovenia, Croatia, Bosnia-Herzegovina.

Serbia, and much of Hungary, were under Ottoman rule. Zagreb built up a number of Croatian national institutions, particularly the cathedral and Roman Catholic religious organizations.

Dinaric Alps

The Dinaric Alps are predominately limestone, surface streams are few, and most drainage is underground. When underground caverns collapse or solution hollows form, *polja* are the result. Most settlement is found in *polja,* where reddish clay soils and moisture can be found. There are few settlements of much size in the Dinaric Alps apart from marketplaces along routeways leading through the region.

Adriatic Coast

The Dinaric Alps drop steeply to the often narrow coastal plain and the offshore islands of the Dalmatian coast. The Adriatic coastlands of Croatia have a Mediterranean climate, the summers hot and sunny; the winters mild and moist, except when the cold *bura* blows down the Adriatic. The *bura* is not confined to the coast and brings snowstorms to the Dinaric Alps.

Adriatic Croatia extends from the Istria peninsula in the northwest via a number of islands and ports to the border with Montenegro in the southeast. Bosnia-Herzegovina has a tiny, ten-mile Adriatic coastline some thirty miles to the northwest of Dubrovnik.

Population and the Distribution of Population

The present population of Croatia is 4.4 million people, down by 400,000 since the Croats voted to leave Yugoslavia in 1991. The decrease is a result of combat deaths and outmigration, including the exodus of 180,000 Serbs from eastern Slavonija. The birth rate has dropped below the death rate and the population is projected to decline in number and to age.

Nearly a quarter of all Croats live in Zagreb on the river Sava. The historic city, with a cathedral dating to 1093 and a university founded in 1669, is a center of manufacturing and commerce. Zagreb was not under Turkish control.

To the east, on the river Drava and the Danubian plains, is the town of Osijek, with 100,000 inhabitants. Osijek is a historic fortress, market, food-processing, and manufacturing center. The place was under Turkish rule from 1526 to 1687, when Austria gained control. The Adriatic coast is rich in historic places, including, from north to south, Pula on the Istria peninsula, Rijeka (168,000), Split (190,000), and Dubrovnik (Ragusa), with 50,000 inhabitants. (Photo 13.8)

Dubrovnik, like Venice, was founded in a remote place by refugees fleeing barbarian invasions with the fall of the Roman Empire. The city came under Byzantine, Venetian, and Hungarian protection, but, as a major trading port on the Adriatic, had much independence. Although shelled by the Serbs as Yugoslavia broke up, the walled city still displays fine medieval streets, monasteries, and other historic buildings. Dubrovnik is a hub of the tourist industry on Croatia's Adriatic coast.

Summary

Croatia had some coal, oil, and gas but few other minerals and the fossil fuels do not supply all needs. Manufacturing includes pharmaceuticals, transport equipment, textiles, brewing, distilling, and chemical industries. EU entry would improve access to export markets and might resolve minority tensions by the guarantee of rights under European law.

▶ SERBIA

In the fourteenth century the Kingdom of Serbia was expanding territorially but in the Danube valley came into conflict with the Ottoman Empire. In 1389 the Serbian army was defeated in Kosovo at the "Field of the Blackbirds" and Serbia came under Turkish

Photo 13.8   The Adriatic coast of Croatia is rugged where the Dinaric Alps come to the sea. Dubrovnik is a strongly fortified medieval town.

Source: Age fotostock/SUPERSTOCK.

rule. Some autonomy was achieved, with Russian help, in 1815. Formal independence was proclaimed in 1882 and Serbia wanted to expand territorially to the Adriatic to gain a coastline. Austria, not wanting the territorial expansion of Serbia, annexed Bosnia in 1908, and then pushed the Serbs further back from the coast in the Balkan war of 1912.

The events leading to World War I started when Archduke Ferdinand was assassinated at Sarajevo in the summer of 1914. In August, Austria and Serbia went to war, dragging all the powers into a conflict that lasted until November 1918. The allies then created what became Yugoslavia with Serbia, and Belgrade, in the dominant position. In the early 1990s Serbian nationalist politicians were prominent in the breakup of Yugoslavia, leading to the reemergence of Serbia.

Physical Environment

We can divide Serbia into two major regions (Figure 13.5):

1. The Danubian plains
2. The Mountainous south, consisting of several distinct ranges

The Danubian Plains

Serbia occupies an extensive area of the Danubian plains, which produce many of the crops we saw growing on the Danubian plains of Hungary, including wheat, corn, and vineyards. The orchards produce famed plums that are distilled into brandy.

Several major tributaries join the Danube near Belgrade, including the Sava flowing from the west. The Sava, with a fertile valley, is navigable to Sisak in Croatia. The

Figure 13.5 Serbia, Macedonia, Montenegro.

Tisa, flowing south through Hungary, joins the Danube between Novi Sad and Belgrade (Photo 13.9). It is navigable upstream to the Hungarian ports of Szeged and Szolnok. The river Timis, arising in the Romanian Carpathians, to the east of Serbia's Danubian plains, joins the Danube downstream of Belgrade. The canalized lower course of the Timis is navigable. Flowing to the Danube from the mountainous south is the Morava river. Running through mountainous country the Morava is of little use for navigation, but the railroad from Belgrade to Thessaloníki, on the Mediterranean coast of Greece, follows the valley.

Photo 13.9   In the Kosovo crisis of 1999 NATO bombed the bridge at Novi Sad, Serbia, blocking the main navigation route on the river Danube and forcing vessels to side channels of the river.

Source: Photograph by Ed Kashi/Aurora Photos.

Belgrade has considerable centrality within the Danube valley and it is surrounded by productive agricultural plains. Upstream of Belgrade is the city of Novi Sad, around which, and stretching to the north, is the heavily populated Vojvodina region producing grains, fruits, vegetables, and livestock. The ethnic population of Vojvodina is mixed, including Serbs, Croats, Hungarians, Romanians, and Slovaks. In the northeast of the country Serbia possesses a part of the Banat, another agriculturally rich region that extends into Romania and Hungary. The Banat is a region with a long history of changing rule and settlement by many ethnic groups.

The Mountainous South

The region is predominately upland and there are few good routeways, with the exception of the Morava valley providing a passage south to Skopje in Macedonia. From Skopje the Vardar valley provides a route to the Aegean Sea. The mountain region produces minerals, timber, and livestock. Population densities are relatively low. The province of Kosovo, which is predominately Albanian and Muslim, wants to detach itself from Serbia. (Photo 13.10)

▶ MONTENEGRO

Ethnically Montenegrans are Serbs speaking a dialect of Serbian. There are Roman Catholics and Muslims in the country, but the majority belong to the Montenegran Orthodox Church.

When the Turks took control of Serbia in the fourteenth century they were never able to subdue Montenegro. By the late eighteenth century the Turks recognized Montenegran independence under a prince-bishop (i.e., a ruler who was both the crowned head of state and patriarch of the church). At the Congress of Berlin (1878) the independence

Photo 13.10 Kosovo is rich in Serbian history, but is now largely populated by ethnic Albanians who are Muslim. Kosovo contains many outstanding orthodox monasteries. The monasteries now have to be protected by UN (KFOR) soldiers as Albanians press for independence from Serbia.

Source: Diomedia/Age Fotostock America, Inc.

of Montenegro was formally recognized, although Austria was responsible for patrolling the Adriatic coastal waters.

In 1900 the country had no railroads and much of the interior was reached on horseback using bridle paths. Livestock raising was the main occupation and there was limited international trade. The ruling prince held most of the power, but the forty tribes, or clans, elected elders who were responsible for law and order in their districts.

When Austria and Serbia started World War I in 1914, Montenegro joined forces with Serbia. At the end of the war a political clique engineered the entry of Montenegro into the land of the South Slavs: Yugoslavia. The power of the prince ceased and the clans divided over the issue of giving up sovereignty.

In July 1941, after Germany invaded Yugoslavia, Montenegro declared independence and neutrality but was quickly occupied by Italy. The clans fought against both the fascists and Tito's communist partisans.

After World War II, Montenegro became part of a reconstituted federal Yugoslavia and received considerable investment from the central government in roads and port facilities. The Yugoslav navy had a base on the Adriatic at the Montenegran port of Kotor and the commercial port, at Bar, became an important Yugoslav seaport.

After the breakup of Yugoslavia, Montenegro remained part of the state of Serbia and Montenegro. Montenegrans, in the minority, felt they were at a disadvantage. Squabbles broke out in 2006, on the issue of whether a Serb or a Montenegran would represent the country in the Eurovision song contest!

In 2006 Montenegro declared independence from Serbia. The Serbian parliament recognized the Montenegro initiative promptly. Serbia wants entry to EU, and another war would alienate Brussels. Montenegro has an Adriatic coast and the loss of the territory land locks Serbia. A major reason for Serbia severing itself from Montenegro is that Serbia retains membership of international organizations. Montenegro will have to apply to join organizations, including FIFA, which organizes the soccer World Cup.

Physical Geography

Montenegro is largely composed of the Dinaric Alps, the mountain range that parallels the east coast of the Adriatic Sea (Figure 13.5). The east coast region of the Adriatic is called Dalmacija and in the medieval era, Venice had a commercial and cultural influence in the coastlands. The Venetians gave Montenegro its name—the black mountain, Crna Gora in Serbian. The coast, in general, is marked by cliffs, islands, and inlets. The Adriatic coast of Montenegro has these characteristics but along the border with Albania, around Lake Skadarsko, is a lowland. The major river, the Morača, flows through the capital, Podgorica (formerly Titograd), and into the lake. Some trade between Montenegro and Albania is conducted across the lake.

Moving inland from the coastal region, and the lowland around Lake Skadarsko, into the mountains, a harsher landscape is encountered. The Dinaric Alps are predominantly limestone and do not support much surface drainage; there are many dry valleys, sink holes, and collapsed caverns. Streams frequently flow underground, except after heavy rain when the underlying channels and caves fill with water.

Climate

The reader will be able to anticipate some features of the climate of a Balkan country with an Adriatic coastline. In the coastal region the climate is Mediterranean but with more rainfall than is typical. In winter, the north wind, the *bora,* can bring cool, blustery weather to the coast and inland precipitation may come as sleet or snow. Summers are warm and dry, but going inland temperatures are modified by altitude.

In areas close to the Adriatic, typical Mediterranean crops are grown. Going into the mountains, farming is predominantly concerned with raising livestock, except in favored valleys where crops can be grown.

► BOSNIA-HERZEGOVINA

Bosnia-Herzegovina is a Slav country composed of the Catholic Croats, Orthodox Serbs, and Bosnian Muslims in an uneasy federation with a central government that conducts foreign policy and regulates monetary policy. The official language is Serbo-Croat. As a result of the Dayton accords (1995), the state consists of a Serb Republic and a Croat-Muslim Federation. On the map, the distribution of Croats is relatively compact and the Croat region lies against Croatia to the west. The Serbs dominate distributions in the north and in the east and south, but are territorially divided by the mainly Muslim central region focused on the capital Sarajevo. The Muslims are themselves divided into the central region and, in the northwest, a Muslim region focused on Bihac (Figure 13.6).

Figure 13.6 Bosnia-Herzegovina Ethnic Regions.

There is a president of the Federation of Bosnia and Herzegovina. The Serb Republic has an elected president and a National Assembly. Bosnia-Herzegovina functions with much outside aid and many international peacekeepers on the ground. That a Slav country, in the Balkans, could contain both Catholics and Orthodox adherents is not difficult to understand. The region lies in the boundary zone between the Church at Rome and the centers of Greek, Eastern, and Russian Orthodox churches. Slavs subscribing to Islam are more difficult to explain and we have to go back to the twelfth century when the Bogomil heresy was established in Bosnia. Bogomils advocated a simple life, rejected the Old Testament, the sacraments, and the organization of the established Christian churches. Neighboring Croatia, Serbia, and Hungary all tried for territorial control of Bosnia and the suppression of the Bogomils.

In 1453 Constantinople fell to the Turks. By 1459 all of Serbia was under Ottoman control and in 1463 Bosnia was invaded. Having been persecuted by the Catholic and the Orthodox church and, having a simple Bogomil religion of their own, many Bosnians

**TABLE 13.3   Population of Bosnia-Herzegovina
by Religious Affiliation, 1910**

Orthodox	825,418	43.5%
Muslim	612,137	32.4%
Roman Catholic	442,197	23.2%
Jews	11,868	.6%
Protestants	6,342	.3%
Total	1,897,962	

found Islam attractive and converted to become a Slav, Muslim elite who looked down on the Christian communities they came to rule. Many Bosnian Muslims occupied administrative positions in the Ottoman Empire, or served in the Turkish armed forces. Below this ruling caste of Bosnian Muslims were Christians who worked the soil, paid taxes, and were liable to have sons drafted into Turkish fighting forces. Bosnian nobles were allies of the Turks in the wars against the Catholic Hungarians in the sixteenth and seventeenth centuries. The Bosnian nobles were conservative and more Turkish than the Turks, resisting Turkish reform efforts in the nineteenth century. When revolts broke out against Turkish rule, either by anti-reform Bosnian Muslims or dissatisfied Christian communities, they were put down by Turkish forces with severity.

The great powers stepped in and at the Congress of Berlin (1878) Bosnia was given to Austria-Hungary to administer while remaining, nominally, under Turkish sovereignty. The annexation of Bosnia-Herzegovina in 1908 by Austria was a factor in the breakdown of diplomatic relations between the European powers before World War I.

Although Austria-Hungary improved the administration of the country, many Catholic officials and settlers were introduced. The Orthodox Serbs, in Bosnia-Herzegovina, most of whom had a low standard of living, looked increasingly toward Serbia. It was at Sarajevo, in Bosnia-Herzegovina, on June 28, 1914, that the heir to the throne of Austria-Hungary, Archduke Ferdinand, and his wife were assassinated, setting off a chain of events leading to World War I, a few weeks later.

After World War I, Bosnia-Herzegovina was made a part of the Kingdom of the Serbs, Croats, and Slovenes (Yugoslavia after 1929). Yugoslavia was invaded by Germany on April 6, 1941, and the country was partitioned. Bosnia-Herzegovina was controlled by the newly formed Croatian state that allied with the Axis powers.

After World War II Bosnia-Herzegovina became a part of Marshall Tito's Communist Yugoslavia. The state came to control most economic activity and owned much land and property, but in general did not make economic progress. In 1992 Bosnia-Herzegovina declared independence and entered a long period of strife between the Serbs, Croats, and Bosnians, which has not been resolved.

Physical Geography

Topographically Bosnia-Herzegovina can be divided into two major regions. In the north is part of the Danubian lands, drained by the river Sava. The more extensive southern region consists of upland, including the Dinaric Alps that parallel the Dalmatian coast of the Adriatic Sea. Most of the coast is part of Croatia, and Bosnia only possesses a short shore on the Adriatic (Figure 13.7).

The upland is largely drained by rivers running in a generally northern direction to join the Sava. The major rivers from west to east are the Vrbas, the Bosna, and the Drina. The rivers are marked by gorges and are not used for navigation. The valleys do broaden into basins where agriculture is practiced. Sarajevo is surrounded by upland but downstream, on the Bosna, there is an extensive basin used for farming. The core of Bosnia-Herzegovina is around Sarajevo and the Bosna basin.

Only one major river, the Neretva, finds a passage through the Dinaric Alps to the Adriatic Sea. The river flows through limestone gorges and is much photographed at Mostar where the famous bridge, built in the reign of Sulaiman the Magnificent (1522–1566), was heavily damaged in the recent civil war, as ethnic groups fought for control of the city. The lower, navigable course of the river runs through Croatia to the sea. (Photo 13.11) (Photo 13.12)

Figure 13.7 Albania and Neighboring States.

Photo 13.11 The famous pedestrian bridge at Mostar was built in 1566 under Turkish administration. The bridge linked the Muslim and Croat communities in the town. In the early 1990s Croat artillery demolished the bridge. A makeshift span served until the restored bridge opened in 2004.

Source: KRT/Presslink/NewsCom.

Photo 13.12 The famous Mostar bridge was reopened in 2004.

Source: Age fotostock/SUPERSTOCK.

Climate

The climate of Bosnia-Herzegovina is continental, with cold winters and warm summers. Winter cold is moderated by the influence of the Mediterranean Sea. As we have seen, the eastern shore of the Adriatic receives more rainfall than is typical in a Mediterranean area. At Sarajevo there is precipitation in all months. Much of the upland region is forested. Timber and wood products are important industries.

Population: Numbers and Cultures

There are 3.9 million people in Bosnia-Herzegovina. The largest ethnic groups are the Muslims, followed by the Serbs and then the Croats. The overall birth rate is nine and the death rate eight. The total fertility rate is 1.2 and the population is projected to slowly shrink.

The major city is Sarajevo, with an estimated population of half a million people. Sarajevo was founded as a fortress in 1263. The place came under Turkish rule in 1429, where it remained until the late nineteenth century, when Austria took charge. The city has a rich heritage of Turkish architecture but much of the city was damaged in World War II and the breakup of Yugoslavia. Since the civil war, the economy of Sarajevo and Bosnia has rebounded, but it may be generations before the scars of civil conflict heal.

▶ MACEDONIA

Landlocked Macedonia, with a population of approximately two million people, is bordered by Serbia and Kosovo to the north, Greece to the south, Bulgaria to the east, and Albania to the west. In the region adjoining Albania there are over 400,000 Albanians. Conflict with Albanians has been quieted by EU negotiators. The Albanians are a large minority, making up a quarter of the population. Other groups include Turks, Serbs, and Romas. There are Macedonians living across the borders with Greece and Bulgaria.

Although Macedonia was a powerful force under the rule of Alexander the Great (336–323 B.C.), there is no modern history of statehood until the country declared independence from Yugoslavia in 1991.

The Macedonian region came under Turkish rule in 1355 and remained there until the Turks were ousted in 1912–1913. Macedonia was then divided between Serbia, Greece, and Bulgaria. After World War I much of the region was incorporated into Yugoslavia. During World War II Bulgaria occupied Macedonia. After the war, in Marshall Tito's Yugoslavia, Macedonia was recognized as autonomous within the federal state.

The boundaries of Macedonia are not settled. Greece and Bulgaria have territorial claims. In the west, the Muslim Albanians are akin ethnically to the population of neighboring Albania. Should Kosovo separate from Serbia, the Albanians in northern Macedonia might want to join the new state. Two-thirds of Macedonians are Orthodox Christians. Thirty percent of the population is Muslim.

Physical Environment

At a general scale, the topography of Macedonia consists of eastern and western mountain masses, separated by the valley of the river Vardar, which follows a line of faults southward to the Aegean Sea (Figure 13.5).

The eastern mountain mass is an extension of the Rhodope Mountains of Bulgaria, consisting largely of metamorphic rocks and underlying granites. Although not as high as the Bulgarian Rhodope, the mountains rise in rugged country to several thousand feet.

The western mountain mass is an extension of the Greek Pindos Mountains and forms the border region with Albania. Here the highest peak, Mount Korab, rises to 9,032 feet. In the mountains, to the southwest of Macedonia, are two large, tectonic lakes. Prespa, at an altitude of 2,798 feet, drains by underground channels to lower Lake Ohrid, at 2,250 feet. Ohrid is clear and deep (1,017 feet) and flows via the Crni Drin to the Adriatic. Both lakes yield freshwater fish and Ohrid possesses resorts and a number of historic buildings.

The greatest part of Macedonia is drained by the Varda and its tributaries, which flow southward out of Macedonia, into the hinterland of Thessaloníki, and joins the Aegean at Thermaïkós Kólpos.

The valley of the Varda falls rapidly through the mountainous country. The river does not provide navigation into the heart of the country but is harnessed for hydroelectric power. The Varda valley is a routeway running from the Aegean northward through Skopje and over the northern upland into the Morava drainage basin that links to the Danube.

Climate

The mean monthly temperatures at Skopje for December, January, and February are all around freezing and in more mountainous areas of the country considerably lower. In winter the influence of cold continental air is frequently felt, with the north wind, *vardarac*, being unpleasant. In summer the mean monthly temperatures for July and August are in the low 70s, less on the uplands.

Macedonia is partly in rain shadow, not receiving the high rainfall totals of the Adriatic coast. Every month receives an inch or more of precipitation, with July, August, and October receiving at least 2 inches. The total rainfall at Skopje is 20 inches, with some higher stations receiving greater amounts.

Skopje

Skopje, on the river Vardar, is the capital of Macedonia, containing approximately half a million people, a quarter of the total population. The Turks developed Skopje as a fortress town and it did possess some good buildings of Turkish origin.

Within Yugoslavia Skopje was developed as an industrial center, becoming one of the larger cities in the federal state. In 1963, a powerful earthquake destroyed most of the city. The historic buildings were lost and replaced, eventually, by modern ferro concrete structures, but years after the quake whole streets were still piles of rubble.

No other urban place in Macedonia has 100,000 inhabitants and most people live in relatively small towns and villages that lack modern services.

Economy and Prospects

Even by Balkan standards Macedonians have low incomes. Only Albania ranks lower. Lacking extensive tracts of arable land, farming is dominated by the production of livestock. Arable crops include grains, vines, root crops, and a wide range of fruits and vegetables, which do well in the warm summer if watered.

Industrial activity is basic: the mining and smelting of metals, textiles, and chemicals. On the surface it seems Macedonia is a long way from having a diversified, competitive economy that could prosper in EU.

Some politicians see that within EU, Macedonia's problems with neighboring countries Albania, Bulgaria, Greece, and Serbia could be contained. The situation is complicated by the fact that Kosovo, to the north in neighboring Serbia, wants independence. In the last Kosovo crisis (1999), many Muslim Kosovans went into Macedonia to avoid Serbian forces. Kosovo is under UN jurisdiction.

We can speculate that within EU, with CAP funds arriving and regional development funds improving infrastructure, living standards in Macedonia would improve, particularly if the country could exploit the possibilities of the routeway running from the Danubian lands to the Aegean Sea.

However, economic development would continue to be slowed by ethnic conflicts, in border areas with neighboring states. Overall 64 percent of the population is Macedonian, 25 percent is Albanian, and just over 10 percent other minorities, including Turks, who make up 4 percent of the population (Johansen, 2004). The Albanians, who are Muslim, have larger households, low female participation in the workforce, a higher rate of unemployment, and a more rapid rate of natural increase. These ethnic differences are masked by averaged national statistics.

The Albanians are concentrated in regions close to Albania and the Serbian province of Kosovo. When the Serbs fought the Albanians in Kosovo in 1999, 300,000 Albanians fled, greatly increasing the number of Albanians in Macedonia. It is likely that Kosovo will become independent of Yugoslavia, as Montenegro has done. But whatever the outcome of the Kosovo independence movement, Macedonia will have relations with Albania and Serbia/Kosovo, complicated by ethnic issues. Internally the existence of a Muslim Albanian population, constituting a majority in several areas, complicates development issues. An important segment of the population has distinctive cultural, demographic, social, and economic characteristics and views development from a different perspective, but will claim discrimination if excluded from plans. Any development program will have to embrace "issues of human capital and ethnic reconciliation" (Johansen, 2004, p. 543).

▶ ALBANIA

The Albanians came under Turkish rule in 1478. Most of the population adopted Islam over time and a number of Albanians rose to positions of power within the Ottoman empire. The emergence of countries like Serbia and Bulgaria from Ottoman rule in the late nineteenth century stirred Albanian nationalism. It was clear that Montenegro, Bulgaria, and Serbia were territorially expansive and wanted to incorporate Albanian lands in efforts to gain frontage on the Adriatic Sea. In addition, Italy aspired to control both coasts of the Adriatic.

In the first Balkan War (1912), while Serbia, Bulgaria, Montenegro, and Greece were attacking Turkish territories in Europe, Albania declared independence from Turkey. After World War I, Albania was recognized as an independent state. A 1925 uprising headed by Ahmed Bey Zogu led to the creation of a republic quickly transformed to a monarchy when Zogu declared himself King Zog in 1928. Economically the country was increasingly linked with Fascist Italy, which sent an invasion force to occupy the country on Easter Sunday, 1939. Mussolini used Albania to attack Greece in 1940. The Italian forces were repulsed but, with the German conquest of Yugoslavia and Greece in 1941, all of the Balkans came under Axis control.

Albanian guerrilla resistance was led by the communist Enver Hoxha. After the war Hoxha established a hard-line Stalinist government that stayed close to Moscow when Yugoslavia broke with the Soviet Union in 1948. After Khruschev denounced Stalin in 1956, Albania broke with Moscow and linked economically and ideologically with Communist China, even allowing a Chinese naval base on the Adriatic coast. Albania broke with China after it made a *rapprochement* with the west in 1978. Hoxha ruled until his death in 1985. A deputy succeeded. The isolation and hard-line Communist policies continued until the collapse of the Soviet Union when non-Communist parties were allowed to contest elections.

Today, Albania has half the workforce in farming and more than half the GDP is derived from agriculture. Manufacturing is limited to food processing and basic consumer products. Albania is one of the few European countries growing in population numbers by natural increase. The birth rate is fourteen per thousand, the death rate five, and the population grows by 0.8 percent per annum. GNI (gross national income), in purchasing power equivalent, is around $5,000 and living standards are obviously low. (Tables 13.4 and 13.5)

Physical Environment

Albania fronts onto the Adriatic Sea and has its back to the mountains of Macedonia (Figure 13.7). The Adriatic coastal plain is not extensive and the mountainous interior rises up to 9,066 feet on the Albania–Macedonia border to the east. The coastal plain produces grains, potatoes, sugar beets, and the Mediterranean summer crops including tomatoes and squashes. Although the country is largely Muslim, there are vineyards and wine production. The hills and mountains are predominately in pasture and forest, producing cattle, sheep, and goats, with pigs being of lesser importance.

In the coastal regions, Albania has a modified Mediterranean climate. Summers are hot, winters are moist. The temperatures at the capital Tirana, not at the coast, average 44°F in January and 75°F in July. The annual rainfall, at over 50 inches, is not typically Mediterranean. The high total results from the large number of active depressions in the Adriatic during the winter months. In the mountains, rainfall is even higher than on the coastal plain and, at altitude, precipitation comes as snow in the colder months. Winters in the mountains of Albania can be severe and snow clogged.

Although well south in the Adriatic, the coast of Albania can still experience the *bora* (north wind) that blows down the sea in winter, bringing cold air and fierce winds to disrupt shipping and chill inhabitants in coastal areas.

Population and Urban Places

The population of Albania is 3.2 million and the number of Albanians grows by just under 1 percent per annum. Less than 50 percent of Albanians live in towns and cities and the country has a rural majority.

TABLE 13.4 Albania Urban Places

Tirana	345,000
Durrës	100,000
Elbasan	88,000
Shkodër	82,000
Vlorë	78,000
Fier	56,000
Korçë	55,000
Berat	40,000

TABLE 13.5 The Balkans: Summary Statistics

	Pop. (in millions)	BR	DR	%NI	IMR	TFR	%Urban	GNI PPP
Albania	3.2	14	6	0.8	8	1.9	45	5,420
Bosnia-Herzegovina	3.9	9	9	0.1	7	1.2	43	7,790
Bulgaria	7.7	9	15	−0.5	10.4	1.3	70	8,630
Croatia	4.4	9	11	−0.2	6.1	1.4	56	12,750
Macedonia	2.0	11	9	−.02	11.3	1.4	59	7,080
Montenegro	0.6	13	9	0.3	8	1.7	NA	5,333
Romania	21.6	10	12	−0.2	16.8	1.3	55	8,940
Serbia	9.5	13	12	0.1	10.0	1.2	51	NA

Source: Population Reference Bureau, 2006.

Albania: EU Prospects

Albania has low per capita incomes and is not a modern country. Many people live on subsistence farms and have a limited cash income. More than half the GDP comes from agriculture and about half the workforce is in farming. Only 80 percent of the population is literate and the literacy rate for females is lower than for males. Economically, socially, and politically Albania is a long way from EU standards and most Albanians have little interest in those standards, seeing modern Europe as an alien world.

▶ SUMMARY

Only one country, Albania, has a total fertility rate that will sustain population numbers. None of the countries in the list will attract migrants and outmigration to the west is likely to reduce population numbers.

Gross national income per capita, expressed in terms of purchasing power parity, is a rough and ready means of comparing the economic performance of countries (Table 13.5). The index takes the total value of domestic production and divides the figure by the number of inhabitants. The figure does not reflect average actual incomes in any country, taking no account of the distribution of wealth. The index makes any country with a subsistence sector appear particularly poor. In the Balkans many live on subsistence holdings, consume what they produce, barter with neighbors, and kill livestock in rotation to provide meat for a village. Subsistence activities do not involve recorded cash payments and do not enter national accounts.

However, although households in the Balkans are adequately fed, there is little disposable income. The hope for aspiring entrants is that the EU will provide the impetus for modernization and economic growth. The experience of Romania and Bulgaria will provide commentary on the prospects of other Balkan states in EU, should they pursue entry.

▶ FURTHER READING

Crampton, R. *The Balkans Since the Second World War*. New York: Longman, 2002.

Crnobrnja, M. *The Yugoslav Drama*. 2nd edition, Montreal and Kingston: McGill-Queen's University Press, 1996.

Darby, H.C. "Bosnia and Herzegovina," in Stephen Clissold (ed.), *A Short History of Yugoslavia*. Cambridge, UK: Cambridge University Press, 1966, pp. 58–72.

Hitchens, K. *Romania 1866–1947*. Oxford, UK: Clarendon Press, 1994.

Holbrooke, R. *To End a War*. New York: Random House, 1998.

Johansen, H. "Ethnic Dimensions of Regional Development in Macedonia: A Research Report." *Eurasian Geography and Economics* 45(7) (2004): 534–544.

Kaplan, R.D. *Balkan Ghosts*. New York: Picador, 1993.

Klemencic, M., and C. Schofield. "An Emerging Borderland in Eastern Slavonia?" in David H. Kaplan and Jouni Häkli (eds.), *Boundaries and Place: European Borderlands in Geographical Context*. Lanham, MD: Rowman and Littlefield, 2002.

Lampe, J.R., and M. Mazower (eds.), *Ideologies and National Identities: The Case of Twentieth Century Southeastern Europe*. Budapest, NY: Central European University Press, 2004.

Lampe, J.R. *Yugoslavia as History: Twice There Was a Country*, 2nd edition, Cambridge, UK: Cambridge University Press, 2000.

Matley, I. *Romania: A Profile*. New York: Praeger, 1970.

Mazower, M. *The Balkans: A Short History*. New York: Modern Library, 2000.

O'Loughlin, J. "Ordering the 'Crush Zone': Geopolitical Games in Post–Cold War Eastern Europe," in Narit Kliot and David Newman (eds.), *Geopolitics at the End of the Twentieth Century*. London: Frank Cass, 2000, pp. 34–56.

Poulton, H. *Who are the Macedonians?* Bloomington: Indiana University Press, 2000.

Rugg, D.S. "Communist Legacies in the Albanian Landscape." *Geographical Review* 84(1) (1994): 59–73.

Sullivan, K. "Out of Darkness: Romania Tries to Shed its Primitive Past for Entry into EU." *The Washington Post*. March 12, 2006, pp. A12–13.

Wilkinson, H.R. *Maps and Politics: A Review of the Ethnographic Cartography of Macedonia*. Liverpool, UK: University of Liverpool Press, 1951. A classic account of the manner in which cartography can be used by competing countries to support territorial claims.

PART

V

REGIONAL SURVEY: RUSSIA AND FORMER SOVIET SOCIALIST REPUBLICS

14

BELARUS, THE UKRAINE, MOLDOVA, GEORGIA, AND RUSSIA

▶ INTRODUCTION

Belarus and Russia are closely linked and show no interest in joining the European Union. Within the Ukraine there are political groups that see membership of EU as a way to fully establish independence from Russia and the former Soviet Union. In the eastern Ukraine, with considerable populations of ethnic Russians, the idea of Europe is less attractive. As Romania joins EU, the Romanian-speaking Moldovans will want membership. In the east of Moldova, in Transdnistria where many Russians and Ukrainians live, the region runs its own affairs.

▶ BELARUS AND THE UKRAINE

Belarusians and Ukrainians are Slavs, speaking related eastern Slavonik languages. Neither cultural group has a history of national independence, for they have been a part of larger entities: the Grandy Duchy of Lithuania, the Kingdom of Poland and, after the partition of Poland, the Russian Empire, and the Empire of Austria-Hungary. Briefly, at the end of World War I, Belarus and the Ukraine were independent before absorption into the Soviet Union.

A Byelorussian National Republic was proclaimed on March 25, 1918, but at the end of the Soviet-Polish war (Treaty of Riga, 1921) eastern Belarus became a Soviet Socialist Republic. Western Belarus was part of Poland.

The Ukraine detached from the Russian Empire in January 1918 and at Brest-Litovsk (March 1918) was recognized as independent by Germany and the Soviets. The western allies insisted that Germany renounce Brest-Litovsk as part of the armistice agreement of November 1918. The Ukraine became involved in the Russian civil war and fought Poland on the delimitation of the boundary between Poland and the Ukraine. By 1922 the Ukraine was a Soviet Socialist Republic.

Belarus and the Ukraine became independent in 1991 but the only system of government that either country had experienced was the Soviet model. The influence of Russia remains strong, particularly in Belarus.

Physical Environment of Belarus and the Ukraine

Belarus

The surface features of Belarus and the Ukraine are a product of the last Ice Age. The Scandinavian ice cap extended across the Baltic and, at maximum extent, had a southern front across the territory now occupied by Belarus. Running through Belarus, from the southwest to the northeast, is a large terminal moraine, rising at Mount Dzyarzhynsk to 1,135 feet, the highest point in the country (Figure 14.1). At the southwest end of the Belarusian Ridge, and extending into Poland, is a remarkable ancient forest, the Belovezskaya, a nature reserve in both Poland and Belarus. The forest was preserved for hunting by Polish kings and then Tzars. The fauna, no longer legally hunted, includes the European bison (Wisent), deer, elk, boar, and smaller game. (Photo 14.1)

To the north and the south of the end moraine forming the Belarusian Ridge are lakes, marshes, forests, and poorly drained ground, which only intermittently can be made into arable land. To the north, the Dvina and the Neman drain to the Baltic. South of the Belarusian Ridge, in the area bordering the Ukraine, is a shallow, ill-drained depression filled with glacial outwash material. These Pripet marshes, through which the river Pripet

Figure 14.1 Belarus.

Photo 14.1   In addition to the Belovezskaya nature reserve that runs from Belarus into Poland, bison are now settled in the exclusion zone around the Chernobyl nuclear reactor which exploded in 1986. The reactor was in the Ukraine, but contamination spread north into Belarus.

Source: REUTERS/Vasily Fedosenko/NewsCom.

flows to join the Dneiper, are an extensive wetland with many lakes and peat bogs. In the second half of the twentieth century, much land was drained for agriculture.

On the Belarusian Ridge, which rises above the lakes and marshes, there is good, well-drained farmland. The capital of Belarus, Minsk, is sited on the ridge–moraine. The main west-to-east route follows the Belarusian Ridge to Smolensk and Moscow. In the summer of 1941 a German Panzer group, coming from Poland to attack the Soviet Union, advanced along the end moraine.

The Ukraine

Much of the Ukraine lies south of the glacial end moraine, the glacial outwash materials, and the Pripet marshes. In the glacial era the Ukraine was south of the ice cap. Although ice free, the Ukraine was a cold, exposed, windswept region, with the winds carrying particles of silt washed out by streams flowing from the ice front. The silt particles were deposited as loess and the loess is the basis of the Black Earths, the rich Chernozem soils of the Steppe, which helped make the Ukraine a major nineteenth-century exporter of grain through the Black Sea ports.

Not all of the Ukraine is treeless steppe. In the west of the country is a part of the Carpathian mountains where peaks attain 6,000 feet.

Rivers of the Ukraine

The greatest part of the Ukraine consists of the valleys and interfluves of rivers that flow, northwest–southeast, to the Black Sea (Figure 14.2). From west to east the major rivers

Figure 14.2   Ukraine.

are the Dnister, the southern Buh (Bug), the Dnieper, and the Donets. In addition, the Danube forms a part of the southwest frontier of the Ukraine with Romania.

The Dnieper is the core of the Ukraine. There are a number of major cities along the river, including the capital of Kiev and the major industrial cities of Dnipropetrovs'k, and Zaporizhzhia. The port of Odesa lies west of the Dnieper estuary. To the east of the estuary is the Crimea peninsula, which contains, in the south, the historic cities of Sevastopol' and Yalta. The Crimea was attached to the Soviet Socialist Republic of the Ukraine in 1954. Russia still bases the Black Sea fleet at Sevastopol' and pays rent to the Ukraine. Most Russians want the strategic Crimea, with the historic towns and subtropical climate, restored to Russia.

Climates of Belarus and the Ukraine

Except along the southern Black Sea margin of the Ukraine where maritime influence is pronounced, the climates of Belarus and the Ukraine are continental. Summers are warm and dusty, thunderstorms are common. Winters are long and cold. The rivers freeze from December to March. Snow covers the landscape for months but the total snowfall is not great. Ground blizzards are common, especially in the Ukraine, when the northeast wind, the *buran,* blows out from the Siberian winter high-pressure system.

Resources of Belarus and Ukraine

Belarus is not rich in resources other than potash, the basis of fertilizer industries. The numerous bogs yield peat, which, when dried and compressed into briquettes, is used as fuel.

By contrast, the Ukraine is rich in resources, including iron ore, coal, manganese, mercury, bauxite, titanium ore, oil, and gas. The well-known deposits of iron ore at Kryvyi Rih, the manganese at Nikopol', and the hard coal seams in the Donets basin are the basis of a major industrial region producing iron, steel, heavy engineering products, nonferrous metals, and chemicals. The industrial cities of the Ukraine include Donets'k and Kharkiv (f. 1654) in the Donets basin and, on the Dnieper, Kiev, Dnipropetrovs'k, and Zaporizhzhia, with a major hydroelectric power plant. (Photo 14.2)

Origins of Independent States

Both Belarus and the Ukraine were briefly independent at the end of World War I, although neither established a functioning state. Under the Soviets anyone associated with independence was suspect. Many were arrested, executed, or deported. In the late 1920s, Stalin determined to collectivize Soviet agriculture. Collectivization was particularly cruel in the Ukraine, a prosperous farming region with many farms above the level of peasant, subsistence units. A whole class of relatively prosperous *Kulaks* resisted having farms

Photo 14.2 Kiev, on the navigable Dnieper river, is the capital of the Ukraine. The city is a center of administration, heavy industry, food processing and grain shipments. Notice at rear the many blocks of housing for workers, a product of the Soviet era.

Source: Age fotostock/SUPERSTOCK.

confiscated and merged into collectives on which they would be laborers. Resisters were shot; villages were destroyed, inhabitants rounded up and deported. Then, with farming disrupted, there was a terrible, Soviet-made famine in the region. Stalin had made a start on the destruction of millions of Ukrainians that took place through collectivization, purges, and NKVD killing units.

Invasion in 1941

Hitler, obsessed with *Lebensraum* (living space) in the east, wanted control of the grain fields of the Ukraine. *Lebensraum* was about space and race. Control the space, displace the Slavic race, introduce Nordic settlers to give the Ukraine a Germanic face.

Barbarossa, launched on the night of June 21–22, 1941, was successful at first. By July, Minsk was under German control, as was the western Ukraine. Many Belarusians and Ukrainians recalled that it had been German policy, in World War I, to establish Belarus and the Ukraine as independent states. At Minsk, largely destroyed by the retreating Red Army, the Archbishop asked for permission to conduct a service of thanksgiving at the cathedral. The Panzer commander, Guderian agreed. In the Ukraine some inhabitants greeted German units with bread and salt, the traditional mark of welcome.

At first the German occupation, as regards civilians, was orderly. Professional military commanders refused to issue orders that suggested soldiers would not be disciplined for violations. The army command suppressed, or delayed, orders that Soviet political organizers, the Commissars, should be immediately shot.

All changed when the SS units and killing squads arrived behind the front lines. Thousands were summarily killed under suspicion of being Communist, or partisans, or just to clear areas of Slav populations. The Jewish population was targeted everywhere, culminating in the Ukraine at the Babi Yar massacre, when thousands of men, women, and children were shot and dumped in a pit.

Nazi actions provoked resistance and when partisans killed German soldiers the army retaliated by indiscriminately killing Ukrainians. Isolated areas within the Pripet marshes, already home to remnants of Red Army units defeated by the *Wehrmacht,* became partisan bases.

Under the 1939 Non-Aggression Pact between Hitler and Stalin, Germany received raw materials and foodstuffs from the Soviet Union. Now the economies of Belarus and the Ukraine were being destroyed by the Red Army. The retreating Soviets blew up hydroelectric plants on the Dnieper and other rivers, dismantled or crippled steel works, and flooded coal mines. The Ukraine could no longer supply grain and raw materials to Germany.

The German army did restart hydroelectric plants, some coal mines, and a few steel works. But the economic return from occupying the Ukraine was tiny, miniscule against the costs of the war.

The war on the eastern front culminated beyond the Ukraine at Stalingrad (Volgograd) on the Volga river, in the winter of 1942–43. From that point the Red Army pushed into Belarus and the Ukraine. As the *Wehrmacht* retreated it destroyed the plant it had restored. The return of the Red Army was not a liberation. The Soviets destroyed Minsk again: Ukrainians and Belarusians suspected of working with Germans were punished.

Postwar Boundaries

The boundaries of the Soviet Socialist Republics of Belarus and the Ukraine altered after World War II. Parts of prewar Poland were detached and added to Belarus and the Ukraine. Polish populations were forced into Poland, which had been moved westward at German expense. Polish settlements were attacked and people killed to expedite migration.

Ruthenia was part of the original Czechoslovakia, but was attached to Hungary after the 1938 Munich agreement. At the end of World War II, Ruthenia became part of the Ukraine. In terms of language, Ruthinian was more akin to Ukrainian than Czech or Slovak, but what Ruthinians thought of being joined to the Soviet Union few dared say.

The Ukraine also received a strip of coastal territory on the Black Sea adjoining Romania. The Ukraine still controls the territory that cuts off another now-independent Soviet Socialist Republic, Moldova, from the sea. The land is particularly valuable as subtropical crops are grown there. The Ukraine received additional subtropical lands when, in 1954, the Crimea peninsula was added. The Donbas region was extended to the east at about the same time. (Photo 14.3)

Postwar Reconstruction

Postwar reconstruction in Belarus and the Ukraine concentrated on heavy industry. The Ukraine came to produce nearly a third of all the iron, steel, and heavy engineering products in the USSR. Investment was sustained in the Donbas region but the capital cities of Minsk and Kiev were also centers of heavy manufacturing, as was Kharkiv in the Ukraine, which had long been a center of armaments production, including tanks.

Photo 14.3   The rivers of the Ukraine and Russia freeze in the winter haulting navigation. At Astrakhan, in Russia, on the delta of the Volga river, at the north end of the Caspian Sea, in early April vessels await the opening of navigation on the river as winter ice breaks up.

Source: ITAR-TASS photo/Vladmir Tyukaev/NewsCom.

All the industrial centers were marked by the construction of large blocks of workers' housing often set close to squares with statues of Lenin and Stalin.

Nuclear power stations were built and one of the plants at Chernobyl', on the Pripet river in northern Ukraine, exploded on April 25, 1986. The Soviet Union was silent. Then sensors in Sweden picked up the radioactive fallout and the disaster became world news. There were many deaths from radioactive exposure in both the Ukraine and Belarus. Crops and livestock in the countries, and many parts of Europe, could not be consumed because of radioactive content. Birth defects rose for many years, but now the afflicted area is returning to normal, although the worst affected zones should never be reinhabited.

Population and Urbanization

Ukraine

The Ukraine is nearly 70 percent urban. The major cities, in rank order by population number are listed in Table 14.1.

On the Crimea peninsula, the Sevastapol' fortress town and naval base has a population of 341,000. The fortifications and the town were heavily damaged in World War II, but after rebuilding became the base of the Soviet Black Sea fleet. Yalta, with less than a 100,000 inhabitants, is a resort town with many hotels, villas, and palaces.

The population of the Ukraine is 47 million people. The birth rate is nine per thousand Ukrainians, the death rate is seventeen, and the total fertility rate is 1.2. Population numbers shrink by three quarters of a percentage point each year. In 1989 there were 51.5 million people living in the Ukraine. Since then, most regions of the country have lost population. Decline is particularly marked in the east with concentrations of ethnic Russians, many of whom returned to Russia. Population loss on the predominantly Crimean peninsula is less marked, partly because the Tartars, expelled by Stalin, are returning (Rowland, 2004).

Belarus

In Belarus over 70 percent of the population lives in urban areas. The largest cities are listed in Table 14.2.

The total population of the country is 9.7 million and falling. The birth rate per thousand is nine, the death rate fifteen, and the population shrinks by half a percent per annum.

TABLE 14.1 Cities of the Ukraine, 2001

Kiev	2,602,000
Kharkiv	1,470,000
Dnipropetrovs'k	1,064,000
Odesa	1,029,000
Donets'k	1,016,000
L'viv	732,000
Kryvyi Rih	667,000
Mykolaïv	514,000
Mariupol'	492,000
Luhans'k	463,000

TABLE 14.2   Cities of Belarus, 1999

Minsk	1,729,000
Homyel'	504,000
Mahilyow	371,000
Vitsyebsk	359,000
Hrodna	309,000
Brest	300,000

Prospects

The Ukraine made approximately a third of Soviet steel. In 1991, the Ukraine produced 45 million tons of steel. By 1995, production had halved but recovered to 38 million tons in 2005, as new markets were found. Nearly 40 percent of Ukrainian GDP is derived from manufacturing.

There are no plans for talks with Belarus or the Ukraine concerning EU entry. However, the Ukraine wants to negotiate an EU–Ukraine free trade agreement. In Belarus and the Ukraine, living standards are low as economies make the transition from smokestack to modern and from central planning to market driven. Nearly 20 percent of the Ukrainian population is Russian and wants to maintain connections with Russia. In the predominately Ukrainian west, the population is interested in joining the EU. Neither Belarus or the Ukraine would be currently competitive in EU economic space. Joining EU would speed outmigration and accelerate population decline.

▶ MOLDOVA

Landlocked Moldova, lying between Romania and the Ukraine, consists largely of the Bessarabia region bounded by the river Prut in the west and the Dniester river in the east (Figure 14.3). Moldova is on the linguistic frontier between the Romance and the Slavic language groups. The majority of Moldovans (75%) speak Romanian but in the east of Moldova are significant minorities of Ukrainian and Russian speakers. Other minorities speak Gagauz and Bulgarian. *De facto,* Transdnistria (Transnistria), where most people speak Russian or Ukrainian, conducts its own affairs within Moldova.

The Orthodox Christian Gagauz are a Turkic people speaking a dialect of Turkish, which incorporates many Romanian and Slavic words. The approximately 200,000 Gagauz live largely in the Prut valley in southern Moldova, with some spillover into adjoining Ukraine. The Gagauz Republic is autonomous within Moldova and should Moldova decide to unite with Romania, the Gagauz have an agreement allowing them to secede. The Gagauzi are noted for vineyards and wine making.

The Russian and Ottoman Empires contested control of Moldova until the end of World War I, when Romania got the region. The Soviet Union never recognized the incorporation and signaled intentions by setting up a small Moldovian Soviet Socialist Republic in the Ukraine, on the Romanian border. In June 1940, as France fell, the Soviets demanded and got Bessarabia and incorporated the territory into the Moldovian Soviet Socialist Republic. In June 1941 Romania, an ally of Germany, joined the attack on the Soviet Union. Bessarabia reverted to Romania until the end of World War II

Figure 14.3 Moldova.

when the Soviets took it back. With the breakup of the Soviet Union, Moldova became independent in August 1991.

Physical Environment

Neighboring Ukraine consists of generally south-flowing rivers separated by interfluves. The pattern continues, on a smaller scale, in Moldova with two rivers, the Prut and the Dneister (called the Nistru in Moldova), being separated by a loess-covered interfluve.

Because of proximity to the Black Sea, which moderates temperatures, Moldova has a milder version of a continental climate than northern Ukraine. Nevertheless, the mean monthly temperature for January at the capital Chişinău, in the south center of the country, is below freezing. The mean monthly temperature for July is 70°F. Rainfall, on average, is 20 inches a year but is highly variable and prolonged dry spells reduce agricultural yields.

As in the adjoining Ukraine, Moldova has a cover of loess, which has developed into rich Black Earth (Chernozem) soil. The loess is deepest in the north and thins to the south. There are woodlands in the north. The south of the country is a grassland (steppe).

Agriculture

Basic crops include winter wheat, corn, tobacco, sugar beets, and sunflowers for seed. The great value of Moldova to the Soviets was the relatively long growing season and the ability to produce fruits, vegetables, and grapes. In the Soviet era, vineyard acreage was expanded and wine production increased. Under the Soviets private farms were taken over for collective and state farm operations. Farmland privatization started in 1995.

Manufacturing

The industrial sector is based on food processing, with agro-processing plants making up nearly half of industrial production. The plants produce sugar, cigarettes, wine, liquor, and package fruits and vegetables. In the Soviet era, Moldova made large exports of wine, tobacco, fruits, and vegetables to other parts of the USSR. Entry into Russian markets is no longer automatic and economic swings in Russia influence demand. Moldova had a food-producing role in the Soviet Union but the agricultural sector would struggle to compete in Europe. Non-agro manufacturing is not well diversified and includes tractors and agricultural equipment, some consumer goods, textiles, and footwear.

Prospects

Moldova, with a population of four million, has a birth rate of eleven and a death rate of twelve per thousand inhabitants. In addition, many young adults choose to emigrate to find higher living standards elsewhere for, even by the averages of eastern Europe, Moldovans have low incomes.

Transdnistria, a legacy from the Soviet era when the Moldovian S.S.R. was set up in 1924, is an issue. The region is undemocratic, administers its own affairs, and is effectively beyond international law. The future of Transdnistria is unclear.

Romanian-speaking Moldovans think of themselves as Europeans and many will want to follow Romania into the EU. Economically, the country is a long way from being able to compete in EU markets.

► GEORGIA

Technically Georgia is not in Europe, but due to its role in linking Caspian and Central Asian oil and gas to the Black Sea, the country is of interest to EU.

Georgia adjoins Russia, Azerbaijan, Armenia, and Turkey. The cultural geography of the country is complex as the Caucasus region is known for a rich intermix of linguistic groups. The Georgia Church is independent within the Eastern Orthodox Church but in the national territory are Muslims, of several sects, and Christians who are Russian Orthodox, or Armenian Orthodox. Of the population of 4.4 million, approximately 70 percent are Georgians. The minorities include Armenians, Russians, Ukrainians, Greeks, Azerbaijanis, Ossetians, Abkhazians, and Ajarians.

The Georgian language is Caucasian and some minorities speak variants of Georgian. Other minorities speak Russian, Greek, Armenian, and Azeri. The complexity of the cultural intermix is indicated by the existence of the autonomous, self-governing regions of Abkhazia, in the northwest, and Ajaria in the southwest, both with Black Sea frontage. South Ossetia is no longer autonomous but has a distinct history and culture. Abkhazia and South Ossetia border Russia.

The Georgian Black Sea coastal region was colonized by Greece and then Rome. The Abkhazians still incorporate Latin elements in their language. Christianity was introduced by the Romans in the fourth century A.D. In the sixteenth century the Georgia region became part of the Turkish Empire until Russia gained control in the early nineteenth century. Georgia was briefly independent at the end of World War I, until invaded by the Red Army and incorporated into the Soviet Union in 1921. The country became independent in 1991.

Physical Geography

A map of the topography of Georgia appears easy to read. A triangular coastal plain, with its base on the Black Sea, is enclosed to the north by the Great Caucasus Mountains and in the south by the Lesser Caucasus (Figure 14.4).

The upland, east of Kutaisi, rises to several thousand feet, but valleys and passes allow passage to the east. The capital, Tbilisi (1,332 feet), is on the line of communication from the Black Sea to the Caspian, at a point where it intersects the north-to-south routes (Table 14.3). Tbilisi is on the Kür (Kura) river, which flows to Azerbaijan and the Caspian Sea.

Figure 14.4 Georgia.

TABLE 14.3   Cities of Georgia

Tbilisi	1,080,000
Kutaisi	186,000
Batumi	122,000
Sukhumi	121,000
Rustavi	116,500
Gori	49,500
Poti	47,000

Population totals are estimates.

North and south of the central upland, on which Tbilisi stands, are the Great Caucasus and Lesser Caucasus Mountains. The highest point in Georgia is in the Great Caucasus at Mt. Kazbek (16,541 feet). In the Great Caucasus range, which trends northwest to southeast from the Black Sea to Caspian, there are several higher mountains including El'brus rising to 18,510 feet, but on the Russian side of the border. The Lesser Caucasus is not as high as the northern range but still has peaks rising to near 10,000 feet.

Climate

The coastal plain, under the influence of the Black Sea, has a subtropical climate. Crops are more Mediterranean than temperate and feature fruits, including citrus, tomatoes, squashes, tobacco, mulberry for silkworms, grapes, and tea. (Photo 14.4)

At Batumi, on the Black Sea coast, the mean monthly temperatures for December, January, and February are 48°F, 43°F, and 44°F, respectively. The means for June through September are all in the 70s, with temperatures being moderated by the Black Sea. Batumi receives over 90 inches of rain a year. All months from September through January

Photo 14.4   In the Black Sea coastal area of Georgia climates are relatively mild and it is possible to grow subtropical crops including tea.

Source: Elisabeth Peters/Alamy Images.

are wet and September receives over 12 inches of rain. On the upland around Tbilisi, climate is influenced by altitude and distance from moderating maritime influence. Mean temperatures in the capital, in the summer months, are in the high 60s and low 70s. In January the monthly mean is around 32°F. Being further inland Tbilisi receives less rainfall, 22 inches, than coastal stations. May and June are the wetter months, each receiving on average 3 inches of rainfall. The Greater Caucasus and Lesser Caucasus mountains have alpine climates with permanent snow and ice fields.

Economic Activities

Georgia is still an underdeveloped country. Nearly half the population lives in the rural areas and approximately a quarter of GDP comes from agriculture. Large areas of the country, in the mountains, are too high or too steep for farming, although providing pasture for livestock. Forty percent of the country is forested.

Georgia produces and exports the steel hardener manganese and a number of other minerals. Hydroelectric power has been developed. The country produces some coal, oil, and gas but gas imports are required to meet energy needs. Manufacturing, including textiles, food processing, metallurgy, oil refining, and chemicals, accounts for less than 10 percent of GDP. Incomes are low and many live at a subsistence level.

Population numbers shrink slowly, with a total fertility rate of 1.6. Estimates suggest that the present populations of 4.5 million will be down to 4 million within 20 years, by which time the population will have aged, unless developments in the oil transit industry results in an influx of young workers.

Strategic Importance

The EU, and all oil-and gas-consuming countries, see the strategic value of Georgia in the context of transporting energy supplies from Central Asia. Azerbaijan, on the west shore of the Caspian Sea, has long exported oil via Georgia to the Black Sea coast. Now a new major pipeline will move oil from Baku, on the Caspian coast, through Georgia to the Turkish Mediterranean port of Ceyhan. There are overland routes for Central Asian oil and gas through Russia but the Azerbaijan–Georgia route is not under Russian control. The President of Georgia, Mikheil Saakashvili, wants to bring Georgia into NATO and the European Union, but Russia resists western influence in the region. In 2006 Georgia arrested Russian officers suspected of spying. The EU intervened and had the officers returned to Russia. Then Russia announced it had been insulted, closed the border with Georgia, targeted Georgian businesses in Moscow for harassment, cut off airline and communication links between the countries, and put up the price of natural gas.

▶ RUSSIA

Introduction

In terms of physical geography, Russia, west of the Urals, is part of Europe. The northern European plain that runs from the North Sea, along the south Baltic shore, continues into Russia, with the climate becoming increasingly continental to the east. Within Russia, Europe, and Asia, there are few topographic barriers to movement. The Ural Mountains, the traditional line of division between Europe and Asia, rise to a few thousand feet and passes offer routeways through the range, which runs from the Arctic Ocean in the north to Kazakhstan in the south (Figure 14.5).

There is no present intention to bring Russia into EU. Apart from political issues, Russia is just too large in terms of population, landmass, and industrial structure to be integrated and the EU, having admitted twelve new members, does not have the structural adjustment funds to correct the many Russian economic problems.

On the political front, one of the forces moving Europe to integration was fear of the Soviet Union. Few in the newly admitted eastern European countries are ready to trust Russia, remembering the ruthless way Stalinist regimes were imposed upon them after World War II. Within Russia there is no history of democratic institutions, the rule of the Czar having been replaced by a state in which the will of the party was enforced by secret police, summary executions, and deportations to labor camps. The movement to create democracy and establish human rights began in the early 1990s. The president of the country, Putin, is a former KGB colonel with autocratic tendencies, who concentrates power in the Kremlin.

The reader might ask, If Russia is unlikely to join EU, why give the country coverage? Russia is so large and important a neighbor that it is a factor in European affairs. Russia has a resource base important to European economies, particularly the oil and gas sent west by pipeline. Russia, west of Urals, is in Europe.

In the Soviet era, Mackinder's Heartland concept was ignored. Today, Mackinder's ideas on the strategic potential of Russia's central position in Eurasia are studied. It is a mistake to assume that the imperial impulse has deserted Russian policymakers (Coones, 2005; O'Loughlin and Talbot, 2005).

Origins and Growth of Russia

By the ninth century A.D. there was a trade route linking the Ladoga region on the Baltic to the Dnieper river and the Black Sea. Goods, including furs, moved along the route, many going, via Constantinople, into Mediterranean trade.

In the north were trading settlements on the river Volkhov, which flows from Lake Il'men' to Lake Ladoga. The river Neva links Lake Ladoga to the Baltic. The preeminent center in the region was Novgorod on Lake Il'men' (Figure 14.5).

In the south, on the Dnieper, the major trade center was Kiev. Vikings came to command the north and brought the Novgorod–Kiev trading region under one rule by 882 A.D. The Vikings were leaders, warriors, and merchants who formed a ruling elite. By 1054 Kievan Rus controlled lands between the Black Sea and the Ladoga region, extending eastward to include the Moscow area and the headwater region of the Volga. Later, power shifted northward to the Novgorod Republic and in 1240 Kiev was taken by the golden horde riding from the east.

After the eclipse of Kiev, on the boundary between the forest and the steppe, more northerly centers developed, partially protected by the forest, which reduced grazing for the hordes of Heartland horsemen. By the late fourteenth century Muscovy was a rising state, close to the headwaters and routeways of the Dnieper, Volga, Dvina, and Don. Muscovy centered on Moscow where the Grand Dukes began the powerful Kremlin fortress in 1367. In 1547 the Grand Duke of Moscow took the title Tzar, from the Latin

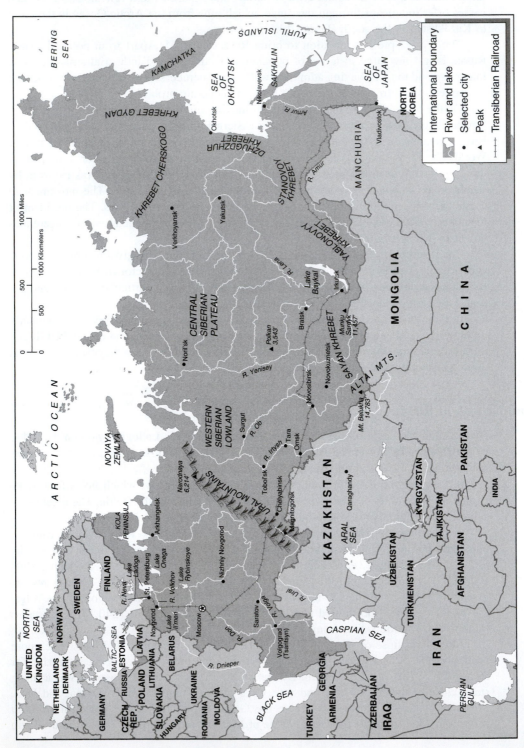

Figure 14.5   Russia.

Photo 14.5 Nizhniy Novgorod was founded as a stronghold to protect eastern Russia. Notice the Kremlin like structure with fortifications enclosing cathedrals and churches. The city was named Gorky from 1932 to 1991 after the writer born in the place.

Source: Roman Shljapnikov/Foto S.A/NewsCom.

Caesar, and the Tzars developed a grand strategy of imperial expansion. To the east, Nizhniy Novgorod was established in 1552 as a base for expansion, and within a few decades forts were built beyond the Urals at Tobol'sk (1587), Tyumenʻ (1586), and Surgut and Tara (1594). (Photo 14.5)

To the south, control of the Volga, flowing to the Caspian, was started with the foundation of Tsaritsyn (1589) (Stalingrad, in the Soviet era, now Volgograd) and Saratov (1590). By the end of the sixteenth century Russia had control of the north shore of the Caspian Sea and, in Siberia, had forts and settlements in the Ob drainage basin.

Siberia was sparsely populated by hunters and gatherers who produced furs as tribute. The indigenous inhabitants were too few in number to offer effective resistance. Expansion along a few routes across Siberia was rapid. Okhotsk on the Pacific was founded in 1649. The southern part of the Western Siberian lowland was not controlled until the eighteenth century when Omsk, in the Ob basin, was founded in 1716 and penetrations were made into Kazakhstan, beginning the annexations of the Central Asian Khanates. The annexations were not completed until the nineteenth century.

In the second half of the seventeenth century southern expansion brought in part of the Ukraine and by the early nineteenth century most of the north shore of the Black Sea was part of Russia, at the expense of the Ottoman Empire.

In the west, territory was ceded to Russia by Sweden in 1721. During the Napoleonic wars Poland was divided and, by 1815, Warsaw was under Russian rule. Finland had been taken from Sweden in 1808, becoming a Russian Grand Duchy in 1809.

In the Far East the Pacific coast and hinterland, between Nikolayevsk and Vladivostok, was annexed 1858–60. The strategic port of Vladivostok was started in the latter year, but ice was a problem in winter.

Much of the imperial expansion was undertaken against lesser forces. Siberians could not resist, the Central Asian Khanates were no match for imperial forces. The declining

Swedish and Ottoman empires ceded territory to expanding Russia. More powerful opponents were prepared to contain Russian expansion. Britain and France landed forces in the Crimea (1854–56), in support of the Ottoman empire, as Russia pressed for control of the Black Sea and the Straits leading to the Mediterranean. Britain resisted Russian expansion into Afghanistan. Persia (Iran) kept the Tzars from the Persian Gulf and the Indian Ocean.

Russia wanted warm water ports and in 1898 leased territory in Manchuria, including Port Arthur (Lüshun in China today). Port Arthur was south of Vladivostok and ice free. Japan had imperial ambitions in Korea and Manchuria. On February 8, 1904, before a declaration of war, Japan crippled the Russian fleet at Port Arthur and, later in the year, destroyed the Baltic fleet that sailed into the Strait of Korea. The Port Arthur lease passed from Russia to Japan by the Treaty of Portsmouth in 1905.

World War I extinguished Tzarist Russia. The Bolsheviks signed a peace agreement with Germany in the spring on 1918 at Brest Litovsk. The western allies insisted the treaty be revoked at the armistice of November 11, 1918, but *de facto* Finland, the Baltics, Poland, Belarus, the Ukraine, and Georgia were independent. The successor Soviet Union was able to bring Belarus, the Ukraine, and Georgia into the USSR after the war, but the territorial losses in World War I were extensive and not accepted by Soviet leaders.

During and after World War II Stalin took control of all the lost lands and more. The Baltics and Moldova were absorbed in 1940. Territory was taken from Finland in the same year. At the end of the war Finland, Poland, Germany, and Czechoslovakia experienced Soviet expansion at their expense. Puppet governments were installed in Poland, East Germany, Czechoslovakia, Hungary, Romania, and Bulgaria. The Soviet empire extended from the Danube and the Elbe to the Pacific, where part of Sakhalin island and the Kuril islands were taken from Japan.

The Soviet Empire was never stable. There were uprisings in East Germany (1953), Hungary (1956), Czechoslovakia (1968), and continual discontent in Poland, where many had fought to win World War II, only to find that Nazi control was replaced by Stalinist rule. Russians referred to eastern Europe as the Near Abroad and when control slipped in the nearby empire the collapse of the Union of Soviet Socialist Republics followed. In 1989, Hungary, and then East Germany, broke away, leading to German unification. In 1990, Latvia and Lithuania declared they had been illegally incorporated in the USSR and announced the intention to leave. The following year all the republics left.

The collapse of the Soviet Union does not mean the end of Russia as a major power. With a population of over 140 million, it is a large state in the European context. The oil, gas, coal, and metal ore resources still exist. In economic terms the loss of republics, many now in the Commonwealth of Independent states, may not be great. Russia retains an extensive resource base and a number of the former Soviet Socialist republics were costly to develop.

Physical Geography

Russia can be divided into four major physiographic regions (Figure 14.5):

1. The European Lowland
2. The Ural Mountains

3. The Siberian lowland and plateau:
   a) The Western Siberian Lowland
   b) The Central Siberian Plateau
4. The Southern and Eastern Mountains

The European Lowland

Lying west of the Urals is a broad continuation of the European plain that runs from the Atlantic to the Urals. The southern part of the region is drained principally by the Volga, running to the Caspian Sea, and the Don and Dnieper, which flow to the Black Sea. The principal rivers draining the northern European lowland of Russia are the Northern Dvina flowing to the White Sea and several shorter streams draining to the Baltic, including the Western Dvina, which passes through Belarus and Latvia to reach the Baltic at the Gulf of Riga.

The northern part of Russia's European lowland was heavily glaciated. There are deep glacial deposits, much ill-drained marshy ground, and many depressions filled with lakes. The southern lowland was south of the great glacier and was covered with wind-blown silt (loess), which, as we have seen in Moldova and the Ukraine, makes fertile Black Earths. The natural vegetation of the northern lowland is forest and marsh. The southern lowland is steppe.

In the northwest is an area of different character close to the borders of Finland and Norway. Here the glaciers scraped off surface deposits, exposing the ancient rocks of the Eurasian shield. On most of the Russian European lowland, high points rarely reach a thousand feet and most land is a few hundred feet above sea level. In the exposed rock areas of the northwest, the old rocks of the Eurasian shield rise to 1,896 feet in Karelia and to nearly 4,000 feet on the Kola peninsula.

The Ural Mountains

The Urals run north to south from the Arctic to the deserts of Kazakhstan. The mountains are highest in the north, rising to 6,214 feet at Mt. Narodnaya. In the south, peaks exceed 5,000 feet. In the lower central section are several passes, one of which takes the Trans-Siberian railroad through the Ural Mountains.

In the cold north, with a short growing season, there is a tundra environment, lacking trees. The central and southern sections of the mountain range are forested. The Urals are well endowed with ferrous and nonferrous mineral ores, including bauxite, copper, lead, nickel, silver, tungsten, and zinc.

The Siberian Lowland and Plateau

Because of the massive mountain ranges to the south, Siberia is drained by rivers flowing north to the Arctic Ocean. From west to east the major rivers are the Ob, the Yenisey, and the Lena, all rising in the mountains to the south. We can subdivide Siberia into the Western Siberian lowlands and the Central Siberian Plateau.

Western Siberian Lowland   The Urals drop abruptly to the Western Siberian lowland, which from west to east extends over a thousand miles to the Yenisey river and the Central Siberian plateau. North to south the Western Siberian lowland extends 1,500 miles from Arctic seas to the foothills of the Altaic mountains. The lowland is drained principally by the river Ob and its tributaries. Most of the plain is low lying, and no more than a few hundred feet above sea level.

The northern part of the lowland is tundra. The growing season is hardly three months, too short for tree growth. Vegetation is grasses, sedges, mosses, lichens, and small, stunted shrubs. For eight months of the year the surface is frozen, but thaws in the short summer. The subsoil is permanently frozen (permafrost) and no water percolates downward, creating extensive marshes and ill-drained areas. Movement is difficult because the land surface is covered in mud and water. When hard fall frosts freeze surfaces again, travel is easier.

South of the tundra are coniferous forests (taiga). Here the growing season, and the frost-free period, is greater but the rivers draining to the longer frozen north cannot discharge for much of the year into the Arctic seas. The forests are a valuable source of timber, but exploitation is hindered by distance and transportation difficulties.

In the south the Western Siberian lowland is higher, better drained, and has a more moderate climate. Deciduous trees appear and southward woodland gives way to steppe. Many temperate crops are grown and processed in towns along the routeways. Along the Trans-Siberian railroad are major cities including Omsk, founded as a fort in 1716, on the route to the east. The town developed into an agricultural service center with industries processing timber. Today, the city has 1.1 million inhabitants, many working in manufacturing industries, including oil refining.

Further east, in the Kuznetsk basin with a huge coalfield, is the industrial center of Novosibirsk, with 1.4 million inhabitants. Started as a stopping place on the Trans-Siberian railroad where it crosses the river Ob, the city grew rapidly in the 1930s when it was decided to build a major industrial region in the heart of Siberia to exploit the coal resources of the Kuznetsk basin. Much coal was trained westward for the Ural metal smelters. On the return journey, trains hauled Ural iron ore to steel works in the Kuznetsk basin.

The Central Siberian Plateau   The Yenisey river roughly demarcates the eastern boundary of the Western Siberian lowland. Beyond the east bank of the river the land rises gently toward the Central plateau. Much of the plateau is over 1,500 feet and in parts exceeds 2,000 feet, the highest point being Polkan, 3,543 feet above sea level. The region is marked by extensive, relatively flat surfaces into which rivers have eroded deep, steep-sided valleys. The rivers on the western side of the plateau drain to the Yenisey. Rivers draining to the east are tributaries of the Lena.

Being higher than the Western Siberian lowland, tundra is more extensive on the plateau. Population is sparser, towns are fewer, and the natural resources are less intensively exploited than in the lowland. The major west-to-east Trans-Siberian routes pass to the south of the region and the plateau is isolated. Yakutsk, founded in 1632, on the Lena river, has a population of less than 200,000 and is a manufacturing, commercial, and administrative center. The Lena is ice free from June to October. During this period timber and minerals can be shipped from Yakutsk and other ports along the river.

The Southern and Eastern Mountains

To the south and east, Siberia is enclosed by mountain ranges, interspersed by high plateau. Starting in the south and moving to the east the ranges include the Altai (or Altay) Mountains, which rise, at Mt. Belukha, to 14,783 feet. The Sayan Khrebet attains 11,457 feet at Munku Sardyk, which lies to the west of Lake Baykal. East of the lake and running in a northeasterly direction is the Yablonovyy Khrebet, rising to over 9,000 feet. The eastward-trending Stanovoy Khrebet runs toward the Pacific shore to link with the Dzhugdzhur range, which trends north along the Pacific coast.

Imbedded in the Southern Mountains is Lake Baykal (Figure 14.7 and Photo 14.6). Faulting and mountain building produced a mile-deep depression, which is filled with the freshwater of ancient Lake Baykal. The lake holds more water than the Great Lakes and contains a rich fauna and flora, including a freshwater seal. Timber and chemical industries, around the lake, have caused environmental degradation and species loss. (Photo 14.7)

Lake Baykal is drained by the Angara river, which is dammed to provide hydroelectric power but eventually falls into the Yenisey drainage system. There are routeways north and south of the lake. The Trans-Siberian railroad runs through Irkutsk (f. 1654), and then along the south shore of Lake Baykal to a pass through the Yablonovyy Khrebet, leading to the Amur valley and eventually, the Pacific port of Vladivostok.

The Irkursk–Bratsk region west of Lake Baykal is a timber-, chemical-, and metal-producing region based on hydroelectric power from the Angara river. Bratsk, founded in 1632, is on a rail line that passes north of Lake Baykal, before rejoining the Trans-Siberian railroad and making a terminus at Vladivostok.

Vladivostok started as a military base in 1860 and grew rapidly after the Trans-Siberian railroad reached the port. Although relatively far south, the sea freezes in winter and ice breakers keep the port open. The mean monthly temperature for January is well below freezing. In July the monthly mean is around 68°F and cool days are common. Strategically, Vladivostok was built up as a large naval base with supporting army and airforce facilities. Although an important industrial center, westerners were kept out of home port of the Soviet Pacific fleet until the collapse of the USSR.

Photo 14.6   The Trans-Siberian railroad passes near the south shore of Lake Baykal on the route between the Urals and the Pacific Ocean.

Source: Age fotostock/SUPERSTOCK.

Photo 14.7 Lake Baykal is a large, deep freshwater lake containing many species of aquatic animals. Fishing is important to smaller communities around the lake. Elsewhere the chemical industry has caused pollution and species depletion.

Source: Brand X/SUPERSTOCK.

Climate

Russia, occupying a large landmass, has a continental climate. Much of Russia is at latitudes comparable to northern Norway and Sweden. Winters are long and cold, with short days and little heat from the sun. Summers are warm, except in the north, and are marked by thundershowers (see Figure 1.4).

Russia's continentality is reinforced by distance from the Atlantic Ocean and the depressions that track off that sea over western and central Europe. To the south the Himalayas, and other mountain ranges, exclude the influence of the warm Indian Ocean. The Eastern mountains shut out the influence of the Pacific and in winter, the large high-pressure system developed over Eurasia has outward winds that export cold and resist penetrations of warmer air.

With the onset of fall the Eurasian landmass cools rapidly, as a large pool of cold air, in a high-pressure system, covers the continent. East of the Urals many places are below freezing for months and in northern tundra environments the land is frozen for eight or nine months of the year. Few parts of Russia, except in the south, near the Black Sea, have a mean monthly temperature for January above 32°F. For many regions the mean for the month is much lower. In northeast Siberia the figure is −40°F.

The transition from winter to summer is rapid, as the snow melts and the sun warms the surface of the earth. Warmer air is less dense and low pressure develops over Eurasia. Summer is the rainy season, with precipitation commonly coming in thunderstorms. In the fall, as the angle of the sun declines, and the income of solar radiation decreases, the land cools rapidly. The onset of winter comes quickly. In most of Russia, spring and fall are short seasons.

Some of the climate tendencies can be illustrated by examining temperatures from stations across Russia. At Petersburg the January mean is 17°F. At Moscow, further inland, the figure is 15°F. In July Petersburg has a mean monthly temperature of 64°F.

Moscow, subject to more continental influence, is slightly warmer in summer. Going north to Arkhangelsk (Archangel), on the White Sea, the January mean is 8°F and the July mean 59°F (see Figure 1.6).

Deep in the Eurasian landmass the coldest temperatures on earth are recorded. Most of Siberia has a January mean below 0°F and at Verkhoyansk, in the northeast, the January mean is −58°F. The Siberian winter is not fearsome all the time. The air is cold but usually calm, dry, clear, and, in the south, sunny. When the *buran* blows, wind chill becomes a threat to life and although snowfall is not heavy, the wind produces ground blizzards, closing routeways and disorientating those caught in the open.

Much of western Russia receives more than 20 inches of precipitation per annum as Atlantic depressions do penetrate into the region. Moving eastward, and north and south, precipitation totals drop. Much of southern Russia, north of the Black Sea, receives less than 20 inches, which is insufficient for farming without irrigation in many areas. On the Western Siberian lowland precipitation is below 20 inches. This is not a constraint on farming, except in the south, as much of the region is swamp or has a growing season too short for arable crops.

The Central Siberian plateau receives slightly more precipitation than the Western Siberian lowland due to altitude, but the region is generally too cold for farming, except in favored locations. Northeastern Siberia is a cold desert, receiving less than 10 inches of precipitation a year.

Vegetation Zones

If we travel through Russia from the Caspian in the south to the Arctic Ocean in the north, we would pass through the following ecological zones: semi-desert, grassland (steppe), forest steppe with broadleaf trees, mixed forest, coniferous forest (taiga), and tundra (see Figure 1.7). The vegetation in the zones has been modified by human action, deserts have been irrigated, grassland ploughed, and forests cut but the zones are recognizable. North of the desert is the steppe, with enough rainfall to support grassland, but not trees, except along the rivers. As we have seen in Moldova and the Ukraine the steppe is underlain by loess, the parent material of the Black Earths (chernozem), the rich soils on which grains yield well.

Further north the steppe becomes partially wooded before giving way to broadleaf forest and then mixed forest as a cooler climate brings more conifers into the woodlands. The broadleaf and mixed forests taper toward the east, due to decreasing precipitation, and are not as well developed east of the Urals.

The coniferous forest (taiga) stretches from Scandinavia through European Russia and the Urals to Siberia. The taiga grows on acidic podsol soils, with acidity increased by decaying pine needles. The coniferous forest is worked to provide timber, pulp, and wood products but, overall, the taiga belt is intact.

Coniferous trees require a minimum growing season of three to four months. Northern Eurasia has a growing season shorter than this and trees do not grow. As we move northward through the taiga, the trees thin to forest tundra and then the treeless tundra. Tundra environments are frozen and snow covered for much of the year. Summers, and growing seasons, are short, although days and sunshine hours are long. Vegetation consists of mosses, lichens, sedges, grasses, and some shrubs. In summer only the surface thaws. The subsoil, in a state of permafrost, stays frozen. Water collects in ponds in which mosquitoes rapidly breed to form clouds of insects seeking blood. The environment has enough vegetation to support reindeer herds and, in general, the fauna is more varied than might be expected, including geese, ptarmigan, moose, bears, wolves, foxes, the Arctic hare, ermine, and numerous species of fish. Hunters and gatherers still operate in the tundra and make a living herding reindeer, collecting pelts, fishing, and, in the short summer, gathering wild berries and mushrooms. In the Soviet era, the herders and hunters were employed by the state. Today, they subsist and make a small income selling furs (Ziker, 2002).

Industrial Development

Russia has an extensive endowment of mineral resources, possessing coal, oil, gas, iron ore, and ores of most of the nonferrous metals, including lead, copper, zinc, bauxite, nickel, and chrome. In addition, gold and diamonds are mined.

The main centers of population, in European Russia, are on lands covered with glacial till. There is a coalfield close to Moscow but when the industrial age arrived in the nineteenth century, the core region of the Russian state was not proximate to major deposits of raw materials. There were valuable mineral deposits elsewhere in the empire. Iron had been mined and smelted in the Urals since the mid-seventeenth century. Early in the eighteenth century copper was discovered in the Altai Mountains and the huge Qaraghandy (Karaganda) coalfield was found in the 1830s in Kazakhstan (then part of the empire). The field was not developed until the 1930s.

Despite the rich raw material base, Russia was a late industrializer. The reasons for late industrialization are debated. Obviously long distances, poor transportation facilities, and high transportation costs made many schemes unprofitable. The state attempted to encourage industrial enterprises but private investment capital was scarce. Large estate owners were wealthy but wealth was tied to land. As land was the basis of an aristocratic lifestyle, and social standing, estate owners were reluctant to sell and preferred the life of a country lord to that of risk-taking entrepreneurs. Banks were relatively few and it was difficult to raise capital using land as security.

Russia was technologically backward but by the 1830s and 1840s all the basic technologies could be bought from companies in Britain, Belgium, France, Germany, and, to a lesser degree, Switzerland. If investors wanted a textile factory, a railroad, or an iron and steel works there were western European companies that could supply and start up the technology.

There was technology transfer to Russia. Around 1840 a young German, Ludwig Knoop, born in Bremen, who had worked in Lancashire went to Russia to represent an English manufacturer of textile machinery. In 1852 Knoop established his own company in St. Petersburg and Moscow. In succeeding decades the company, representing British manufacturers of machinery, equipped over 150 factories with modern textile equipment. Knoop was able to obtain credit from the manufacturers of machines and frequently the Knoop Company took a financial stake in new factories.

The textile industry was a foundation of European industrialization, but in the Russian case there was no takeoff, after 1850, into more broadly based economic activity. Demand for goods was limited by the low incomes of most of the population. Lack of technical education meant that there were few engineers with the skills to develop industrial processes and new technologies in a Russian context.

Examination of the output of coal and steel shows that growth in production remained slow after 1850 (Figure 14.6). The mining of minerals and coal was still done on a small scale. The Urals continued to smelt iron using the old charcoal technology.

Demand for iron and steel rose in Russia as railroads were built and modern armaments bought. To satisfy demand, imports grew and now traditionalists understood that imports were a drain on national wealth, and that manufacturing should be promoted.

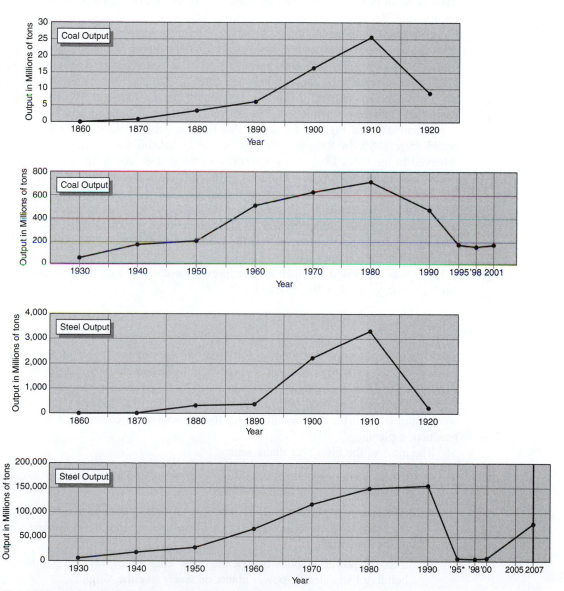

*The 1995 figure reflects the loss of steel making capacity in Ukraine, etc.

Figure 14.6 Production of Steel and Coal.

Tariffs were raised on imports. Foreigners were given inducements to produce in Russia. The Welsh iron master John Hughes, for example, established the New Russia company to smelt iron ore in the Donets' Valley starting in 1872. Hughes brought miners and technicians from Wales and built a company town, Hughesovka. The town became Stalino in 1924 and Donets'k in 1961 and remains a major center of iron, steel, engineering, and heavy engineering industries (Figure 14.2). Other foreigners came to use the ore and coal resources of the Donets' region. The Swedish Nobel brothers developed Caspian oil wells at Baku in the 1870s. The Transcaucasus railroad, linking Baku to the Black Sea, opened in 1882.

Policies of state invention, to encourage foreign investment, and direct state investment in enterprises were further developed by Count Sergei Witte, Minister of Finance from 1892 to 1903. By the 1890s railroad track was being laid in Russia at the rate of 1,500 miles a year. Most lines ran between cities in European Russia but Witte financed the Trans-Siberian railroad to link St. Petersburg on the Baltic to Vladivostok on the Pacific and opened economic space in Siberia to settlement and industrial development.

Examination of Figure 14.6 shows that Russian production of coal and steel accelerated after 1880. In Rostow's view (Chapter 5), takeoff into economic growth was achieved in the years 1890–1914. Certainly in the first decade of the twentieth century Russia was industrializing at a rapid pace. Takeoff was followed by collapse.

World War I stimulated industrial production, but military defeat, revolution, and the fall of Tzarist Russia brought economic collapse. In 1913 Russia produced over four million tons of steel. During the war, production declined slowly until 1917–18, when it fell rapidly. Immediately after the war the country produced only 200,000 tons of steel. The 1913 production level was not attained again until 1929. The experience of the steel industry was replicated in other sectors: prewar production figures were not regained until a dozen years after World War I.

Five Year Plans

The Communist regime that took over Russia and created the Union of Soviet Socialist Republics wanted to implement Marxist–Leninist economic planning with the state controlling and funding economic development. As we have seen, state economic intervention had been practiced before World War I in Russia. Stalin's Five Year Plans were ruthless, with the state commanding the economy and centralizing all aspects of production planning.

The aims of the Five Year Plans were:

1. Create a heavy industrial base; iron, coal, steel, and the smelting of nonferrous metals.

2. Improve infrastructure with the building of more railroads, roads, bridges, and telephone systems to link all regions of the country into the command structure of the economy.

3. Improve power supplies by opening new coal mines, oil and gas fields, and building hydroelectric power plants on rivers like the Volga.

4. Consolidate agricultural land into collectives and state farms. Those who owned land in estates or small farms had property confiscated by the state. The aim

was to increase food production, make agriculture more efficient, and release labor from the land to work in the new heavy industry projects.

Stalin wanted to utilize the raw material base to make the Soviet Union self-sufficient is all aspects of economic life. In addition, Stalin had a strategically driven spatial view. In the Five Year Plans, Stalin was preparing for a war he knew would come. Stalin had read Hitler's *Mein Kampf,* which talks of Germany needing living space in the east. The *Lebensraum* theme—we want the grain fields of the Ukraine, the ores of the Urals, and the forests of Siberia—was publicly declared by Hitler at the 1936 Nuremberg rally.

Stalin wanted to develop manufacturing capacity on the Volga, in the Urals, and Siberia, areas that would be difficult for an enemy to attack (Figure 14.7). Of course, the existing manufacturing zones in European Russia claimed a full share of the state resources, but Stalin did get new manufacturing regions (Stone, 2005).

The Volga river was harnessed to provide hydroelectricity. Large dams were built and reservoirs filled with water, drowning towns, villages, and farmland. Locks were constructed around dams to maintain Volga navigation in the ice-free period from April through November. Volga valley oil and gas fields were exploited by petrochemical industries. In response to development, new industries including shipbuilding, heavy engineering, tractor manufacturing, and food processing were established. Irrigation works and navigation canals were started, including a waterway to link navigation on the Volga with the River Don and the Black Sea.

Towns were enlarged in size or were built on green field sites to house workers at new plants. Stalingrad (now Volgograd) was expanded on the nucleus of the old frontier town Tsaritsyn, established in 1589. Numerous other urban places were enlarged or founded close to the Volga river. Western visitors, being shown what the regime wanted to show, and having little idea of the terror apparatus of the state, were impressed with Soviet economic development, particularly observers from countries that suffered in the Great Depression, with the closure of shipyards, steel works, and engineering plants. It seemed to many commentators that Soviet Socialist planning could overcome the economic cycle of growth and recession to provide full employment and expanding economic activity.

In the Urals, the resources of iron ore and nonferrous metals were exploited in plants built in new industrial centers. Magnitogorsk, on the Ural river, was started in 1929 close to an iron ore deposit. The city became a major iron and steel center with Siberian coal brought in by rail. Metal working and chemical industries were established. Workers were housed in new accommodation blocks, which quickly became unsightly as a result of high levels of pollution generated by the metal smelters. (Photo 14.8)

The development of the Kuznetsk basin, in Siberia, was on the basis of a large coalfield. The major city in the industrial region is Novosibirsk on the Trans-Siberian railroad. Today, the region is diversified and not solely dependent upon heavy industry. Close to Novosibirsk is Akademgorodok, a research city founded in 1959, containing many scientific institutions and during the Cold War, secure armament-related projects.

An independently calculated index of Soviet industrial production showed that output increased from 37 in 1928 to 111 in 1939. Soviet propaganda projected an image of a peace-loving worker's paradise, but armament production contributed to the greatly increased industrial output. Several "tractor" factories produced tanks of advanced design that were effective in World War II. The aircraft industry produced fighters and bombers.

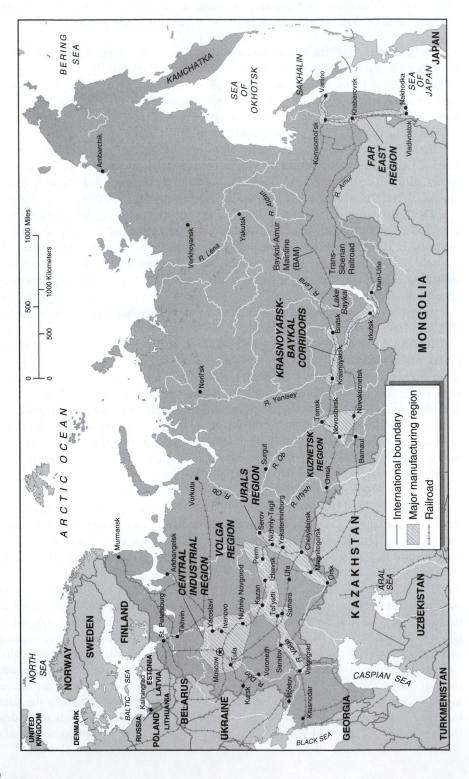

Figure 14.7 Russia: Industrial Regions.

Source: H. J. de Blij, A. B. Murphy, E. H. Fouberg, and John Wiley & Sons, Inc.

Photo 14.8 In the Soviet era, many industrial towns were developed in the Urals together with hydro-electricity plants to provide energy. Industries including metal smelting and chemicals led to environmental degradation.

Source: ITAR-TASS/Anatoly Semekhin/NewsCom.

Collectivization of Agriculture

The collectivization and modernization of agriculture was more difficult to achieve than the creation of heavy industries and the supporting infrastructure of roads, power plants, electricity transmission systems, and improved railroads. In farming the land was already worked by peasants and small farmers who lacked education and did not see the benefits of an agricultural modernization program designed to produce cheap food for urban dwellers, while reducing the number of farmers.

In the Ukraine the *Kulaks* resisted the state takeover. Dissidents were shot or deported. Farm-to-farm searches seized food stocks. In the early years the attempt to collectivize farming created food shortages. The old patterns of cultivation were uprooted. It was not easy to take peasant farmers and, with the aid of tractors and other new agricultural equipment, turn them into modern, industrialized farm workers. Some production increases were achieved. Between 1930 and 1939 wheat output doubled. Advances were not so great for other crops. The production of pigs was increased on the state units. Cattle numbers went down, partly because the owners of livestock killed and consumed animals rather than turn them over to a collective.

World War II

Stalin failed to read Hitler's intentions in the spring of 1941 (Murphy, 2005). Warnings came from many sources, to be dismissed as misintelligence designed to undermine the Non-Aggression pact with Germany. Soviet forces were ordered to give no provocation, even though air space was violated by reconnaissance aircraft and forward Red Army units knew of the presence of German divisions. When on midsummer night June 21–22, 1941 the German attack was launched, there was little resistance at first. Front-line Soviet positions were overrun; aircraft were bombed on the ground. Then armies were enveloped and annihilated. By November 1941 German forces were besieging Leningrad, threatening Moscow, and occupying the Ukraine.

Soviet industrial production declined by at least a third. The grain, coal, iron, steel, and manganese output of the Ukraine was lost. However, the Volga, the Urals, and Siberian regions increased production and industrial capacity was moved from European Russia to the east. As the *Wehrmacht* advanced, factories were dismantled and put on trains with the workers, to be reassembled and restarted a few months later, in an eastern industrial region. Armaments output did not decline at the same rate as overall industrial production.

In 1942, while maintaining the siege of Leningrad and keeping pressure on the Moscow front, the *Wehrmacht* advanced from the Ukraine into the Volga region and sent forces south to capture the oil fields on the north side of the Caucasus. Hitler insisted that Stalingrad be taken, although German tank divisions were designed for mobile war, not street fighting. The Red Army was able to bring fresh troops into the city from the east bank of the Volga and when German forces were cut off they were forced to surrender early in 1943. The siege of Stalingrad (today Volgograd) was the turning point on the eastern front. Soviet armies pushed into eastern Europe, taking control of the region and occupying East Germany.

Victory was costly for the Soviet Union (Harrison, 1996). In 1940 the Soviet industrial production index was 116. In 1945, with the war over and some capacity restored, the index was at 77. The number of Soviet dead was approximately 20 million people. In European Russia, west of Moscow, towns and cities were ruins. Industrial centers like Leningrad, Minsk (now in Belarus), Kiev (now in Ukraine), Rostov (Rostov-na-Donu), and Stalingrad had to be rebuilt. Not all the damage was done by German armies. The Red Army in retreat, in the early part of the war, destroyed everything it could to deny the *Wehrmacht* resources and shelter. Later in the war, as the Soviets pushed west, many Soviet citizens, in what had been occupied territory, were shot for supposed cooperation with Germans.

Postwar

Soviet production was at prewar levels by 1948, and in 1960 industrial output was four times the 1939 level. The Five Year Development plans were continued, with much the same emphasis on heavy industry, infrastructure, agriculture, and armaments. The Red Army remained large and well equipped. The Soviets developed thermonuclear weapons, a navy including nuclear submarines and, for a time, led in the development of missiles. The Soviets put the first satellite (*Sputnik*) into orbit in 1957. Yuri Gagarin, in *Vostok,* was the first astronaut on April 12, 1961, and in 1966 the USSR landed the first probe on the moon.

The Five Year Plans were a success in increasing industrial output and developing distant regions in the Urals and Siberia, but they failed in major ways. First, agricultural production, in spite of state farms, mechanization, hybrid crops and the ploughing up of new lands, could not keep pace with demand. In the 1970s the Soviet Union admitted failure and started to buy grain on world markets, which helped farmers in North America receive a better price.

Second, living standards for workers never kept pace with the increases in industrial output. Workers were provided with housing, but it was often in unattractive, prefabricated, ferro-concrete blocks in towns that symbolized socialist urban landscapes. Food was scarce and consumer goods in short supply. People spent long hours in line waiting to buy a cabbage or other basic food items. There was no starvation, just a continual struggle to acquire foodstuffs. Eventually many people, aware of living standards in western Europe, began to ask when the state would deliver the rewards that had been promised at the end of decades of sacrifice. The plans were poor at responding to consumer demand and failed to produce consumer goods in sufficient quantity.

The Five Year Plans were primarily concerned with meeting production targets. Little thought was given to environmental issues. Plant managers, struggling to deliver a quota, were given no incentive to protect environments within and around production facilities. Ordinary citizens raising pollution issues were ignored or silenced.

Political power was concentrated in the Kremlin. There was no room for initiative or debate in a search for solutions to local and regional problems. George Kennan, writing in 1947, foresaw the downfall of a system in which there was no local government and power was tightly held at the center. Kennan declared, early in the Cold War, that the Soviet system carried within it "the seeds of its own decay" and the sprouting of the seeds was well advanced.

When Mikhail Gorbachev, speaking of *glasnost* (openness) and *perestroika* (restructuring), attempted to decentralize government and introduce elements of a market economy, the Soviet Union lost coherence. "The traditional Soviet system depended on a chain of command and Gorbachev destroyed that chain of command" in "an unintended economic side-effect of policies pursued in the cause of political reform" (Hanson, 2003, p. 253). Constituent Soviet Socialist Republics declared independence. The USSR was dissolved in December 1991 and Russia became a founding member of a Commonwealth of Independent States, along with Belarus and Ukraine. Nine other former Soviet Socialist republics became members. The Baltics did not and are now part of EU.

The Post-Soviet Economy

The centrally planned Soviet economy was not designed to evolve into a market economy. When the USSR dissolved, the old Soviet economy of Russia failed. Quickly production units were working below capacity and the payment of wages fell months in arrears. People found themselves in several activities to make a subsistence. The end of the socialist state meant a drop in living standards and hardship for many who sold their few personal possessions in street markets. In the mid-1990s industrial production was half what it had been in 1990. The state sold off (privatized) industries, but many of the sales were suspect. Party insiders were able to buy assets at a fraction of their real value and became billionaires, while many workers lost jobs at the sold-off plants.

Output has recovered in recent years and GDP growth averaged 6 percent between 1999 and 2004. In particular, oil and gas exports have helped the economy recover and are the largest exports, by value, from Russia. The traditional products—metals and metal ores, timber, pulp, and wood products—are significant exports. Steel production has recovered as export markets have been found for the relatively cheap Russian product. The home production of cars, refrigerators, and televisions is poor, partially because of low wages for the majority. The affluent minority indulges every taste and invests overseas, including purchasing leading soccer clubs, such as Chelsea in the English Premier League. In some respects Russia has reverted to the maldistribution of wealth that characterized the rule of the Tzars.

Population Density and Distribution

The greatest part of the population lives in European Russia. Population densities are below those found in western Europe and reduce rapidly in the northern Arctic region. European Russia contains the major cities of the country: Moscow, with a population of over 10.4 million inhabitants, and St. Petersburg, with over 4.7 million people.

The Ural Mountains may not be a barrier, but they are a divide. East of the Urals, beyond the string of industrial towns established to exploit the mineral resources of the range, the density of population decreases markedly, except along the line of the Trans-Siberian railroad. The railroad follows a southern route across Siberia, passing through lands where arable farming is possible and linking the early towns of conquest and colonization. New towns, to exploit resources, were added along the routeway in the Soviet era. Now, as the grand design of state-planned and -supported production has ceased, many communities have lost inhabitants, as people have left for European Russia and birth rates have dropped.

All across the northern periphery of Russia from Magadan on the Pacific coast through the tundra region of Siberia and the northern Urals to the lands draining to the White Sea, the sparse population is being reduced. Subsidies have ceased and people move to work elsewhere. Only in the northern gas fields and mining towns is employment attractive.

Population Problems

When the Soviet Union broke up, the population of Russia stood at just below 150 million. Today, the population of Russia is 143 million. In the 1980s the population of Russia increased slowly, but, with the Soviet collapse, the birth rate declined and the death rate increased. People lost jobs and benefits, wages were cut, and payment delayed. Harsh times made children unwelcome and the rate of abortions increased. In 2004 Russia recorded 1.5 million births and 1.6 million abortions.

Since the Soviet breakup, life expectancy for females has dropped from 74.3 to 72 years. For males the fall has been from 63.5 to 59 years! The male–female life expectancy gap is now 13 years! The large fall in male life expectancy is a result of alcoholism, cardiovascular disease, industrial accidents, and military conflict. Two hundred thousand people die of unnatural causes every year. Suicides are common amongst those demoralized by the trauma of economic collapse.

When the Soviet Union broke up there was return migration of Russians moving from former Soviet Socialist republics back to Russia. The trend has slowed, although significant populations of Russians remain in the Baltics and the Ukraine. Official policy remains: encourage Russians to return to Russia.

In 1987 the Russian total fertility rate was 2.2. By 1999 the figure had nearly halved to 1.2. As the Russian economy stabilized, fertility recovered a little. Today, the average woman has 1.3 children. President Putin describes the lack of children as a national problem and declared that Russia needs to "create conditions that will encourage people to give birth and raise children." But the government has been slow to improve child allowances and other benefits that might enable people with limited incomes to raise more children.

Rural depopulation is a major issue in Russia. Rural communities are losing population to outmigration and remaining inhabitants are in older age groups. Numerous smaller places have shrinking, aging populations.

In aggregate, Russian agriculture is in decline. Farms in isolated areas, distant from urban markets, and farms in marginal environments, are unprofitable now that Soviet subsidies have ceased. Rural depopulation and village abandonment is marked (Ioffe and Nefedova, 2004). However, there are areas where farmers have expanded production to serve urban markets. Von Thünen's ideas regarding agricultural land use around markets are asserting themselves.

Urban Places

Nearly 75 percent of Russians live in urban places. The country does not have a dense network of towns and cities and in the northern lands there is a lack of service centers.

During the Soviet era many new towns were started or small, older places were expanded according to a grand plan to develop the resources of the state. People were given inducements, or forced to live in new places in isolated regions. Movement to major cities like Moscow was discouraged.

Within towns and cities accommodation for workers was carefully planned. Clusters of apartment blocks were provided with basic services, including food stores, to which inhabitants could walk. In location terms, the clusters of blocks were usually away from the city center. Such apartment block clusters were used by planners in many European cities; you can see them at Stockholm, Sheffield, and Paris.

When the Soviet Union collapsed, urban areas began to restructure. People with the means wanted to live closer to the amenities of the city center and left the outlying apartment complexes, many of which have decayed. Others wanted single-family homes and viable Russian cities are developing suburbs. Retailing, which used to be conducted in government stores, convenient to housing complexes, is locating more in city centers and on lines of communication. Outlying malls are appearing. (Photos 14.9 and 14.10)

The urban network of Russia is changing. Smaller towns and cities are losing population and many have ceased to function as urban places. Bigger places, with viable economies, attract economic activity and migrants. Such places are reshaping their urban structure with the growth of suburbs and conveniently situated retail outlets.

Photo 14.9 Central Moscow and the area around the Kremlin is being rapidly redeveloped with high value real estate projects and prestige buildings for businesses.

Source: George Simhoni/Masterfile.

Moscow, St. Petersburg, and Novosibirsk

The three cities were established in different phases of Russian history. Moscow is at the center of the medieval origins of Muscovy and Russia. St. Petersburg represents the spatial expansion of the state in the seventeenth and eighteenth centuries. Novosibirsk, founded as a stop on the Trans-Siberian railroad, grew rapidly in the Soviet era, as the center of a major industrial region in Siberia.

The Kremlin

Moscow contains the classic morphological elements of a European city, a powerful fortress, the Kremlin, beside a river, with the church enclosed in the center of power. Around the triangular Kremlin compound is an area kept free of buildings that might give attackers cover. Red Square and the Alexander Gardens form the space on the landward fronts. The Moscow river, flowing to the Volga, protects the south flank. Arterial roads ran from the Kremlin into surrounding regions. Later, as the city spread, radial roads linked the outward arteries.

Within the heavily fortified Kremlin were the major institutions of social control: palaces for the rulers, barracks, an arsenal, the Court of Justice, and Cathedrals of the Russian Orthodox Church. With imperial power concentrated in Moscow, as the empire grew to the Baltic, the Black Sea, the Caspian, and the Pacific, it is not surprising that commerce, manufacturing, and bureaucrats congregated there too.

St. Petersburg—The Capital, 1712–1918

Peter the Great understood that Moscow was relatively isolated and he tried to connect Russia more effectively with the West. St. Petersburg was started in 1703 on a site where

Photo 14.10 Moscow's central shopping streets have been refurbished and contain many fashionable shops.

Source: Grigory Dukor/Reuters/Landov LLC.

the river Neva, flowing from Lake Ladoga, enters the Gulf of Finland. St. Petersburg became the capital in 1712. If Moscow is the medieval fortress city, St. Petersburg, in the fabric of the place, proclaims it is a city built by an absolutist ruler of the eighteenth century. The squares, parade grounds, palaces, boulevards, and administrative buildings radiate elegance but project power. Even the gardens and canals are geometrical and ordered. The citizen was to admire, and be awed by, a state that could plan, build, and maintain such well-controlled spaces. Peter could not control the winter and although St. Petersburg winters are less cold than at Moscow, the port and naval base freeze for several months in the winter and shipping ceases. Nevertheless, St. Petersburg gave Russia a Baltic port. Previously the main port had been Archangel on the White Sea. (Photo 14.11) (Photo 14.12)

St. Petersburg, although the capital from 1712 to 1918, never displaced Moscow as the largest place and center of the Orthodox Church. The cities were linked by rail in 1851. Moscow remained the imperial center, on the routeways to the Volga, the Caspian, the Urals, Siberia, and the Pacific. The Trans-Siberian railroad, started in 1891, reinforced Moscow's position at the center of the communications system. St. Petersburg was at one

Photo 14.11 Alexander Column and General Staff building on Dvortsovaya Place, St. Petersburg. St. Petersburg has many squares well suited to project the power of Czars and Soviet leaders.

Source: Age fotostock/SUPERSTOCK.

Photo 14.12 Canal, St. Petersburg. The city beside the Gulf of Finland and on the river Neva has numerous artificial waterways.

Source: ITAR-TASS/Boris Kavashkin/NewsCom.

end of the line. St. Petersburg, with museums, the Hermitage Art Gallery, a fine urban fabric, grand parks, prestigious universities, and the largest port in Russia, has attracted investment and visitors. Being close to Finland, the Baltics, and the economic space of the European Union are advantages. In 2003 St. Petersburg celebrated the tercentenary of foundation. The city was cleaned, special events were organized, and large numbers of visitors came to the celebrations. The city will prosper but not threaten the preeminence of Moscow.

Moscow: Transformed Communist Capital

The Bolshevik Revolution damaged Russia economically and demographically but after the Communist takeover, Moscow, less exposed to the west, regained capital status. By 1925 population was growing rapidly and with power concentrated in the city, growth outran the ability of the regime to provide accommodation. Houses were made of wood, families crowded into one room, as the growth of manufacturing, bureaucracy, and military organizations drew migrants to Moscow. Tower blocks of apartments were erected by the state, but, in the Communist era, the authorities could not meet housing demand and attempted to restrict in-migration. The regime did provide public transport and the Moscow underground rail system, with palace-like stations, was admired by visitors. Living standards for most inhabitants were basic.

With the end of the Soviet Union Moscow has been transformed. At first, as state-subsidized activities failed, there was hardship and desperate searches for any paid employment. Then as outside investment arrived, to exploit the huge potential market, there was an economic revival and a reordering of urban structure in response to market forces:

> ... capitalism exploded through Moscow, which has attracted more foreign investment than any other Russian city. New foreign and Russian capital built business centers, set up real estate companies, and established a new retail sector. Once empty avenues filled with cars seemingly overnight. TVs, VCRs, designer clothes, furniture—Muscovites bought voraciously and then bought some more. In the early 1990s, kiosks appeared everywhere, stocked with an improbable mix of everything from candy bars to vodka, socks, and toys. Five years later, stores selling liquor, food, clothing and toys, as well as every other possible consumer good, largely replaced the kiosks. Now, malls are beginning to replace hastily built and remodeled stores. (Mitchneck and Hamilton, 2003, pp. 242–243)

The central streets of Moscow are lined with expensive stores, supermarkets, gallerias, and offices, however, "the streetscapes of Moscow's outer zones have hardly changed from Soviet times" (O'Loughlin and Kolossov, 2002, p. 166).

Novosibirsk

Novosibirsk started in 1893 as the New Village (*Novaya Derevnya*) close to the Ob where a bridge was built to carry the Trans-Siberian railroad across the river. As the town grew it was named in honor of Czars. The name became Novosibirsk, New Siberia, in 1925. (Photo 14.13)

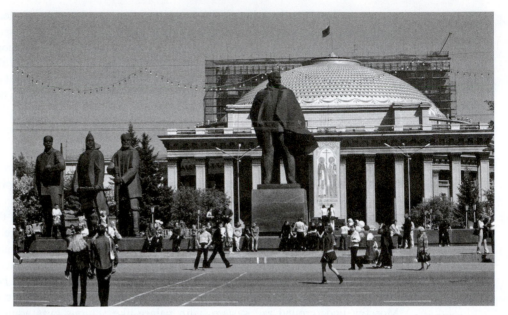

Photo 14.13 Novosibirsk is characterized by broad streets, spacious squares and ambitious public buildings. In the picture, the largest Opera House in the world is guarded by heroes of the Russian Revolution.

Source: Topham/The Image Works.

Having Siberia in the city name does not project a positive image to outsiders. However, Novosibirsk is in southern Siberia, in the wooded-steppe region, where a range of grains, root crops, vegetables, and small fruits can be produced along with livestock and poultry. When the railroad arrived there were agricultural and timber products to transport. The river Ob is ice free and navigable for part of the year and the new town acquired both port and railroad functions.

When, in the first Five Year development plan, the Kuznetsk basin began large-scale growth, Novosibirsk was already a substantial transportation center, with agricultural and timber-processing industries. Transportation functions were enhanced with the completion, in 1931, of the Turkistan–Siberian railroad. The line runs from Novosibirsk to link with the Trans-Caspian railroad near Tashkent in Central Asia.

As Siberia industrialized in the 1930s, the early emphasis was on heavy industry. World War II led to rapid growth and diversification. Novosibirsk was far from the fighting front, beyond the range of German bombers. Industrial capacity increased and many plants were moved into Siberia from threatened European Russia.

Postwar diversification continued until the city possessed most types of engineering works and research facilities. Akademgorodok was laid out about fifteen miles south of the center of Novosibirsk, containing research institutes together with accommodation for scientists, technicians, and families. Many found life in the Novosibirsk area relatively attractive. Living standards tended to be higher and security less oppressive than in the large cities of European Russia. Goods were more readily available, as part of a Soviet program to attract people to the Siberian city region.

By 1970 Novosibirsk had over a million inhabitants and had acquired museums, a university, art galleries, theaters, and libraries. A metro system opened in 1986 and the city spread out, generously creating a Soviet landscape of large public buildings, broad streets, memorials to the Great Patriotic War, and block complexes housing workers.

Because industry and research was state funded, the collapse of the Soviet Union reduced some activities, but the city still functions and remains attractive to many, even if it is 2000 miles east of Moscow and temperatures do not rise above freezing until April.

► SUMMARY

As the Soviet Union broke up, many state agencies and industries ceased to provide secure employment and pay wages. In recent years, with increased demand for oil, gas, and basic materials like steel, the Russian economy has grown. High energy prices have helped provide the government with income and some of the privatized industries, like steel, have benefited from improved worldwide demand.

Many people found themselves displaced from jobs, which they thought secure, and have had difficulty finding work or are in jobs unrelated to previous careers. As discussed, disruption led to a drop in the birth rate, increased alcoholism, more abortions, and a high suicide rate. The worst may now be past, but, for many, life will never be the same again. Nostalgia for the past regime is powerful. Many still think of Stalin as a powerful and effective Soviet leader (Mendelson and Gerber, 2006). Khruschev, who denounced Stalin and made many reforms, has a negative image. The present Russian leadership, including President Putin, sees the breakup of the Soviet Union as a disaster. It is not surprising ordinary people with secure jobs in the Soviet era share that view.

In spite of prosperity in some Russian cities and regions, the effectively used national territory is shrinking. Marginal farming areas, which survived on Soviet subsidies, have ceased to produce, small isolated towns are depopulated, and unprofitable factories have shut. The formerly integrated Soviet economy has shrunk spatially, with many areas being allowed to wither. "Geographically, economically and socially Russia . . . is an archipelago" with prosperity confined to major urban agglomerations and resource–extracting centers. (Dienes, 2004, p. 443)

Within Russia there are twenty republics, which are homes to ethnic minorities. Although the republics have considerable autonomy, several want to leave the Russian Federation. On the north side of Caucasus are non-Russian ethnic groups that wish to be free of Moscow control. The best known group are the Chechens, with their major city of Groznyy. Chechnya is a member state of the Russian Federation but wants independence. To the east, with frontage onto the Caspian Sea, is Dagestan, briefly freed from the Russian Empire at the end of World War I. To the west, the Karachai and Balkar peoples are Sunni Muslims, speaking a Turkic language. The Karachai are numerous in the Kuban valley. The Balkar are found in the Baksan valley. The peoples attempted to achieve autonomy in 1917–1919 and, as Muslims, were strongly anti-Communist. The Red Army arrived in August 1920, and a Communist regime held power until August 1942, when German forces supported the independence of Karachai-Balkarai. Independence was short. When the Soviets returned, tens of thousands were deported to central Asia, where many died. Karachai-Balkaria wants independence from Russia, but this has not been recognized.

The whole North Caucasus region is non-Russian and the national groups desire independence. Chechnya has been the most forceful in its demands. The Russians fear that allowing an independent Chechnya would result in a tier of independent states running from the Caspian to the Black Sea along the north flank of the Caucasus. The sensible policy would be to let the peoples go, but this is unlikely.

The desire to exert control from Moscow still affects many aspects of life in Russia. President Putin concentrates control in the Kremlin. The state increasingly wants to control the industries that were privatized in the early post-Soviet years. Supplies of oil and gas are tampered with to let the Ukraine and Georgia know they cannot incur the displeasure of Russia. Contracts that western oil and gas companies entered into are being revised or revoked. Russia is determined to project major power status on neighboring states.

▶ FURTHER READING

Áhrend, R. "Russia's Post-Crisis Growth: Its Sources and Prospects for Continuation." *Europe-Asia Studies* 58(1) (2006): 1–24.

Argenbright, R. "Remaking Moscow: New Places, New Selves." *Geographical Review* 89(1) (1999): 1–22.

Baxter, J.H. "Privatization in Moscow." *Geographical Review* 84(2) (1994): 201–215.

Billington, J.H. *Russia in Search of Itself*. Baltimore: Johns Hopkins University Press, 2004.

Bradshaw, M., and J. Prendergrast. "The Russian Heartland Revisted: An Assessment of Russia's Transformation." *Eurasian Geography and Economics* 46(2) (2005): 83–122.

Coones, P. "The Heartland in Russian History," in Brian W. Blouet (ed.), *Global Geostrategy: Mackinder and the Defence of the West*. London: Frank Cass, 2005.

Davies, R.W., and S.G. Wheatcroft. *The Years of Hunger: Soviet Agriculture, 1931–1933*. New York: Palgrave Macmillan, 2004.

Dienes, L. "Observations on the Problematic Potential of Russian Oil and the Complexities of Siberia." *Eurasian Geography and Economics* 45(5) (2004): 319–345.

Hanson, P. *The Rise and Fall of the Soviet Economy*. New York: Longman, 2003.

Harrison, M. *Accounting for War: Soviet Production, Employment and the Defence Burden 1940–1945*. Cambridge, UK: Cambridge University Press, 1996.

Hill, F., and C. Gaddy. *The Siberian Curse: How Communist Planners Left Russia Out in the Cold*. Washington, DC: Brookings Institution Press, 2003.

Hooson, D.J.M. *A New Soviet Heartland?* Princeton, NJ: Van Nostrand, 1964. The classic account of the development of Soviet industry in remoter regions.

Hosking, G. *Russia and the Russians: A History from Rus to the Russian Federation*. New York: Penguin Press, 2001.

Ioffe, G. "Understanding Belarus: Economy and Political Landscape." *Europe-Asia Studies* 56(1) (January 2004): 85–118.

Ioffe, G., and T. Nefedova. "Marginal Farmland in European Russia." *Eurasian Geography and Economics* 45(1) (2004): 45–59.

Ioffe, G., T. Nefedova, and I. Zaslavsky. *The End of Pesantry? The Disintegration of Rural Russia*. Pittsburgh, PA: University of Pittsburgh Press, 2006.

Jaimoukha, A. *The Circassians: A Handbook*. New York: Palgrave, 2001.

Jordan, B.B., and T.G.B. Jordan. *Siberian Village: Land and Life in the Sakha Republic*. Minneapolis: University of Minnesota Press, 2001.

King, C. *The Black Sea: A History*. Oxford, UK: Oxford University Press, 2004.

Lewis, A. (ed.). *The EU and Moldova: On a Fault-line of Europe*. London, Federal Trust, 2004.

Meier, A. "Endangered Revolution" (Ukraine). *National Geographic* 209(3) (March 2006): 32–59.

Mendelson, S., and T. P. Gerber. "Failing the Stalin Test, Russians and Their Dictator." *Foreign Affairs*. 85(1) (January/February 2006): 2–8.

Minahan, J. *The Former Soviet Union's Diverse Peoples: A Reference Sourcebook*. Santa Barbara, CA: ABC-C110, 2004.

Mitchneck, B.A., and E. Hamilton. "Cities of Russia," in Stan Brunn, Jack Williams, and Donald Ziegler, *Cities of the World*. Lanham, MD: Rowman and Littlefield, 2003.

Murphy, D.E. *What Stalin Knew: The Enigma of Barbarossa*. New Haven, CT: Yale University Press, 2005.

O'Loughlin, J., and V. Kolossov. "Moscow: Post-Soviet Developments and Challenges." *Eurasian Geography and Economics* 43(3) (2002): 161–169.

O'Loughlin, J., and P.F. Talbot. "Where in the World is Russia? Geopolitical Perceptions and Preferences of Ordinary Russians." *Eurasian Geography and Economics* 46(1) (2005): 23–50.

Overy, R. *Russian's War*. New York: Penguin Books, 1999.

"The President of Russia on Population." *Population and Development*. 31(2) (2005): 400–403.

Round, J. "Rescaling Russia's Geography: The Challenges of Depopulating the Northern Periphery." *Europe-Asia Studies* 57(5) (July 2005): 705–727.

Rowland, R. "National and Regional Population Trends in Ukraine." *Eurasian Geography and Economics* 47(7) (2004): 491–514.

Sellar, C., and J. Pickles. "Where will Europe End? Ukraine and the Limits of European Integration." *Eurasian Geography and Economics* 43(2) (2002): 123–142.

Stone, D.R. "The First Five-Year Plan and the Geography of Soviet Defence Industry." *Europe-Asia Studies* 57(7) (November 2005): 1047–1063.

Williams, B.G. "Caucasus Belli: New Perspectives on Russia's Quagmire." *Russian Review*. 64 (October 2005): 680–688.

Ziker, J.P. *Peoples of the Tundra*. Prospect Heights, IL: Waveland Press, 2002.

PART

VI

THE MEDITERRANEAN FRINGE

15

TURKEY, CYPRUS, MALTA, AND GIBRALTAR

▶ INTRODUCTION

Malta and the Greek part of Cyprus joined EU in 2004. Northern, Turkish, Cyprus is a separate republic. Few recognize the country except Turkey. In 2005 Turkey started talks with the EU seeking accession (Figure 15.1).

With a population of 75 million inhabitants Turkey would become the second largest country in EU, after Germany. Currently, Turkish living standards are well below European averages; life expectancy (71) is low, and infant mortality, 25 deaths per thousand live births, is high. The demands Turkey would make on the CAP and EU structural readjustment funds would be considerable. Some doubt that the EU could absorb such a large new member. Others take the view that the inclusion of Turkey would strengthen EU and provide "greater global geopolitical equilibrium" (Cohen, 2004, p. 582).

▶ TURKEY

Turkey is a bridge between Europe and the Middle East. Istanbul, the largest Turkish conurbation, lies on either side of the Bosporus waterway flowing from the Black Sea to the Aegean, dividing Europe from Asia Minor. A short ferry ride takes visitors from Europe to Asia; the journey over the Bosporus bridge into Asia Minor is a few minutes. (Photo 15.1)

Physical Geography

Turkey can be divided into four major physical regions (Figure 15.2):

1. The Black Sea coast of Anatolia

2. The Anatolian plateau and mountains

3. The Mediterranean coastlands of Anatolia

4. Western Anatolia, the Bosporus, the Sea of Marmara, and European Turkey

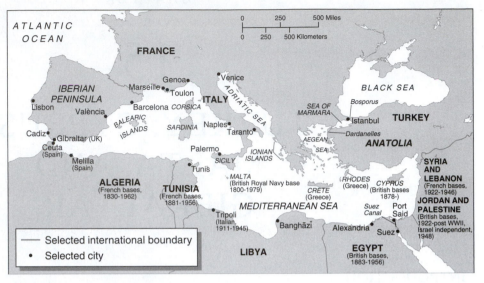

Figure 15.1 Mediterranean sea.

The Black Sea Coast of Anatolia

The Pontic Mountains of Anatolia run close to the Black Sea coast and the areas of coastal plain are limited. The climate of the coast and the north face of the Pontic Mountains is influenced by the Black Sea. At the coast, the climate is of a Mediterranean type, but with more rainfall in summer than is typical. Samsun receives nearly 30 inches of

Photo 15.1 The Bosporus bridge at Istanbul links Europe and Asia.

Source: José Fuste Raga/SUPERSTOCK.

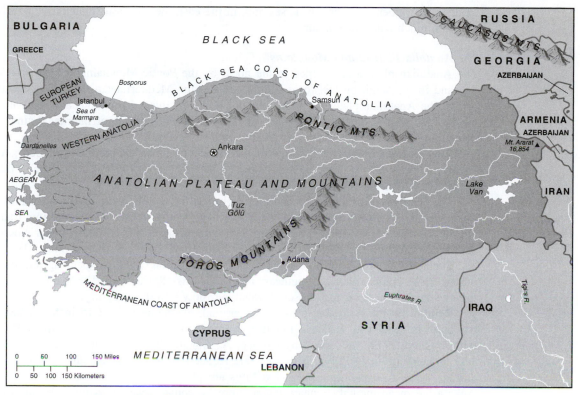

Figure 15.2 Physical regions of Turkey.

rain during the year and although June, July, and August receive less than 2 inches of precipitation, every month receives significant rainfall (Table 15.1).

Moving inland and up the north flank of the Pontic Mountains, annual precipitation increases and hillsides are covered in forests of deciduous and coniferous trees. In winter, at high elevations, precipitation comes as snow, and slopes are snow covered for months. As precipitation-bearing depressions pass, colder air from lands to the north of the Black Sea, including the steppes of the Ukraine, lowers temperatures. Samsun, on Turkey's Black Sea coast, can experience frost, but the mean monthly temperature for January is in the low 40s. The relative winter mildness on the north coast of Turkey results from the

TABLE 15.1 Turkey Weather Data

	mm Jan°F	mm July°F	Rainfall total in inches
Samsun	44°	72°	29.1
Ankara	31°	72°	13.6
Adana	48°	82°	24.3
Istanbul	40°	73°	31.5

moderating influence of the Black Sea and, in the east, the Caucasus mountains, which keep out much cold Siberian air.

The Anatolia Plateau and Mountains

The Anatolian plateau is bounded in the north by the Pontic Mountains and the Black Sea and in the south by the Toros Mountains and the Mediterranean Sea. The central part of the Anatolian plateau is characterized by downfaulted depressions and intervening uplifted blocks (horsts). The depressions are frequently zones of internal drainage containing salt lakes, as at Lake Tuz Gölü, south of Ankara.

Anatolia rises from west to east, with the highest peaks on the borders with Georgia, Armenia, Iran, and Iraq. The preeminent peak is Mount Ararat (16,854 feet) close to the border with Armenia and Iran.

The headwaters of the Tigris and the Euphrates rivers rise in the southeast of Anatolia. The Tigris flows to Iraq and the Euphrates through Syria before entering Iraq. On the borderlands of Turkey, adjoining Syria, are alluvial plains of the Tigris where summers are hotter than the Anatolia plateau. Dams have been constructed on the Tigris and Euphrates, to provide hydroelectric power and irrigation water to expand agricultural output on the alluvial plains of southeast Turkey (Harris, 2002, 2006). The Syrians worry that the Turkish projects may lessen downstream flow into their country.

Rainfall on the Anatolian plateau is marginal, averaging around 15 inches a year. Irrigation is needed to sustain agriculture. Summer on the plateau is hot and rainless, but temperatures are moderated by altitude. At Ankara (2,825 feet) the mean monthly July temperature is 74°F, but daily highs are frequently over 90°F. At sundown, due to altitude, temperatures drop rapidly. Evenings are pleasant and cool. Winter is cold. At Ankara the mean monthly January temperature is around freezing. Istanbul, beside the sea, is 10 degrees warmer. In higher, eastern Anatolia winters are longer, snow cover is prolonged, and glaciers are found in the higher valleys.

All of Turkey, lying in the zone where the African and Eurasian plates converge, is prone to earthquakes and the faulted plateau experiences the hazard frequently.

The Mediterranean Coastlands

The Toros Mountains (*Toros Dağlari*) run down to the Mediterranean coast where there are several agricultural lowlands including the lands around Adana and Antalya (Figure 15.3). The climate is typically Mediterranean with a long, hot, dry summer and a generally mild, wet winter (see Adana, Table 15.1). Where irrigation water is available from streams flowing from the Toros Mountains, agriculture is sustained in the summer months. The port of Ceyhan is the Mediterranean terminal of the pipeline running from Baku on the Caspian, through Georgia, to Turkey. Medieval crusader castles dot the coast. (Photo 15.2)

Western Anatolia, the Bosporus, the Sea of Marmara, and European Turkey

Western Anatolia is drained by rivers that flow to the Aegean. There is still plenty of upland but the topography is lower than further east in Anatolia. On the coastal plain the climate is Mediterranean and extends inland up the broader valleys. At Izmir (former Smyrna) the mean monthly temperature for July is 82°F, with the record high exceeding one hundred degrees. In January the city has a mean monthly temperature of 47°F, although colder air can fall to the coast down the valleys running from the interior.

Figure 15.3   Turkey: Major Urban areas

On the west bank of the sea of Marmara and the Bosporus we are in European Turkey. For the most part the land running to the borders with Greece and Bulgaria is a low-lying, rolling region producing livestock and Mediterranean crops. In the north of the region, close to Bulgaria, the Istranca Dağlari rises to several thousand feet.

At Istanbul, on the Bosporus, the mean for January is 42°F. In July the weather is warm and sunny, with a monthly mean of 74°F. Istanbul on the Black Sea has a more

Photo 15.2   The Mediterranean coast of turkey possesses the remains of several crusader castles. Anamur castle is of Roman origin and rebuilt by Crusaders.

Source: Age fotostock/SUPERSTOCK.

moderate summer than Adana (July mean 82°F) on the Aegean, as the Black Sea is cooler than the Aegean in summer. (Photo 15.3)

The physical origins of the Bosporus waterway are geologically recent. At the end of the Ice Age an enclosed freshwater lake (Pontian Lake) occupied the central region of the depression now holding the Black Sea. The lake and the Aegean Sea were separated by higher ground.

In the postglacial sea level rise, water levels in the Mediterranean rose until, in 5,600 B.C., the sea overflowed from the Aegean into the Pontian lake, rapidly cutting the great notch of the Bosporus. The inhabited coastal plains around the lake were flooded, setting off movement, which *may* have stimulated the migration of peoples and the diffusion of cultural traits. The freshwater Pontian lake became the more extensive salt water Black Sea.

Distribution of Population, Urban Areas, and Economic Activity

The demographic statistics of Turkey are un-European. The birth rate is 21 babies per annum for every thousand Turks, more than double the German statistic. The relatively high birth rate produces a youthful population and the death rate, at seven for every thousand Turks, is one of the lowest in the region. The rate of Turkish natural increase, at 1.4 percent per annum, is high by European standards as is the total fertility rate at 2.4 children per woman. If Turkey's rate of natural increase were sustained over the next decade, it would overtake Germany in population numbers. Other vital statistics are also un-European. For every thousand live births, 25 children do not reach their first birthday. Twenty years ago the figure was over 100. Life expectancy is ten years less than in Northern and Western Europe. Incomes compare with Belarus and the Ukraine rather than Southern Europe.

Photo 15.3 The Blue Mosque and Saint Sophia overlook the Bosporus waterway, Istanbul.

Source: Age fotostock/SUPERSTOCK.

Turkey has many families dependent on farming for a living, and in general rural population distribution reflects the agricultural potential of Turkish environments. Rural population densities are higher on coastal plains, decrease on mountain sides, and are sparse in poorly watered parts of the Anatolian plateau. Apart from subsistence crops, Turkey is a large producer of olives, cotton, tobacco, citrus, fruits (particularly apricots), nuts, and the summer Mediterranean crops of tomatoes, peppers, and squash. Forests cover over 10 percent of the country and are exploited for cut timber but, in general, forestry does not support high population densities.

Higher population densities are found along the Black Sea coast, around the Bosporus, the Sea of Marmara, and the valleys running to the Aegean coast of Turkey. Population densities are generally low in Anatolia, except where water and agricultural land are found in depressions; in favorable well-watered areas, ancient cities thrived. Today approximately half of Turkey's top dozen cities are in Anatolia, several on the overland routeway that runs from the Bosporus up to Eskişehir and on to Ankara, and Erzurum (population .37 million), high in the Anatolian Mountains (Figure 15.3). The routeway continues to Armenia, Georgia, and Azerbaijan.

On the Black Sea coast the major cities are, from east to west, Trabzon (.22 million), Samsun (.36 million), and the heavy industry center of Zonguldak. Trabzon and Samsun are linked by a coastal road, but the mountains then meet the coast and there is no seashore road to Zonguldak from Samsun.

On the Bosporus the Istanbul conurbation is the largest concentration of population in Turkey and there are numerous towns and cities around the Sea of Marmara and the Dardanelles. On Turkey's Aegean coast the largest place is Izmir (former Smyrna), with over two million inhabitants. The south coast of Turkey, fronting onto the Mediterranean, has contrasting population densities. Where coastal plains connect with valleys running into the Anatolian hinterland, there are large cities including Mersin and Adana, with high surrounding rural population densities. Where the upland comes to the coast, population numbers are low.

Turkey still has a large rural population, but now approximately two-thirds of all Turks live in the rapidly growing cities, up from one-third, thirty years ago. Rural to urban migration has been rapid and the cities have been unable to absorb all the newcomers. The result has been the growth of shanty towns, *Gecekondu* in Turkish. Another term is self-help housing, for the units are built by the inhabitants, with their own labor, using materials purchased over a period of time. At first the self-help units lack services but as these are provided and houses acquire additional stories, the shanty towns evolve into working-class suburban areas (Tas and Lightfoot, 2005, pp. 263–271).

Turkey possesses five cities with over a million inhabitants and around 30 urban places with populations exceeding 200,000 (See Table 15.2).

Istanbul

Istanbul is an excellent city to illustrate the concept of site and situation. Istanbul is on the narrow waterway that separates Europe from Asia and links the Black and Mediterranean Seas. The city is situated in a position "to control overland trade between Europe and Asia and the shipping lanes between the Mediterranean and Black Seas (Zeigler, 2003, p. 283).

TABLE 15.2   Major Turkish Cities 500,000 or more

Istanbul	8,830,000
Ankara	3,200,000
Izmir	2,250,000
Bursa	1,180,000
Adana	1,130,000
Gaziantep	860,000
Konya	760,000
Antalya	610,000
Diyarbakir	550,000
Mersin	540,000
Kayseri	520,000
Eskişehir	490,000

The site of Istanbul is on the European shore of the Bosporus where it joins the Sea of Marmara. Here a broad inlet, the Golden Horn, provides a safe harbor for shipping and creates a defensible peninsula on higher ground. There has been a city on the site since 657 B.C. The emperor Constantine chose the site and situation for his capital of the Roman empire in 330 A.D. When Rome fell in the fifth century A.D. Constantinople remained as the capital of the eastern empire, Byzantium, until captured by the Turks (1453) to become the capital of the Ottoman empire. The capital of Turkey was transferred to Ankara in 1923 and Constantinople was renamed Istanbul in 1930 (Sarkis, 2006). The city is one of the largest in Europe, a cultural, commercial, financial, and industrial center, housing many ethnicities and religions. Istanbul is also a major port and shipbuilding center.

Ankara

When Turkey established itself as a secular state it was decided to move the capital from Istanbul, which was associated with the Sultan and the Caliphate, to Ankara (Angora). In 1923, when the capital change was made, Ankara had a population of under 30,000 people. The city had a central location in Turkey. The agricultural hinterland of Ankara is based on irrigation farming and mohair production from the Angora goat, which gave the city its name. Ankara grew rapidly as an administrative, cultural, communication, commercial, and industrial center. There are numerous museums in Ankara, together with the mausoleum for Atatürk, the man who ruled dictatorilly as president from 1923 to 1938 but started the modernization and industrialization of Turkey. (Photo 15.4)

▶ TURKEY, THE MIDDLE EAST, AND EUROPE

At maximum extent, in the seventeenth century, the Ottoman empire controlled the Danube valley, including most of the Hungarian lands and Budapest. The Turks also controlled the Black Sea, Anatolia, Mesopotamia, the Levant, the west shore of the Persian Gulf, Egypt, and North Africa to the Strait of Gibraltar and the Atlantic.

By the nineteenth century the empire was contracting. Greek independence was recognized in 1829. The Serbs (1879), Romania (1877), and Bulgaria (1878) became

Photo 15.4 Atatürk's mausoleum, Ankara. Atatürk rebuilt, ruled, and modernized Turkey in the years after World War I.

Source: Age fotostock/SUPERSTOCK.

independent, although Bulgaria, nominally, was under Turkish sovereignty. It became fashionable to refer to Turkey as "the sick man of Europe." Today, those who do not want Turkey in EU reject the idea that Turkey had ever been in Europe. Certainly Anatolia is in Asia Minor, but Istanbul is in Europe and for centuries Turkey has been a factor in European affairs. As the Ottoman Empire declined, Britain and France wanted to maintain Turkey as a container of Russia. In 1854–56 Britain and France joined the Crimean War to support Turkey in resistance to a Russian attempt to control the sea routes running from the Black Sea to the Aegean via the Dardanelles.

In the 1890s a German company received a concession from the Ottoman Empire to develop a Berlin-to-Baghdad railroad. The European track was largely in existence and the rail line would be extended from Istanbul to Baghdad in Mesopotamia and down to the head of the Persian Gulf. In 1899 Britain, fearing that Germany might threaten the empire of India, via the Gulf, went to the Gulf Sheikdoms, technically part of the Ottoman empire, with a deal. Do not cede, sell, or otherwise give into occupation any part of your territory to a foreign power and Britain would provide protection. *De facto* the Gulf Coast Sheikdoms, including Kuwait, Bahrain, Qatar, the eventual United Arab Emirates, Muscat, and Oman, became independent.

The "young Turks" started to take power and modernize the Ottoman empire, beginning in 1908. The country that wanted to invest most in Turkey was Germany and unfortunately, for everyone, this led to the Ottoman empire allying with the Central Powers in World War I. Once the war started, Britain annexed Cyprus, took over Egypt, put an army in Iraq, which was defeated and tried disastrously, with the French navy, to force battleships up the Dardanelles toward Istanbul. More successfully, militarily, Arabic-speaking British army personnel went to Arabia to ally with the Sherif of Mecca, Lord Feisal, against the Ottoman empire. The uprisings were successful (see the film *Lawrence of Arabia*), but, long term, there is no peace. After World War I, the League of Nations created the Mandated Territories of Palestine (including a national home for the Jews), Trans-Jordan and Iraq, under British administration. Syria and Lebanon were administered by France. As we know there has been little stability in the states created after World War I by the League. The Ottomans with detached, indirect, administrative techniques had kept the Middle East relatively quiet in their empire.

After World War I the Ottoman empire had gone. The state of Turkey appeared, no longer controlling non-Turkish populations, except the Kurds in eastern Turkey. Alone among the defeated Central Powers, the Turks rejected the treaty dictated to them after the war (Treaty of Sèvres, 1920). Turkey refused to be bullied by the Greeks, who wanted the city of Smyrna, now Izmir, and the surrounding territory on the Aegean coast of Anatolia. Smyrna had a large Greek population. There was war. The Treaty of Lausanne (1923) established boundaries between Greece and Turkey. Further the countries agreed to exchange ethnic populations. Greeks in multicultural Constantinople were exempt but Smyrna (Izmir) lost its Greek population and Thessaloníki (Salonika) its Turks and other nationalities. Many Greeks coming from Turkey resettled on the plain of Drama, near Thessaloníki, in government-sponsored agricultural settlements. The losers at Lausanne were the Kurds who remained part of Turkey. Sèvres had suggested a separate Kurdish state.

Turkey underwent rapid change in the 1920s under Mustafa Kemel Atatürk (1880–1938). In 1923 Turkey became a republic, and Ankara was made the capital. In 1928 Islam ceased to be the state religion, women received the vote, Turkish was written in the Roman alphabet rather than Arabic script, and people got surnames. Atatürk's well educated wife had an influence on the developments (Callslar, 2006).

In World War II Turkey stayed neutral until early 1945, joined NATO in 1951, and became an important ally in the containment of the Soviet Union. As the West German economy expanded in the 1950s and 1960s, Turks moved to work in Germany. Often young Turks lived together in neighborhoods that became ghettos. They did not assimilate, nor were they encouraged to do so. Sons and daughters of Turkish migrants, born in Germany, were not given German nationality, as they were not of "German blood."

The Turkish economy could integrate with EU. Turkey has democratic institutions but the military plays a political role, taking power in 1960 and 1980 and forcing a prime minister from power in 1997. The Kurds in Turkey want independence or autonomy. Armed aspects of this struggle have led to warfare, atrocities, and human rights violations. However, Kurds and Turks are moving to a political settlement because EU does not like importing disputes with expansion. Within EU the Kurds would have the protection of European human rights legislation.

The greatest obstacle to Turkish EU membership is the attitude of some EU states toward Turkey (Dahlman, 2004, Figure 3). Historically, the Austrians have resisted Turkish advances up the Danube valley. Today, Austrian resistance to full membership for Turkey reflects the view of right-wing political parties rather than fear of another Turkish siege of Vienna. Some Germans worry that existing Turkish communities will reestablish migration chains from Turkey to central Europe, but others understand that Germany is Turkey's largest trade partner. France, with unassimilated Muslim communities, is fearful of more immigrants from anywhere. French leaders say there will be a referendum in France on Turkish membership if Turkey meets EU qualifications. A *non* vote from the French electorate would veto Turkish entry.

In Turkey, a growing number question the benefits of EU membership and the costs of entry. Turkish industries would face heavy competition within European Union economic space. On the other hand, CAP payments and structural readjustment funds would stimulate economic activity.

► CYPRUS, MALTA, AND GIBRALTAR

Cyprus, Malta, and Gibraltar have different cultures, but they were all British military bases on the searoute through the Mediterranean to Suez, India, and the Far East (Figure 15.1). Cyprus and Gibraltar still have a British military presence. All three territories were British Crown colonies. Gibraltar, with an elected legislature and internal self-government, still has a Governor, who as commander in chief has the power to overrule in matters of security, defense, and foreign affairs.

► CYPRUS

Cyprus is an island lying off the south coast of Turkey and the west coast of Syria, divided between a Greek majority and a Turkish minority (Figure 15.4). The Greeks are Orthodox. The Turks are Sunni Muslims. In 2000 the Greeks numbered some 650,000, the Turks less than 90,000. Turkish numbers have declined in recent decades. The total population of Cyprus is approximately 800,000. In theory Cyprus joined EU in 2004. In practice the Greek-dominated Republic of Cyprus joined EU and Turkish Cyprus was excluded. UN attempts to find a federal formula to reunite Cyprus have failed.

Physical Environment

Cyprus consists of the northern Kyrenia Mountains, the southern Troodos Mountains, and between the two ranges is the Plain of Mesaoria (Figure 15.5). The Kyrenia range is an Alpine range and consists of upthrust sedimentary rocks rising to 3,360 feet. The Troodos Mountains are largely igneous rocks pushed up from the floor of the Sea of Tethys during the Alpine orogeny. The Troodos range rises to 6,401 feet at Mount Olympus, the highest point in Cyprus. The igneous rocks contain minerals including copper, the metal from which Cyprus takes its name. Today, copper exports are in decline.

The Troodos mountains contain extensive forests that have been well managed and conserved. In winter the higher elevations are snow covered and provide skiing. The lower slopes are famed for vineyards and orchards.

The Mesaoria, between the mountain ranges, runs from Morphou Bay in the west, past the capital Nicosia, to the bay of Famagusta in the east. The major river crossing the plain is the Pedhieos, which rises in the Troodos Mountains. The river is heavily used for irrigation as the plain is hot, dry, and dusty in summer, but where water is available crops yield well. In the frost-free winters the temperate crops, including potatoes and grains, are raised on the plain. The Greeks and the Turks have differing crop emphases. The Greeks specialize in deciduous fruits and wine making. Turkish farmers grow more tobacco and grapes to be consumed as fruit.

Figure 15.4 Cyprus.

The climate of Cyprus is Mediterranean, with altitude and relief lowering temperatures and inducing rainfall in the Kyrenia and Troodos Mountains. The Mesaoria is mild and moist in winter, but hot, dry, and brown in summer. At the capital, Nicosia, close to the center of the island, summer daytime highs are in the 90s and frequently exceed 100°F. At night temperatures rarely drop below 70°F. In winter the Nicosia highs are around 60°F. It is unusual for lows to fall below 40°F.

In coastal areas temperatures are moderated by onshore breezes in the summer and by the relatively warm sea in winter. (Photo 15.5)

Origins of the Multiethnic Island

Cyprus was a valued part of the Roman empire, being a source of copper, timber, and dyes. When the empire split, the island was part of Byzantium, with the Greek population worshiping in the Greek Orthodox Church. During the crusades, the strategic position of Cyprus, off the coast of Syria, was of value to western Europeans attacking the Holy Land. Starting in the late twelfth century, crusaders acquired estates and serfs on Cyprus and built castles, churches, and cathedrals. The majority of inhabitants remained Greek Orthodox but the Church of Cyprus was placed under a Roman Catholic archbishop. Besides crusading, western Europeans used Cyprus for trade and the island came under Venetian control in 1489, where it remained until taken by the Turks in 1571.

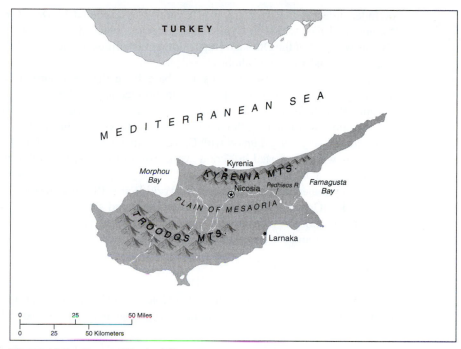

Figure 15.5 Cyprus: Physical features.

Cyprus under the Turks

The arrival of the Turks improved life for many Cypriots. Serfdom was abolished. The Roman Catholic clergy were expelled. The position of the Orthodox Church of Cyprus improved as the Orthodox Archbishop was recognized by the Turks as the Ethnach, the leader of the Greek community in religious and civil affairs. The recognition had disadvantages. When the mainland Greeks revolted against the Turks, in the Greek War

Photo 15.5 The small port of Kyrenia on the north coast of Cyprus still possesses architectural features from the age of the Crusaders.

Source: Age fotostock/SUPERSTOCK.

of Independence, the Archbishop of Cyprus was executed in 1821 for his sympathies. Numerous Turks settled in Cyprus as a minority, which did not feel threatened because Cyprus was part of the Ottoman empire. Other minorities on the island include Armenians, Marionites, and Roman Catholics.

In 1878 Britain leased Cyprus as a base from the Ottoman empire to support the Turks against the Russian empire and to protect approaches to the Suez Canal. The arriving British forces were generally welcomed by the Greek majority who, knowing that Britain had turned the Ionian islands over to Greece in 1864, thought the British connection a step toward union with Greece. This view ignored the facts that sovereignty of Cyprus was still vested in Turkey, with Britain the lessee, and that the Ionian islands did not have a Turkish minority.

Turkey joined the central powers in World War I. Britain annexed Cyprus and offered the island to Greece if it joined the war on the allied side. The offer was rejected by the Greek king who was married to a sister of the Kaiser. Greece did not join World War I until Germany and Turkey were close to defeat.

British Cyprus

In 1925 Cyprus became a British Crown Colony and Greek Cypriot pressure for *Enosis,* union with Greece, intensified. The *Enosis* concept is deeply rooted in Hellenism.

> *From childhood the Greek is conditioned in church and school to believe that...all Greek speaking areas must be united within the frontiers of the motherland. Greeks outside the narrow circles of the most intellectually sophisticated still day-dream of the return of Constantinople and the "lost lands" of Anatolia. (Crawshaw, 1978, p. 18)*

The Turks were criticized by commentators for insisting on an exchange of populations when peace was made between Turkey and Greece in the Treaty of Lausanne in 1923. The Turks understood, however, that the Greeks in Smyrna (Izmir) would never be content in Turkey and any Turks in Greece would be second-class citizens. The Greeks, too, wanted to expel Serbs, Bulgarians, and Turks.

In Crown Colony Cyprus, British administrators, taught at school to see ancient Greece as the basis of western civilization, perhaps gave the impression that they sympathized with *Enosis* aspirations, and this agitated Turkish Cypriots.

In general, the administration of the colony of Cyprus was conventional. Local revenues were used to improve roads, irrigation schemes, and harbor works. Forestry management was introduced and a forestry college established, with students coming from Cyprus and other parts of the empire/commonwealth to study. The college still operates.

There was unrest in the 1930s as the Greek Cypriots agitated for union with Greece and the Turkish community demonstrated against becoming part of Greece. In the 1950s EOKA terrorists pushed for the Greek connection violently. Bombs were planted and Cypriot policemen and civil servants were targeted for killing, as EOKA saw them as collaborating with Britain.

Archbishop Makarios, a leading advocate of *Enosis,* was exiled by Britain to the Seychelles in the Indian Ocean. The Archbishop was the leader of the Church in Cyprus. He did not report to a higher patriarch in the Orthodox Church and going back to the

Turkish era, Makarios was the Ethnarc in charge of the Greek Cypriot community. The Archbishop was a spiritual leader with a political role. There could be no settlement of the Cyprus problems without Makarios.

Other factors were in play. The Turkish Cypriots did not want to become part of Greece. Britain had valuable strategic bases in Cyprus during the Cold War. From Cyprus airfields, Royal Air Force bombers could fly north over the airspace of NATO allies Greece and Turkey, cross the Black Sea, and attack targets in the Soviet Union or eastern Europe.

An apparently sensible settlement was reached in 1959. The Greek and Turkish Cypriots agreed to an independent Cyprus. Britain would retain sovereignty over the bases but cease to administer the country. Archbishop Makarios was elected president. A Turkish Cypriot became vice president. In June 1960 Cyprus became independent.

For the most part Greeks and Turks knew how to do business with each other, although there was little social mixing and virtually no intermarriage. Villages in which Turks and Greeks lived had a church and a mosque in the respective sectors and each community had its own rhythms of life. Left to themselves the islanders could have made an independent Cyprus work. (Photo 15.6)

The EOKA terrorists, many from mainland Greece, who felt they had been denied *Enosis,* tried to assassinate Makarios. Greece and Turkey repeatedly threatened each other, giving militants, on either side in Cyprus, hopes of powerful outside support. Over Christmas in 1963 ethnic violence resulted in more than 300 deaths, with Turks in predominately Greek southern Cyprus suffering badly. Turkey mobilized; Greece prepared for war.

The U.S., fearing a clash of NATO countries close to the border of the Soviet Union, intervened, diplomatically. George Ball and Dean Acheson of the State Department, with

Photo 15.6   Many communities in Cyprus have both a Greek Orthodox church and a mosque to serve inhabitants of differing faiths.

Source: Andrew Holt/Alamy Images.

the aid of a forthright note from President Lyndon Johnson, quieted the Turks, but Ball found Makarios inflexible (Bill, 1997). The Archbishop believed that as there were more Greeks than Turks on Cyprus, the former should decide the issues.

In March 1964 a UN force arrived to keep the peace. Archeson and Ball worked on a plan to divide Cyprus into eight cantons, six Greek and two Turkish. The plan was not accepted, but looking back the arrangement would have been better than the bitter split that exists today. The division of Cyprus, in the sense of the two communities being unable to work together, dates from 1963–1964. Lawrence Durrell's *Bitter Lemons*, written in the 1950s, suggests that a different outcome could have been achieved.

In 1967 Greece came under the military rule of the Colonels. In July 1974, encouraged by the Colonels, Greek-born officers staged a military coup in Cyprus. In response, Turkish forces invaded and large numbers of Greeks fled northern Cyprus, abandoning houses, shops, and farms. Many properties are still Greek owned, although occupied by others. It will take years to resolve disputes. No solution to the partition of Cyprus is in sight.

Beginning in 2002, the UN Secretary General, Kofi Annan, produced four versions of a plan to reunite Cyprus and exclude union of any part of the island with Greece or Turkey. At a referendum in April 2004, 80 percent of Greek Cypriots participating voted no to the federal solution (Berg, 2006). The Turkish Cypriots voted in favor of the plan. Later in the year Greek Cyprus joined EU; Turkish Cyprus was excluded. Previously, EU had wanted boundary disputes settled before countries entered. Germany and Poland and Hungary and Romania settled boundary issues before Poland and Hungary joined. With the entry of Greek Cyprus, EU imported a long-running dispute.

Divided Cyprus

The Turkish Republic of Northern Cyprus (proclaimed in 1983) covers about a third of the island and contains a population of around 200,000, not all of Turkish descent.

Prior to the division of Cyprus, the Kyrenia region was attractive to tourists: the sea breezes cooled the north coast in summer, the region contained Byzantine and crusader castles, the Kyrenia Mountains were scenic. After partition it was difficult to reach Turkish Cyprus, except through Istanbul. The Turkish government deploys troops in northern Cyprus and generally subsidizes the region. Outside investment is deterred by political uncertainty. Outmigration has reduced Turkish Cypriot numbers.

In Greek Cyprus population grows by over 1 percent per annum. Nearly half of the exports are agricultural products. In the manufacturing sector pharmaceuticals are a major part of exports, along with assembly goods using local labor. Greek Cyprus is enjoying an influx of economic energy from EU entry. The whole island would gain economically in every sector, particularly tourism, by reintegrating, but a unified Cyprus is unlikely in the near future. A divided Cyprus is an issue hindering Turkey in negotiations for EU entry.

▶ MALTA

The Malta archipelago, Malta, Gozo, and Comino, has a land area of 124 square miles and a population of 400,000 people. The population density is over 3,000 people per square mile!

Physical Environment

Malta is a limestone island. At the top and bottom of the geological succession are crusts of tough Coralline limestone, which does not weather into a deep soil and supports low-growing shrubs, herbs, wild flowers, and goats. The filling between hard Coralline limestone is dominated by the golden Globigerina limestone, which weathers to a workable soil and cuts easily into a building stone used in most of the towns and villages. Farm land on the Globigerina is divided into small fields surrounded by rubble limestone walls. Farms on the hard Coralline limestone are few and confined to depressions where the detritus of erosion has accumulated.

The islands tilt to the east. The west coast is emergent, with steep cliffs rising up to 829 feet at Dingli, the highest point in the islands. In the east, faulting and tilting has allowed the sea to penetrate, creating large bays and harbors including St. Paul's Bay, Marsaxlokk, Marsamxett, and the Grand Harbour, the latter providing a major base for the British Royal Navy from 1800 to 1979 (Figure 15.6). (Photo 15.7)

Malta is one of the most southerly places in Europe and enjoys a well-developed Mediterranean climate. The major characteristics of the climate of the Maltese islands

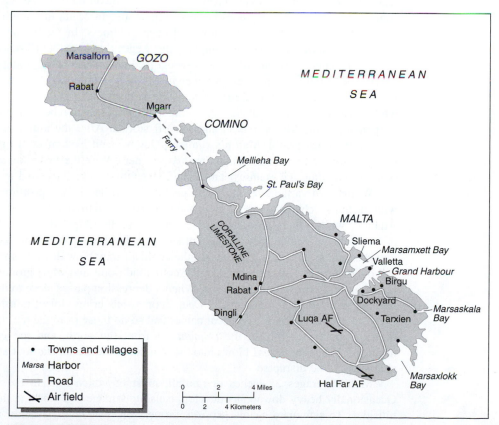

Figure 15.6 Malta.

Photo 15.7 A view from Valletta across The Grand Harbour to Fort St. Angelo and Vittoriosa (Birgu). Fort St. Angelo was utilized by the Knights of St. John and then the Royal Navy. The Royal Naval dockyard is now a commercial operation.

Source: Age fotostock/SUPERSTOCK.

can be summed up as summer drought, and mild, rainy winters with, for a large part of the year, blue skies and a high sunlight intensity.

In summer the Mediterranean region is dominated by high pressure, and in Malta the mean monthly temperatures of June, July, August, and September are in excess of 70°F (20°C). Fortunately, the heat of summer is tempered by breezes blowing from the sea. Occasionally summer climate is overheated by an air current from the Sahara. The wind is hot and humid, having picked up moisture traveling over the sea. In hot, humid conditions the human body does not cool efficiently. In Malta this south wind is called the *xlokk* and it produces enervation and even giddiness. In Sicily and Italy, the wind is known as *scirocco*. The wind impacts the Aegean and southern France. In southeast Spain it is named *leveche,* where it is recognized as a cause of irrational behavior.

In the summer months the weather in Malta is settled, with sunny days and little rainfall, for thunderstorms are rare. Plant life is dormant in the summer drought except where irrigation water is available. Summer breaks in September when cooler Atlantic air penetrates the Mediterranean and sets off storms. After the initial rains the weather cools and calms into St. Martin's summer. October and November are pleasant, warm, sunny, showery months and busy times for farmers. With the first showers, plants grow rapidly, including the grains, potatoes, and vegetables to be harvested in spring.

Winter is a variable season, with weather controlled by the positions of air masses with differing characteristics. Depressions penetrate into the Mediterranean from the Atlantic, while others are generated in the lee of the Alps. If a low-pressure system passes north of Malta, then warm southerly winds blow up from the Sahara. Should a low-pressure system pass to the south, then cold air may blow down from a high-pressure cell over Europe. The north winds bring cold conditions, and while ground frost at Malta is unknown, flurries of snow are not. When a depression passes close to the islands, mild weather with cloud and rain result. Deep depressions bring violent rainstorms.

The southerly winds have local names and so do those from the north. The *majjistral* blows from the northwest, the *tramuntana* from the north, and the *grigal* from the northeast. When the *grigal* blows into the mouths of the Grand Harbour and Marsamxett, shipping can be disrupted.

Malta averages 22 inches of rainfall, most of which falls in the winter months. Occasionally heavy downpours develop, causing damage and washing the soil from the hillsides. To safeguard against erosion, farmers on the Maltese islands, and throughout the Mediterranean, terraced hillsides, but intense rainfall will wash away retaining walls.

Landscape

The landscape of Malta is heavy with history. Around 5,200 B.C. Neolithic farmers coming from Sicily settled the islands, cleared land for agriculture, fished, and until the game was gone, hunted. In basic economy the Neolithic communities were similar to many in southern Europe. Then around 4,000 B.C. a remarkable Temple Culture began, culminating in the Ġgantija temple complex (3,000 B.C.) on Gozo and the Tarxien temples (2,500 B.C.) overlooking the Grand Harbour in Malta. There are over thirty temple structures surviving mostly on, and built of, the tough Coralline limestone. There may have been temples on the softer Globigerina limestone, lost to weathering, intensive agriculture, and demands for building stone. The temples were associated with a fertility cult and contained statues of large goddesses. The temples are older than the pyramids and Stonehenge. In form they employ the megalithic (big stone) construction seen at Stonehenge and many other European sites.

Commentators debate Malta's Neolithic temples. How could the population of a small group of islands build and support so many structures? What form of social system existed? Why did the temple culture collapse around 2,500 B.C.? What is the significance of the temples in European cultural development? Some see the temples as a culture florescence peculiar to the Maltese islands. Others view the large goddesses and the rituals of life and death as part of a wider European cult of the Great Goddess, which flourished in the Neolithic and possibly emanated from Malta (Gimbutas, 1999).

The Temple Culture ended around 2,500 B.C. Theories accounting for collapse include drought, overpopulation, and an eruption of Mt. Etna in Sicily, killing crops with volcanic ash.

By 800 B.C. the Phoenicians were established at Malta, apparently coexisting with people of a bronze age culture. The Phoenicians put down deep roots in Malta, using it as a trading station and a base on the way to Carthage in Tunisia. They held the islands until 216 B.C. when Rome took over, but Punic culture remained for centuries.

Arab Malta

With the fall of Rome, Malta became part of the Eastern empire and in 870 A.D. the Byzantine garrison was defeated by an Arab invasion force. Both Sicily and Malta came under Arab rule. Malta was reconquered from Sicily by the Normans in 1091.

Arab cultural influences did not cease with the reconquest, for the inhabitants now spoke a dialect of Arabic, which evolved, picking up words from European languages, to become modern Maltese. Most place-names in Malta have an Arabic root. Other than the names of Malta and Gozo, which came to us via Latin, there are no place-names that predate the Arab period. This is unusual. In western Europe, as we saw in Chapter 2, Celtic place-names survive in France and England, long after the Celts were dispossessed of territory. In the United States and Canada the landscape is covered with names derived from Native American words for hills, rivers, and regions. In Malta no valley, stream, or cliff has a name that predates the Arab conquest. Possibly there was a period after the Arab conquest when the island was virtually uninhabited. When resettled, the new population generated place-names for every physical feature and inhabited place. An alternative but unfashionable theory is that the place-names of Malta, in 870 A.D., were still derived from Phoenician. As both Arabic and Phoenician were Semitic languages, transliteration into Arabic was possible.

Maltese remains, basically, an Arabic language and place-names are derived from Arabic words: *marsa*; harbor; *gebel*; hill; *triq*; street (from the Arabic *tariq,* way); *dar*; house; *bir*; well; *għajn*; spring; *ras*; headland; and *ramla*; sandy beach. Look at a topographic map of the Arab Middle East and you will see these toponyms are common descriptors of physical features. The old capital Mdina (ar. *Madinah*) has a suburb, Rabat, and all the old villages have Arabic names. Although devout Roman Catholics, the Maltese refer to God as *Alla* (Werner, 2004). Maltese is written in a modified Roman alphabet.

After Roger the Norman's reconquest of Malta in 1091, the islands were brought back into Europe politically and economically. Eventually the islands became part of the Kingdom of Aragon and then Aragon and Castile.

Malta under the Crusading Order of St. John, 1530–1798

In 1522 the crusading Order of St. John was forced out of Rhodes (Chapter 9). The Knights of St. John, under the Grand Master, spent time in Italy and Sicily petitioning European rulers for a new home. The union of Aragon and Castile had placed Malta in the Empire of Charles V and he offered the islands to the Order, in fief, with the Order paying an annual fee of one falcon, providing the origin for Dashiell Hammett's *Maltese Falcon* (1930). The Order accepted the fief, to the annoyance of the Maltese nobility who wanted the islands to be directly under the crown. The knights also got Tripoli in North Africa, with the obligation to garrison the town.

The Order arrived in Malta in 1530 and settled in the small port of Birgu on a peninsula jutting into the Grand Harbour. The knights built a hospital, fortifications, and *auberges* to house members drawn from various parts of Europe. In July 1551 a Turkish fleet arrived at Malta, pillaged the villages, and took most of the inhabitants of the northern island of Gozo into slavery. Then the Turks captured the fortress at Tripoli.

The Order built additional fortresses around the Grand Harbour and prepared to resist the next attack. The Turks built up armaments with the Sultan, Sulaiman the Magnificent, ordering a fleet big enough to carry an expeditionary force of over 20,000 men. Commentators noticed many new galleys being built along the shores of the Golden Horn and wondered who was to be attacked. The target was Malta, with the Turkish force arriving off the island in May 1565. The Siege of Malta lasted until early September, when the Mediterranean weather breaks and oared galleys had to make for home port. Under Grand Master Valette, the knights, the Maltese, and the Spanish troops, sent by the Viceroy of Sicily, resisted the siege but when the Turks were driven off, the island was devastated.

The Order had fortifications repaired and laid out a new fortress capital on a high peninsula, which commanded both the Grand Harbour and adjoining Marsamxett. The fortress city, Valletta, is a fine example of sixteenth-century military engineering and town planning (Photo 4.10). Within powerful walls the Order built *auberges,* the hospital, the Cathedral of St. John, a palace for the Grand Master, and other structures to house the administration. (Photo 15.8)

In the eighteenth century beyond the landfront walls, the fortified suburb of Floriana was laid out with broad boulevards, a parade ground, and a triumphal archway. Valletta

Photo 15.8 The Knights of St. John were drawn from across Europe and lived in residences (*auberges*) representing the regions from which knights came. Above is the 18th century Auberge de Castille, Valletta.

Source: Steve Vidler/SUPERSTOCK.

and Floriana display the urban fabric of the sixteenth through the eighteenth centuries as designed by military engineers, under the direction of absolutist rulers. By the end of the Order's rule, in 1798, 40 percent of the population of Malta was living in the towns and fortresses the knights had built or rebuilt around the Grand Harbour.

Out in the villages the Maltese worked as farmers, craftsmen, and traders. Maltese farming was typically Mediterranean. In the mild, moist winter months the cereals and temperate crops were grown and harvested in the spring. Cotton was then sown and harvested in the dry summer. Irrigated crops were grown in *wieds* (valleys) served by a spring (*għajn*). Raw cotton was a major export in the eighteenth century, but some was woven into cloth locally.

As a result of the spending and employment provided by the Order, the Maltese had a better standard of living than Sicilians, Calabrians, or Neapolitans. All the villages built large baroque parish churches, often decorated by Italian artists brought in by the Order to execute commissions in Valletta. Mattia Preti, a well-known seventeenth-century Italian painter, spent years in Malta decorating the ceiling of St. John's Cathedral in Valletta. His paintings are found in rural parish churches.

The Order had a fleet used to combat corsairs and raid Muslim shipping. It paid for its fleet, soldiers, and fortifications with the income from estates all across Europe. The largest group of knights, and the most income, came from France. With the French revolution an aristocratic Order became a target and the estates in France were confiscated in 1792. Then in 1798 Napoleon, on his way to Egypt with an expeditionary force, took Malta, forced the Order out, reorganized government on revolutionary principles, and attacked the position of the Church in Malta. The Maltese rebelled. The French withdrew into the fortified places around the Grand Harbour and were besieged for two years until starved out by the Maltese, aided by the Royal Navy.

Malta and Britain

Malta became a British base in 1800 and by the end of the nineteenth century a major Royal Naval dockyard had been constructed in the Grand Harbour, providing the home port for the Mediterranean fleet. The British military built barracks, docks, hospitals, and eventually airfields, radar stations, and flying boat bases. Numerous Maltese were employed on military establishments and as staff in army, navy, and airforce headquarters. Maltese joined the British armed services and units (the Royal Malta Artillery and the King's Own Malta Regiment) were locally raised and officered. The major business of Malta became looking after British military installations. When World War II arrived, the Maltese more than fulfilled their side of the bargain with courage and determination.

Malta in World War II

In World War II Malta was under attack from June 1940 to early 1943. For a time it could only be supplied by sea at great cost. The Maltese lived on meager rations (1,200 calories) and a daily meal at "Victory Kitchens," where unappetizing goat stew was cooked on fires fueled by timber from bombed buildings. No place in Europe was as heavily bombed as Malta, with the Regia Aeronautica and the Luftwaffe flying the short distance from Sicily to attack airfields and naval facilities. Many of the bombs fell on civilians in the harborside towns. Large numbers of Maltese were evacuated from the harbor towns; those who stayed were forced to sleep in air raid shelters cut into the limestone, which saved many lives. (Photo 15.9)

Malta lay on the Axis supply lines to North Africa. The Malta submarines and torpedo bombers attacked the convoys, taking supplies to the Italian army and General Rommel's *Afrika Corps,* which were trying to capture Egypt and the Suez Canal. If Malta fell, winning the war in North Africa would have taken another year. D-Day would have been delayed. The Red Army would have come to the Rhine.

Maltese towns and villages are stone built and incendiary bombs had little impact. There was no burning-out of towns as happened in western Europe. Evacuation, and in the rock, air raid shelters, kept the bombing death total remarkably low, at around 1,500 civilians. Approximately 11,000 buildings were destroyed, including some of the major structures erected by the Order. By 1943 Malta had fought off all the Axis attacks and became a base for the liberation of Europe, with the invasion of Sicily and Italy being partly launched from the islands.

As World War II ended, Britain provided a war damage fund to finance rebuilding of the towns and the economy. In the early Cold War the Malta bases were fully utilized

Photo 15.9 The Valletta Opera House was heavily damaged by Axis bombing in 1942. The building had to be demolished and the site sits empty to the present.

Source: ©AP/Wide World Photos.

and the economy prospered. On the constitutional front Britain and Malta could never get the balance right. From the mid-nineteenth century the intention was to have a council or legislature that ran the affairs of Malta, with Britain represented by a Governor, being responsible for the military installations, defense, and foreign affairs. The division of responsibility never seemed to be settled for long.

In the mid-1950s the Prime Minister of Malta, Dom Mintoff, a Rhodes Scholar, proposed that Malta and Britain integrate. Former French colonies in the Caribbean have become departments of France. Malta was to become part of the United Kingdom and elect members to the Parliament at Westminster. Britain undertook to bring living standards up to those in the United Kindom. The project stumbled on religious questions. Britain was largely Protestant; Malta was Roman Catholic. Britain would give formal guarantees that the Church in Malta would retain status and independence. The Prime Minister of Malta, leader of the Malta Labour Party, was trying to modernize Malta by reducing the power of traditional institutions. The Archbishop of Malta never got the guarantees he wanted from the government of Malta and the project faded away. If it had succeeded, Malta would have probably joined Europe as part of the United Kindom in 1973.

Independence

Malta became independent in 1964 with a unicameral legislature. The independent country, like a number of former colonies, recognized the Queen as the constitutional head of state, with the sovereign represented by a Governor General of Maltese nationality. Independence required a transformation from an economy dependent upon military spending to a modern, diversified economy. Britain handed the naval dockyard to Malta to be used as a commercial shipyard. Britain signed an agreement of Mutual Assistance and Defence and made grants and loans available to improve infrastructure, create industrial parks, and attract manufacturing industry. Tourism began to grow rapidly as part of the general trend of northern Europeans taking vacations in the sunny south. Most Maltese speak English and the British connection ensured that many visitors came from the United Kindom, still the largest source of tourists today.

In the 1970s Britain paid a rent to use the bases until 1979, when all were closed. Mr. Mintoff was back in power. He turned Malta into a neutral republic and the military installations had to go, although after an interval, vessels of the Royal Navy make courtesy calls and the U.S. Navy uses the Grand Harbour to give crews a short shore leave.

Summary

Since the Neolithic the Maltese islands have usually contained more people than the land could support with locally grown foodstuffs. Trade and the provision of services have been an important part of life in the archipelago since the emergence of the Temple Culture starting in 4,000 B.C.

The Order of St. John, a multinational European defense force—with members from Iberia, France, Italy, and until the Reformation, England—had the resources, drawn from estates in Europe, to build a military base in the central Mediterranean on the channel linking the east and west basins of the sea. The well-fortified base attracted the Royal Navy and from 1800 until the 1960s Malta lived on British military spending.

The islands attract a million visitors a year and have become an offshore banking and manufacturing center. Maltese labor is no longer cheap and new manufacturing investment is difficult to attract. The cultural and historical resources of the islands are well known and the tourist sector grows. The Royal Naval Dockyard became a commercial shipyard, but never made money and receives governmental support.

As with much of Mediterranean agriculture, small terraced fields are going out of cultivation. On better land, higher value crops for export are grown and the area under vineyards is increasing. Visitors find Maltese wines pleasant and attractively priced.

The educated inhabitants of the island speak three languages: Maltese (*Malti*), English, and Italian. Broadcast programs are available in all three. With strong links to Britain and Italy, together with a history of trade and entrepreneurial activities, Malta prospers in EU.

▶ GIBRALTAR

Gibraltar is a United Kingdom Overseas Territory and Gibraltarians are UK citizens. Gibraltar joined the European Community with the United Kindom in 1973 and enjoys a special status being exempt from EU fiscal policy. Presently there are no plans to change status.

Gibraltar is a rock of Jurassic limestone rising up to 1,396 feet (Figure 15.7) (Photo 15.10). Although not the most southerly point on the Iberian peninsula, the Rock is positioned to exercise control over the Strait of Gibraltar in wartime. There is little flat land around the Rock, apart from the sandy spit running to La Línea in Spain, largely occupied by the airport, partially built out over the sea.

The 30,000 Gibralterians are fed with imported foodstuffs and watered by seawater distillation plants and huge concrete catchments, constructed on the east side of the Rock. The Rock is full of military installations excavated into the limestone.

Climate

Gibraltar has a modified Mediterranean climate, being close to the junction of the Atlantic Ocean and the Mediterranean Sea. In the winter months Atlantic depressions deliver approximately 30 inches of rainfall, a relatively high figure for a Mediterranean climate station. The cooler, less saline Atlantic water, which flows into the Strait, moderates summer temperatures. July daytime highs are in the 80s, rather than the 90s experienced on the Mediterranean coast of Spain. All year round colder Atlantic water, at depth, is kept out of the Mediterranean basin by an underwater sill.

The Rock produces a number of peculiar weather features. Frequently there is a plume of cloud streaming from the peak as the Rock uplifts and cools passing airstreams. The Rock produces notorious wind eddies, which were feared by aviators before the introduction of modern powerful aircraft.

Site and Situation

The town of Gibraltar is sited on the less steep western flank of the Rock. There is no natural harbor and harbor works have been built out into the sea on the west. Spain has

Photo 15.10 Gibraltar with the Spanish town of La Línea in the foreground.

Source: Miles Ertman/Masterfile.

far better sites for ports and, militarily, Gibraltar was not much used when the Rock was under Spanish control. From an English perspective the position of Gibraltar had immense potential. It was worth the cost of constructing a naval base from which warships could be deployed into the Atlantic or the Mediterranean.

Strategic Value

In the Arab-Berber invasion of Spain starting in 711 A.D., Tarik-ibn-Zeyad captured Gibraltar and fortified the place, creating a base for campaigns inland. The name of the Rock commemorates the conqueror. Gibraltar is the hill (*Jebel*) of Tarik. The Rock was reconquered in 1462, becoming a part of the Kingdom of Aragon and Castile in 1501, and developing into a prosperous free trade port on a major searoute.

In the seventeenth century England expanded her overseas possessions. Jamaica, in the Caribbean, was taken from Spain starting in 1655. In the Mediterranean a small fleet was stationed at the North African port of Tangier from 1662 to 1684. In 1703 a treaty with Portugal allowed Royal Navy vessels to use Lisbon and in 1704 Gibraltar was taken from Spain, giving England a port near the Strait of Gibraltar. Gibraltar could be used to keep apart both the French and the Spanish Atlantic and Mediterranean fleets, besides protecting a commercial seaway of growing importance to English merchants.

By the Treaty of Utrecht (1713) British possession of Gibraltar and Minorca (in the Balearic islands) was recognized by Spain with conditions. Catholics were guaranteed freedom of worship. Moors and Jews, groups expelled from Spain, were not to be allowed into Gibraltar. Should Britain leave the Rock, it had to be offered to Spain. The boundary between Gibraltar and Spain was not defined. Smuggling into Spanish ports was prohibited. In practice, smuggling was not suppressed, and Jews and Moors used the port.

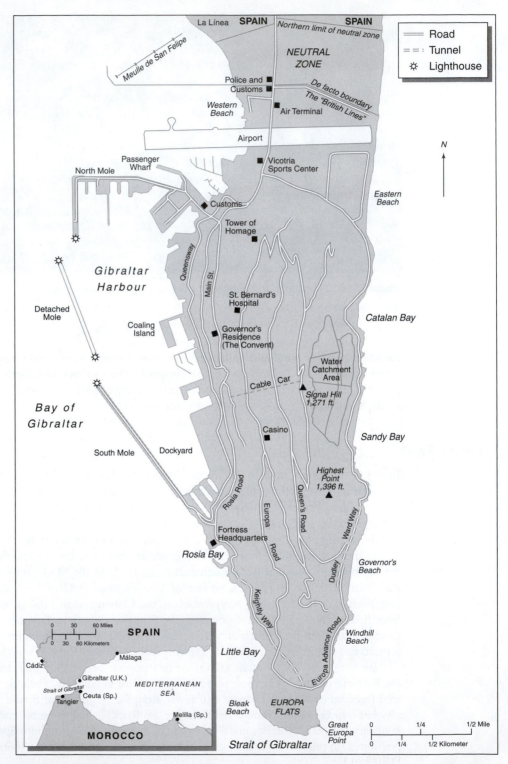

Figure 15.7 Gibraltar.

Spain has never accepted the loss of Gibraltar. Armed attempts were made to recover the territory, most famously in the Siege of 1779–1784 when Britain was involved in a North American colonial war. The Spanish siege of Gibraltar was unsuccessful and, in the Napoleonic wars, Spain was invaded by France, setting off a chain of events leading to Mexico and Spain's South American colonies becoming independent.

During the Napoleonic wars, Gibraltar was a valuable commercial port and naval base. Napoleon instituted a Continental System to exclude British goods from European markets. Gibraltar became a key commercial base running goods to Mediterranean ports. When Nelson was blockading the French Mediterranean fleet in Toulon and chasing Napoleon's expedition to Egypt, destroying that fleet near Alexandria, Gibraltar was an important base, as it was when Lord Nelson, off Cape Trafalgar, on Spain's Atlantic coast, defeated the larger combined fleets of France and Spain in 1805.

Under British rule, merchants from Portugal, Genoa, North Africa, and British ports established businesses at Gibraltar. Some Spanish merchants stayed on, under suspicion from the British and their countrymen, who despised them as Rock Rats making money under a foreign flag. To do business with British regiments, the Royal Navy, and the dockyard, local merchants needed English and that language became dominant, although Spanish, Italian, and Portuguese were in use.

By the end of the Napoleonic wars Britain had the base at Malta and gave up Minorca. The Royal Navy now had a base on the Strait of Gibraltar and another in the central Mediterranean on the relatively narrow passage between Sicily and North Africa. As France developed North African imperial ambitions, starting with Algeria in 1830, the naval bases at Gibraltar and Malta acquired enhanced strategic importance.

The trade and strategic situation in the Mediterranean was transformed by the opening of the Suez Canal. The Canal cut the journey to India, South East Asia, Japan, and the Far East by thousands of miles, as it was no longer necessary to sail around Africa and the Cape of Good Hope. From the beginning, more than half the merchant vessels going through the Canal were British registered. The Mediterranean seaway, in strategic terms, became the passage to India, Australia, New Zealand, Singapore, Hong Kong, and Shanghai. The Suez Canal opened in 1869, the same year that the transcontinental railroad was completed across North America. The world was globalizing, as Jules Verne suggested in *Around the World in Eighty Days* (1873).

With the rise of a German navy in the early twentieth century, large investments were made in the Gibraltar dockyard and naval base. When World War I came, Malta and Gibraltar were on the supply routes for allied forces fighting Turkey. Neither base suffered an attack. However, in presonar World War I, U boats moved through the Strait without detection.

In World War II Gibraltar was bombed by the French airforce, in retaliation after the Royal Navy shelled the French fleet in North Africa to prevent it being used by Germany. The German attack on Gibraltar (Operation Felix) never took place, as General Franco refused to ally with Nazi Germany. Gibraltar was a key base in the antisubmarine Battle of the Atlantic and the North African campaigns. General Dwight Eisenhower established his headquarters for Operation "Torch," the invasion of North Africa, in 1942.

During the Cold War Gibraltar was a major position within the NATO command structure. Today, the base is maintained by approximately 1,000 British military and a slightly greater number of Gibralterian civilian employees. NATO keeps a command operation on the Rock and operates surveillance and communication facilities.

The economy of Gibraltar depends largely on services and port facilities. Although tourist accommodation, and space for second homes and apartments, is limited, many visitors come to spend a day on the Rock from cruise boats. The cruisers shop the main street duty-free outlets, and ascend the Rock by cable car to enjoy views of Iberia and Africa. Financial services are a growth area in the economy.

Constitution

The Gibraltar House of Assembly consists of a Speaker and 15 elected members, with the Attorney General and the Financial Secretary being *ex officio* members. The leader of the Assembly, with the title Chief Minister, presides over a Council of Ministers, which runs domestic affairs. The Governor, appointed by the British Government, is the commander in chief, with responsibility for defense, foreign affairs, and internal security. In most matters the Governor acts on advice from the Chief Minister but has large powers in any time of emergency, for Gibraltar remains a military base.

Constitutionally Gibraltar is in a *cul de sac*. Spain still insists that it should have sovereignty over Gibraltar. Britain suggested a joint sovereignty formula. In 2002 Gibraltarians rejected this idea in a referendum in which 99 percent of the votes cast were against joint sovereignty, which Gibraltarians saw as a step toward Spanish control. In 2006 Britain recognized Gibraltar's right to self-determination, but Spain immediately insisted that under the Treaty of Utrecht (1713), if Britain leaves, the Rock reverts to Madrid.

In 1969 Spain closed the border with Gibraltar. Movement was restored in the 1980s, but with tight border controls. It remained difficult to dial Spain from Gibraltar and the Spanish telephone system did not recognize Gibraltar's international calling code. Most serious were the airspace restrictions. There were no flights from Spain to Gibraltar, or in the opposite direction. Spain would allow no overflights from European cities. Planes coming from Britain had to avoid Spanish airspace on takeoff and landing. The problems will now ease. In 2006 representatives from Gibraltar, Britain, and Spain negotiated a reduction of Spanish restrictions on border crossings, telecommunications, and airspace use. Now Spain will allow direct flights to Gibraltar from Spanish cities and elsewhere in Europe. Spain still claims sovereignty over the Rock.

The EU in Africa

Across the Strait of Gibraltar, in North Africa, are two Spanish provinces, Ceuta and Melilla, which Spain acquired in 1580 and 1497. The exclaves of Spain are surrounded by Morocco, which has the same view of Ceuta and Melilla as Spain has on Gibraltar. Spain has detailed legal arguments to explain why Ceuta and Gibraltar are dissimilar cases. Morocco has difficulty understanding how Spain can argue to acquire Gibraltar while retaining Ceuta and Melilla. (Photo 15.11)

Ceuta and Melilla are part of EU. The provinces are attractive targets for Africans seeking refugee or asylum status in Europe. The provinces are fenced but ladders and battering rams are used to break through and, once in, the trespassers have to be processed. Many are allowed to stay under the rules of asylum.

Photo 15.11   A powerful, well-guarded fence with razor wire separates Spanish Ceuta from Morocco and immigrants who cross the Sahara desert in an effort to enter Europe.

Source: Anton Meres/Reuters/Landov LLC.

Islas Canarias

Spanish sovereignty over the Canary Islands was recognized by Portugal in 1479. The islands, comprised of two Spanish provinces, lie off the west coast of Morocco—an archipelago composed of volcanic rocks rising to 12,162 feet, higher than any peak in the Iberian peninsula. The volcanic rocks break down into fertile soils and, in the subtropical climate, good crops of bananas, grapes, and tomatoes are produced. The mild climate, good beaches, and fine fishing opportunities attract tourists.

In recent years another type of traveler has been arriving in large numbers—the illegal immigrant. Boatloads of migrants seeking entry to Europe arrive everyday from Mauritania, Senegal, and more distant sources. Currently over 20,000 people are housed in camps awaiting determination of status. The EU is establishing joint surveillance projects to curtail the arrival of boat people. (Photo 3.10)

In 1987 Morocco applied for membership of the European Community and was rejected on the grounds that the Treaty of Rome only allows European states to be members. The Maastricht Treaty, which established EU, carries a similar Europe-only provision. Presently there is no likelihood that the EU will take in countries from North Africa.

▶ FURTHER READING

TURKEY

Callslar, I. *Latife Hanim*. Istanbul: Doğan Kitapcilik, 2006.

Cohen, S.B. "The Geopolitics of Turkey's Accession to the European Union." *Eurasian Geography and Economics* 45(8) (2004): 575–582.

Dalhman, C. "Turkey's Accession to the European Union: The Geopolitics of Enlargement." *Eurasian Geography and Economics* 45(8) (2004): 553–574.

Finkel, C. *Osman's Dream: The Story of the Ottoman Empire 1300–1923*. London: John Murray, 2005.

Freely, J. *Istanbul: The Imperial City*. New York: Viking, 1996.

Harris, L.M. "Irrigation, Gender and Social Geographies of the Changing Waterscapes of Southeastern Anatolia." *Environment and Planning D: Society and Space* 24(2) (2006): 187–213.

Harris, L.M. "Water and Conflict Geographies of the Southeastern Anatolia Project." *Society and Natural Resources* 15 (2002): 743–759.

Lahlos, S. (ed.). *Turkey: Current Issues and Background*. New York: Nova, 2003.

Murphy, A.B. "Turkey's Place in the Europe of the 21st Century." *Eurasian Geography and Economics* 45(8) (2004): 583–587.

Ryan, W. and W. Pitman. *Noah's Flood*. New York: Simon and Schuster, 1998. For a description of the creation of the Bosporus.

Sarkis, H. *A Turkish Triangle: Ankara, Istanbul and Izmir at the Gates of Europe*. Cambridge, MA: Harvard University, Graduate School of Design, 2006.

Tas, H.I., and D.R. Lightfoot. "Gecekondu Settlements in Turkey." *Journal of Geography* 104(6) (2005): 263–271.

Zeigler, D.J. "Cities of the Greater Middle East," in Stanley D. Brunn, Jack F. Williams, and Donald J. Zeigler, *Cities of the World*. Lanham, MD: Rowman and Littlefield, 2003.

CYPRUS

Berg, E. "Pooling Sovereignty, Losing Territoriality? Making Peace in Cyprus and Moldova." *Tijdschrift Voor Economische en Sociale Geografie* 97(3) (2006): 222–236.

Bill, J.A. *George Ball*. New Haven CT: Yale University Press, 1997. For an account of the U.S. effort to solve the Cyprus problem in 1964.

Crawshaw, N. *The Cyprus Revolt: An Account of the Struggle for Union with Greece*. London: George Allen, 1978.

Durrell, L. *Bitter Lemons*. New York: Dutton, 1957. Durrell, the novelist, spoke Greek, and spent time in Greece before living in Cyprus, writing an observant account of the island during the *Enosis* crisis of the 1950s.

Solsten, E. (ed.). *Cyprus: A Country Study*. Washington, DC: Library of Congress, 1993.

MALTA

Blouet, B.W. *The Story of Malta*. Fifth Edition, Valletta, Malta: Progress Press, 2007.

Gimbutas, Marija. *The Living Goddess*. Berkeley: University of California Press, 1999.

Werner, Louis. "Europe's New Arabic Connection." *Saudi Aramoco World* (June 2004).

Wettinger, G. *Place Names of the Maltese Islands ca 1300–1800*. Malta: PEG, 2000.

GIBRALTAR

Gold, P. *A Stone in Spain's Shoe: The Search for a Solution to the Problem of Gilbraltar*. Liverpool UK: Liverpool University Press, 1994.

Harvey, M. *Gibraltar*. Staplehurst, Kent: Spellmount Publishers, 2000.

Hughes, Q., and A. Migros. *Strong as the Rock of Gibraltar*. Gibraltar: Exchange Publications, 1995.

Morris, D.S., and R.H. Haigh. *Britain, Spain and Gibraltar 1945–1990: The Eternal Triangle*. New York: Routledge, 1992.

Winchester, S. *Outposts: Journeys to the Surviving Relics of the British Empire*. London: Penguin, 2003.

16

THE EUROPEAN UNION AND THE WIDER WORLD

▶ INTRODUCTION

An age of European Union expansion is coming to an end. The number of countries outside EU markedly decreased with the enlargements of 2004 and 2007. Although Croatia and Serbia are likely to be admitted to EU, in an effort to bring political stability to the Balkans, other possible candidates for membership, including Belarus, Russia, and the Ukraine, are not yet free of corruption nor do they possess secure democracies. The campaign to include Turkey is presently stalling with several EU members, including France and Austria, resisting full membership for the secular, but Islamic country. The Turks have not helped their cause by prosecuting writers expressing unpopular views. There has been a change of mood among European voters. The failure of the 2005 constitution reflected suspicion of the faulty document and reservations concerning further expansion. Europe is in a phase of expansion fatigue.

▶ WIDENING

We can anticipate the completion of the spatial widening of the EU in the next decade. Bulgaria and Romania joined in 2007. Croatia is commencing talks on entry and if successful, the remaining Balkan states are likely to follow, with the exception of Albania. Further east in Europe, Moldova will want to join EU as its western neighbor, Romania, integrates with Europe. Absorbing the Ukraine would be more difficult for it is a large country and, in the east, the ethnic Russians are reluctant Europeans.

Russia is too large for the EU to take in without straining the resources available to restructure industry, improve agriculture, and help backward regions. Russia has many poor regions, often in remote areas, like northern Siberia, which are no longer economically viable now that the subsidies, provided by the Soviet Union, have gone. Beyond the size issues, Russia has governmental and human rights problems that are a long way from meeting EU standards. Few EU members want to give Russians the power to shape European policy by pressuring eastern European countries to vote with them.

Turkey has talked to the EU about entry. Opponents of Turkish entry point to the size of a country that, in terms of population, will be as large as Germany by 2015 and fear large numbers of Turks coming west. Turkey has, by European standards, a high rate of population growth (1.4% p.a.) and low per capita incomes. There are reservations concerning the volume of resources needed to bring Turkey to the average EU economic standards.

▶ DEEPENING

Deepening refers to the creation of European institutions that merge additional aspects of sovereignty and bind the members of EU more closely together. Advocates of deepening suggest that harmonizing more aspects of life will be another step toward creating common economic space in which corporations can operate in a multinational environment. The adoption of the Euro in 1999 deepened the financial ties between members (except Denmark, Sweden, and the UK, which did not adopt the Euro) and made it easier to conduct business across Europe, as differing exchange rates, and the costs of exchanging currencies, were abolished.

The rejection of the EU constitution in 2005 was a disappointment for the deepeners. The document was too complex (see Chapter 6) and in countries like France, with high unemployment, a significant segment of voters were disillusioned with economic performance within EU.

▶ EU EXTERNAL RELATIONS

The EU does not have common foreign and defense policies. There are advocates of creating unified policies. Those against it point out that most of western and eastern Europe are NATO members and creating another set of defense policies will cause confusion. There are commentators who believe that NATO should take a broader role in world affairs. For example, NATO could participate in efforts to prevent Iran acquiring nuclear weapons. Here, however, European countries are already involved.

In foreign affairs, countries like Britain and Spain, with wide overseas connections, legacies from imperial pasts, are more committed to inclusive outreach than France and Germany. A common foreign policy is not likely in a few years. The existing foreign offices in all European countries would be reluctant to give up responsibilities and expertise to a unified, European department of state.

For the EU, common external policies are largely concerned with trade. Europe has a harmonized system of tariffs and makes regulations on what can and cannot be traded into and out of EU economic space. Europe, like the U.S., enforces generally accepted restrictions on trade in endangered species. In other areas there is conflict. The EU does not want to import meat products where the animals have been fed hormones to promote growth. The EU is wary of genetically modified (GM) crops. Citizens' groups make it difficult to grow GM crops and imports are under restriction. The fear is that GM crops may have unintended ecological effects and contaminate existing crops, as has happened with rice varieties in Louisiana. (Photo 16.1)

Hormones and GM crops have led to minor trade wars between the U.S. and EU. Fortunately, the conflicts have usually been referred to the WTO for resolution, which slows everything down and allows emotions to subside and events to move on. The U.S. was imposing special tariffs on steel imports from Europe and elsewhere. EU went to

Photo 16.1 All across Europe there are sustained protests against the growing of genetically modified crops. In this Spanish example, the protest is against GM corn.

Source: Pedro Armestre/Reuters/Landov LLC.

the WTO and got a favorable ruling. In the meantime worldwide demand for steel, led by Asian markets, increased and the need for tariffs to protect U.S. steel makers was reduced.

▶ THE COMMON AGRICULTURAL POLICY IN WORLD AFFAIRS

The strengths and weaknesses of the Common Agricultural Policy (CAP) were discussed in Chapter 5. In terms of EU external relations the CAP has a number of disadvantages. The CAP, by the use of farm subsidies, encouraged farmers to overproduce crops and be paid prices in excess of the world market price. Further, the EU market is protected against cheap imports of agricultural products by high import tariffs.

As discussed previously, the CAP has helped rationalize and modernize farming in EU, but it has disadvantages for world trade and EU relations with less developed countries. For example, subsidized sugar beets are grown in Europe and the price of sugar is higher than it would be if the sugar cane farmers of Brazil, the Dominican Republic, and Central America were allowed low tariff entry into Europe. Powerful farming interests argue for the continued protection of high-cost sugar beet cultivation within EU. By removing protection, the price of sugar in Europe could be cut and the economies of several developing countries helped. The EU spends 42 percent of the budget on the CAP and less than 5 percent on overseas aid.

By the way, the U.S. protects sugar beet farmers by imposing quotas on sugar imports from the Caribbean, Central America, and elsewhere. We too have sugar prices that are at least a third too high in order to protect sugar beet producers. Sugar, of course, is a raw material in many products and high sugar prices inflate food prices in both Europe and the United States.

▶ IMPERIAL PASTS

European countries have imperial pasts—some in recent times, others much earlier. France, The Netherlands, and Britain had large overseas empires. Sweden, Denmark, Italy, and Germany had overseas possessions. Sweden had settlements in New Jersey. Denmark was the former owner of the U.S. Virgin Islands. Germany had African and Asian colonies until the end of World War I. Italy held Tripolitanea and Cyrenaica (Libya) and Abyssinia (Ethiopia) until the end of World War II.

Sweden had a Baltic empire in the seventeenth and eighteenth centuries. Swedish influence remains strong in Finland. The Kingdoms of Poland and Lithuania ruled large parts of eastern Europe in the sixteenth century. Going back to Classical times, Greece and Rome controlled segments of Europe and the Mediterranean world. The former widespread influence of the Italian peninsula is still reflected in the Roman Catholic Church centered upon Vatican City. The extent of ancient Greece was a driving force in the territorial expansion of modern Greece and many Greeks resented the appearance of Macedonia on territory perceived to be part of their ancient world.

Austria, Germany, Hungary, and Turkey had extensive territorial empires and the lines of old boundaries still echo in the modern world. The postwar boundary between Germany and Poland has taken a long time to settle.

▶ DECOLONIZATION

After World War II, decolonization for France was traumatic. French Indo-China (modern Vietnam) was lost in 1954 after an unsuccessful effort to reassert French colonial control after Japanese occupation. Violence also developed in North Africa and General de Gaulle asked colonies to make a choice: remain part of France or become fully independent. The African territories chose independence, the Caribbean islands remained with France, becoming departments with legislatures at home and representatives elected to the *parlement* in Paris. Martinique and Guadaloupe are Western-hemisphere segments of the EU.

As we saw in Chapter 15, Britain was prepared to integrate the Mediterranean colony of Malta into the United Kingdom in 1955. The inability of the government of Malta to provide guarantees to the Roman Catholic Church stalled the integration idea and everyone had second thoughts. Commentators in Britain pointed out that over forty colonies were moving to independence. If a significant number wanted integration, and representation in the Parliament at Westminster, the British two-party system might be radically altered. A large number of members of Parliament from former colonies would be voting in London on issues that were not of much direct importance in West Africa, for example.

Britain decolonized conventionally. Sovereignty was transferred to successor states, most of which stayed in the Commonwealth. Some, such as Trinidad, became republics. Others, like Barbados, retain the Queen as the constitutional head of state, represented by a Governor General of Barbadian nationality. The Governor General signs into law bills passed by the Parliament in Bridgetown and acts on advice from the Prime Minister of Barbados. Barbadians still receive knighthoods and other orders of recognition.

In addition to Gibraltar (Chapter 15), Britain has fourteen Overseas Territories, including Bermuda, the British Virgin Islands, Cayman Islands, Montserrat, and the Falkland Islands. Most of the territories have elected legislatures, but UK-appointed governors are responsible for defense, security, and foreign affairs. Citizens of the overseas territories have UK citizenship and can move to Britain if they wish. *De facto,* citizens of Britain's Overseas Territories can reside in EU countries.

The Netherlands are represented in the Caribbean by Aruba (an autonomous part of the Kingdom of The Netherlands), and the self-governing Netherlands Antilles. In both Aruba and the Antilles, The Netherlands is represented by a governor.

Spain and Portugal had large overseas empires. Portuguese is the language of Brazil and Spanish is used in much of Latin America, providing EU with cultural and business links to Mexico and Central and South America. The Atlantic archipelagoes of the Azores (Portugal), the Canary Islands (Spain), and Madeira (Portugal) are integral parts of the Iberian countries.

The EU does provide aid for former colonies, but where aid has taken the form of advantageous access to European markets, the WTO has decided the arrangements are outside the rules. For example, the EU gave favorable terms of entry to Caribbean and African banana producers. Multinational corporations, wanting to sell Central American bananas, claimed that the aid was unfair to them.

► THE EUROPEAN UNION AND NAFTA

It might appear that a free trade agreement between Europe and NAFTA would be the next logical step to promote North Atlantic trade. Agreement is unlikely. NAFTA, though called a free trade agreement, has many complexities and provisions that would not mesh with the European institutions. The U.S. is under pressure to exclude products from NAFTA trade partners on the grounds that the competition is unfair. Trade associations claim that wheat farmers and timber producers in Canada enjoy subsidies or that Mexican tomato growers have the unfair advantage of a longer growing season! Sometimes the U.S. imposes penalties on imports. If there is to be tariff-free North Atlantic trade, a new North Atlantic Free Trade Treaty would have to be negotiated. Countries in Europe with slow growth economies and high unemployment, including France and Germany, would be wary of more imports and would, at present, resist a major new free trade agreement. Britain would welcome freer trade with North America.

► TURKEY, THE MIDDLE EAST, AND CENTRAL ASIA

The United States would like Turkey to become part of EU, but several European countries are reluctant to welcome the country as a member. It is wrong to assume that admitting Turkey will improve relations with adjoining, Arab, Middle Eastern countries. Turkey is a secular state but most Turks subscribe to Islam. Several Middle Eastern states are Islamic fundamentalist. Turkey recognizes and works with Israel. There was a time when Baghdad and Damascus looked to Turkey as a center of culture, education, and influence. The prestige of Istanbul and Ankara diminished with the rise of Arab nationalism in the post World War II era.

The hope of better relations between Europe and the Arab Middle East was revealed to be unrealistic, early in 2006, with the publication of cartoons depicting Mohammed in European newspapers. The cartoons were originally published in Denmark in the fall of 2005. Republication in French newspapers, and elsewhere, led to rioting in much of the Muslim world. European embassies were attacked, with authorities making little effort to maintain law and order. In the west it was generally felt that a number of Middle Eastern governments had encouraged riots to promote anti-European sentiments. There is little immediate hope for greatly improved relations between Europe and the Arab Middle East.

Turkey has developed good relations with a number of the Central Asian, Turkic republics. The republics have oil and gas resources and Turkey can be connected to the transportation routes coming from the Caspian Sea, via Azerbaijan and Georgia, to the Black Sea. Turkey is becoming important to European energy supplies.

Turkey has strained relations with Armenia over the treatment of Armenians in World War I. Armenia has poor relations with Azerbaijan over Nagorno Karabakh, an exclave of Armenia in Azerbaijan.

▶ THE EU AND THE UNITED STATES: A PARABLE OF PORTS

The U.S. trades heavily with China, Japan, and other parts of Asia. In terms of investment the EU is more important to U.S. interests. Many multinational corporations have interlocking operations in Europe and North America. EU investment rules make it easy for corporations to move capital between EU countries. In economic and cultural terms Europe and North America have strong and long-established connections. In the foreign policy arena there are divisions. Some European countries supported U.S. Iraq policy with troops. Other countries, including France, denounced U.S. interventions.

Europe and North America are heavily linked by the forces of globalization. The term *globalization* is a cliché of our times, but the process goes back centuries and produces complex interconnections. For example, the mid-nineteenth century a British steamship company offered a passage to India via the Mediterranean Sea, passengers being transported overland, via Cairo, from Alexandria to the Red Sea. Then in 1869 the Suez Canal opened and the Peninsular and Oriental Steam Navigation company could take passengers, without transshipment, from Europe, through the canal, to India, southeast Asia, and the Far East.

In the 1890s, with rising Russian and German power, Britain feared foreign presence on the Persian Gulf, that might allow expanding powers to threaten predominance in the Indian Ocean. Britain signed defense agreements with Kuwait, Bahrain, Qatar, and several other small Sheikdoms, including Dubai, (Dubayy) that eventually became the United Arab Emirates. The defense agreements made the Persian Gulf territories *de facto* independent and prevented them being absorbed by Iraq, Saudi Arabia, and Iran when those states emerged in the twentieth century. After World War II oil was exploited in the Gulf Sheikdoms. Great wealth was generated, which was used to diversify economies and to buy investments overseas.

The P&O shipping line, in recent decades, diversified into containers, ferries, and port management and its shares prospered on the London Stock Exchange. In the United

Photo 16.2 Europe is a large importer of oil. Rotterdam is a major terminal and refining center.

Source: Imageshop/Age Fotostock America, Inc.

Arab Emirate of Dubai, (Dubayy) the ports were modernized and the new expertise used to acquire and run facilities in Europe, Asia, and Latin America (Photo 16.2). The state-owned Dubai World Ports made a bid for the shares of P&O. The shareholders accepted. Just another aspect of globalization, with capital moving from the Middle East to London to buy a publically quoted company to the benefit of the financial services industry. However, among the ports P&O ran were facilities in New Orleans, Miami, Baltimore, Philadelphia, and New York. Of course, all the health, safety, security, and immigration rules, mandated by the U.S. government, would remain in place, but this did not stop protests against a state corporation, from an Arab country, making an investment in the U.S. economy.

The regulatory authorities in the United States were aware of the takeover and treated it as a routine transaction, but in the political arena there was turmoil, which the Dubai (Dubayy) company resolved by selling the P&O port operations in the United States.

▶ THE EURO

Most countries in EU adopted the Euro. Now many countries, outside of Europe, use the currency as a vehicle to store reserves. Typically, countries will have foreign currency reserves in the dollar, the Euro, the Yen, and British Pound, and some Swiss Francs. The rise of the Euro has allowed countries to diversify holdings, often shifting part of reserves to the Euro, from the dollar. When the Euro was launched it was weak against the dollar. This is no longer the case.

▶ SUMMARY

The European Union has been successful in integrating the economies of European countries and in harmonizing the numerous legal and regulatory systems. The full economic benefits of the absorption of eastern Europe have yet to appear but the countries admitted in 2004 and 2007 are restructuring economies. Restructuring is painful, particularly when we remember that the collapse of Communism resulted in economic recession a little more than a decade ago. Now opening economies to competition from western Europe is resulting in both new investment and the closing of older industries. Many people lost jobs and unemployment rates in eastern Europe remain high.

In the "old EU," the economies of France, Germany, and Italy have slow growth and high unemployment. All three countries are reluctant to abandon the statist policies of the past. By contrast, the Benelux countries—Belgium, The Netherlands, and Luxembourg—have embraced globalization and even as relatively small countries are home to many transnational corporations and international organizations.

GLOSSARY

Bold type within a definition indicates a term referred to elsewhere in the glossary.

Agricultural Revolution In the seventeenth century the Dutch introduced crop rotation, manuring, and fodder crops to create a system of mixed farming that produced arable crops and livestock products. As a result of manuring and the planting of nitrogen-fixing fodder crops like clover, land could be kept in continuous cultivation and output rose. The industrial revolutions of the nineteenth century initiated the mechanization of farming as labor migrated from rural areas to cities.

Altitude Height above sea level, expressed in feet or meters. One meter is equivalent to 3,281 feet (see **Metric system**).

Anticyclone A high-pressure atmospheric system with stable air. In summer the Azores high brings warm, sunny weather to western Europe. In winter the Siberian high-pressure system becomes established, bringing cold, clear conditions to eastern and central Europe.

Arable land Cultivated land. Literally "plow land."

Autarky Economic self-sufficiency.

Balkanized A region broken in small, competing, hostile states.

Balkans Peninsula projecting into the Mediterranean Sea containing Bosnia-Herzegovina, Montenegro, Serbia, Croatia, Macedonia, Albania, Greece, and parts of Bulgaria, Turkey, and Romania. Literally, a Turkish term meaning "wooded mountains."

Birth rate, crude The number of live births per year, per thousand of a population.

Bosporus Waterway connecting the Black Sea to the Sea of Marmara and the Aegean.

Buffer state An independent country separating more powerful states.

Built environment A term used synonymously with **cultural landscape**, but which usually refers to urban places.

Central place A settlement (village, town, city) that provides goods and services to a surrounding trade area.

Chernozem Rich, black earth soil derived from loess.

Child mortality The number of children per thousand live births who do not reach their fifth birthday.

Cholera Disease of the intestine caused by the bacterium *Vibro cholerae,* spread by infected drinking water or foodstuffs, including shellfish.

Common Market A group of countries promoting the free movement of goods, capital, and labor between their national territories; establishes a common external tariff on goods entering the market from nonmember states and works to create common business conditions. The European Common Market, the European Economic Community, was established in 1957 by France, Germany, Italy, and the Benelux countries.

Continental Climate In temperate lands, a climate with large, seasonal temperature variations (see **Continental Polar**). Much of interior North America has a continental climate, as does central and eastern Europe.

Continental Polar (cP) Very cold, dry air originating in winter over Siberia and the Eurasian landmass. In North America cP air develops over the Canadian north.

Continental Tropical (cT) Hot, dry tropical air originating over deserts, including the Sahara.

Conurbation A cluster of towns and cities that have merged to create a continuous urban area.

Convective rainfall Precipitation resulting from the lifting of air from the earth's surface, often in a thunderstorm system.

Cultural landscape The features constructed by humans that form part of the visible landscape (see **Natural environment** and **Built environment**).

Culture The shared beliefs, values, traits, and technologies that bind a people together.

Customs Union Countries that agree to establish free trade among themselves and impose a common external tariff on goods entering from nonmembers. The free movement of labor and capital is not required, as in a **Common Market**.

Cyclonic rainfall Precipitation that results from the passage of a **depression** or low-pressure system.

Death rate, crude The number of deaths per year, per thousand of a population.

Decolonization The process by which a colonial power withdraws from an overseas territory and transfers political control to an independent successor state.

Demographic Transition Model A series of decreases in birth rates and death rates. As death rates fall sooner in the transition than birth rates, there is a surge in numbers before population growth rates slow.

Demography/demographic The study of human populations, especially size, density, distribution, and vital statistics.

Depression A low-pressure weather system producing clouds and precipitation.

Devolution The transfer of some powers from a central government to regional authorities.

Direct migrant A person moving directly from place of origin to destination without residing at an intervening place.

Doubling time The number of years it takes a population to double in size. Calculated by dividing the percentage rate of increase into 70. A population consistently growing at 2 percent annually doubles in 35 years, if the percentage rate of growth remains constant.

Environmental determinism The view that the physical environment controls, or determines, human activity.

Esker A deposit of sand and gravel left by a stream flowing under a glacier, often high enough to provide a routeway across marshes or lakes in postglacial times.

Estado Novo Literally, new state. The corporatist state imposed upon Portugal by the economist Dr. Salazar in the 1930s.

European Economic Area (EEA) Established by the Treaty of Oporto in 1992. A free trade area containing EU and EFTA members.

European Free Trade Association Established by the Stockholm Convention in 1960 to promote free trade among member states. Via EEA, EFTA has free trade with EU members.

European Parliament Composed of members of the European Parliament, or MEPs, elected in EU-wide elections every five years. The Parliament meets in Strasbourg and in Brussels.

European Union (EU) The European Union was created by the Maastricht Treaty of 1992, ratified by all members in 1993. It evolved from the European Economic Community (EEC) or Common Market, established in 1957.

Exclave A part of a country separated from the main state by the territory of another country (e.g., Kaliningrad is separated from Russia).

Fallow Plow land left unsown to restore fertility.

Fertility rate, total The average number of children born to a woman in her lifetime.

Forced migrant A migrant forced from place of residence.

Free trade area Created when two or more states remove tariffs and other restrictions on trade between them. Members can trade with nonmembers on terms of their own choosing.

Globalization A process by which regions of the globe are increasingly interconnected.

Gross domestic product The total value of goods and services produced in monetary transactions within the national territory of a country. Gross national product (GNP) includes monies earned outside the country such as repatriated profits. Both GDP and GNP are often expressed on a per capita basis. Neither GDP nor GNP includes the value of subsistence activities, and exaggerate wealth differences between cash and subsistence economies.

Gulf stream An ocean current that exits the Gulf of Mexico via the Florida Strait and flows to northwest Europe, where it is known as the North Atlantic Drift.

Horst A fault-formed, upstanding block of rock.

Iberia/Iberian peninsula The peninsula occupied by Spain and Portugal.

Indigenous Belonging naturally in an environment or region.

Infant mortality The number of children per thousand live births who do not reach their first birthday.

Insolation Solar radiation received at the earth's surface.

Interfluve Higher ground lying between the flood plains of adjoining rivers.

Internal drainage A region that drains to a central depression, lake, or sea with no outlet to the sea.

International Monetary Fund (IMF) The formation of the IMF was agreed upon at Bretton Woods in 1944. It became a UN agency in 1947.

Isoline Line joining points of equal value. Isobar is a line joining stations with the same atmospheric pressure. Greek *isis* (equal).

Isomorphic plain A plain with common characteristics of terrain, climate, water supply, and soil fertility. Von Thünen and Christaller constructed models on hypothetical isomorphic plains to emphasize the importance of distance to markets and service centers.

Karst Limestone country characterized by sink holes, underground drainage, caves, and caverns.

Life expectancy The average number of years a newborn child is expected to live.

Loess Fine, wind-borne sediment.

Maritime climate Climate displaying oceanic influences including high humidity, much cloud, plentiful precipitation, and moderate temperatures. See **Maritime Polar** and **Maritime Tropical**.

Maritime Polar (mP) A cool, moist air mass acquiring characteristics over Arctic or Antarctic seas. For Europe, the air mass source region is the Arctic seas of the North Atlantic.

Maritime Tropical (mT) A warm, moist air mass acquiring characteristics over tropical seas. For Europe, the air mass source region is the Caribbean and the Gulf of Mexico.

Market town Small town providing services to a surrounding rural area and having open-air markets on one or more days each week.

Mediterranean climate A climate with hot, dry summers and mild, moist winters.

Mercantilism An economic system involving government regulation of business, industry, trade, and shipping, designed to minimize imports.

Metric system Decimal system of measurement. The U.S. currency is metric, but otherwise the United States retains miles, feet, inches, Fahrenheit, pints, and gallons but not the larger Canadian imperial gallon. Canada is phasing in a metric system. Some common approximate conversions:

2 inches	=	5 centimeters
10 feet	=	3 meters
5 miles	=	8 kilometers
1 meter	=	3.281 feet
32°F	=	0°C
70°F	=	21.1°C
80°F	=	26.7°C
90°F	=	32.2°C
1 hectare	=	2.471 acres
1 acre	=	.405 of a hectare
1 square mile	=	2.5899 square kilometers

Migration chain The social mechanisms by which migration patterns are established and maintained along migration channels.

Migration channel The movement of many migrants along a common routeway from a source region to a destination.

Migration field The area from which a city, town, or village draws migrants. The larger the place, the wider the migration field.

Mobility transition A change in migratory behavior that occurs as traditional societies enter the modern world and begin to urbanize rapidly, resulting in strong rural-to-urban migration.

Monoculture The growing of a single crop over a wide area (see **Polyculture**).

Moraine Deposit consisting of clay and rocks, left by a melted glacier.

Morphology The study of form and structure.

Nationalization The takeover of an industry by a government.

Natural environment The physical elements of the earth, such as landforms, climate, vegetation, and soils (see **Cultural Landscape**).

Natural increase The difference between the birth rate and the death rate expressed as a percentage. A population with a crude birth rate of 20 per thousand and a crude death rate of 10 per thousand has an annual 1 percent rate of natural increase.

Natural vegetation The vegetation of a region prior to modification by human activities such as burning, logging, or agriculture.

Neoliberalism Policies to reduce the role of government in economies by lowering subsidies, privatizing nationalized industries, and reducing tariffs.

Orogeny Mountain-building episode including the Caledonian, Hercynian, and Alpine orogenies.

Orographic rainfall Precipitation resulting when moisture-bearing air rises over higher ground.

Permafrost Permanently frozen subsoil found in Arctic and sub-Arctic regions.

Polder Low-lying drained land reclaimed from the sea or a river. Dutch.

Polyculture The practice of growing several crops simultaneously within a small area (see **Monoculture**).

Population density The number of persons living in a square mile or square kilometer of earth space.

Population pyramid A diagram used to illustrate the age and gender distribution of a population. The population is divided into 5-year age cohorts: 0–4, 5–9, 10–14, 15–19, etc. If the lower age cohorts are larger than those above, the population is increasing. If the 0–4 age group is smaller than the 5–9 cohort, the rate of **natural increase** is slowing.

Primary sector of the economy Economic activities that directly exploit natural resources: hunting, gathering, forestry, fishing, farming, mining, and quarrying.

Primate city A major city whose population is many times larger than that of any other place in a country. Examples in Europe include London, Moscow, Paris, and Istanbul, but not Rome or Brussels.

Privatization The selling off of state-owned companies to the private sector.

Rain shadow Area of relatively low rainfall on the lee side of uplands over which rain-bearing winds have passed.

Rank–size rule A distribution of city size populations in which the largest city is twice the size of the second largest place, and three times as big as the third-ranked city, etc.

Refugee A person displaced from home by famine, warfare, civil unrest, ethnic conflict, or government edict who crosses an international boundary.

Return migrant A migrant who, having moved from a source area to a destination, returns to the source area.

Seasonal migrant A migrant who moves on a seasonal basis between places of residence.

Secondary sector of the economy Manufacturing and construction activity.

Sierra Mountain range.

Source area The region from which migrants originate (see **Migration field**).

Stagflation An economic condition consisting of inflation, little economic growth, and persistent unemployment.

Steppe Grassland frequently underlain by loess soils.

Step-wise migration Stepwise migration occurs when a migrant resides in a number of places before reaching a final destination.

Subsistence farming Agricultural activity in which farming families consume most of what they produce.

Sustainable development Resource use that passes to future generations an environment undegraded by the extinction of species, soil erosion, pollution, or any other means.

Taiga Coniferous forest lying between the **tundra** and the **steppe**.

Tectonic forces Pressures in the earth's crust that produce faulting, earthquakes, and mountain-building forces.

Tees–Exe Line A line of division, devised by Halford Mackinder, to separate highland and lowland Britain.

Tenant farmer A cultivator who rents land by paying cash, by sharing the crop, or by providing labor services to the landlord.

Tertiary sector of the economy The service sector of the economy includes wholesaling, retailing, transportation, financial services, education, recreation, administration, medical services, and tourism.

Three-field system of agriculture An agricultural practice in which arable land of a community is divided into three open fields, with two fields being cultivated and one field, in rotation, left fallow.

Threshold of demand The minimum number of purchases required to make it possible to offer a good or service.

Toponym The name of a place; a place name.

Tsunami High waves produced by an undersea earthquake.

Tundra Far northern region with a short growing season and permanently frozen subsoil.

Urban hierarchy The ranking of urban places on the basis of population numbers and functions.

Urban majority A country in which over half the population lives in urban places.

Urban morphology The form and internal structure of a town or city.

Village Rural nucleated settlement providing basic services, including store, bar, post office, and church.

Voluntary migrant A migrant who chooses to move residence.

Westerly wind system A wind system that transports mild, moist, maritime tropical air from the Caribbean/Gulf of Mexico to western Europe.

World Bank The International Bank for Reconstruction and Development, founded in 1947 as the world's development bank.

World city A city that provides services on a global scale (e.g., New York, London, Paris, Geneva).